Biodegradation Handbook

Edited by **William Chang**

R CALLISTO REFERENCE

New York

Published by Callisto Reference,
106 Park Avenue, Suite 200,
New York, NY 10016, USA
www.callistoreference.com

Biodegradation Handbook
Edited by William Chang

International Standard Book Number: 978-1-63239-088-2 (Hardback)

Contents

Preface

This book, written for readers with interest in the process of biodegradation, serves as a good source of information. It is a compilation of various biodegradation research procedures which lead to various biological processes. It deals with biodegradation with respect to polymers and surfactants and also takes into account the microbial behavior.

This book is the end result of constructive efforts and intensive research done by experts in this field. The aim of this book is to enlighten the readers with recent information in this area of research. The information provided in this profound book would serve as a valuable reference to students and researchers in this field.

At the end, I would like to thank all the authors for devoting their precious time and providing their valuable contribution to this book. I would also like to express my gratitude to my fellow colleagues who encouraged me throughout the process.

Editor

Biodegradation of Polymers and Surfactants

Biodegradation of Medical Purpose Polymeric Materials and Their Impact on Biocompatibility

Elisa Tamariz and Ariadna Rios-Ramírez

Additional information is available at the end of the chapter

1. Introduction

The use of polymeric materials in medical devices and pharmaceutical applications has been extended in the last decades. Biodegradable implantable polymers for tissue engineering and drug release have the advantage to avoid a permanent and chronic immune response, and to avoid removal surgery; moreover the versatility of polymeric materials aloud the design of specific biodegradable characteristics to control drug release, to develop resorbable devices, and to improve cell integration.

Biodegradation is a term used to describe the process of break down a material by nature; however in the case of medical purpose biomaterials, biodegradation is focus in the biological processes inside the body that cause a gradual breakdown of the material.

Biomaterials degradation is a very important aspect to consider when they are used for medical purpose, since their ability to function for a certain application depends on the length of time that it is necessary to keep them in the body.

Polymers biodegradation process and rate within an organism is related to the polymer characteristics and the place in the body where will be exposed. This chapter intended to offer an overview of the mechanisms that influences the biodegradation of polymeric materials used for medical purposes, with special emphasis in the immunological mechanisms that modulates biodegradation rates and biocompatibility, and in the features that implies their use in the central nervous system (CNS). It will be also focused in the importance of modulate the biodegradation for some biomedical application, and how the on purpose control of biodegradation could be a relevant aspect to design biomaterials with a more interactive and efficient role in medicine.

2. Polymeric material for biomedical applications

The use of polymers in biomedical applications is now widely accepted and they are termed with the generic name of polymeric biomaterials. A biomaterial can be defined by their function as a material in contact with living tissue, used to the treatment of disease or injury, and to improve human health by restoring the function of tissue and organs in the body [1]. The 1982 Consensus Development Conference Statement of the National Institute of Health (NIH) defines a biomaterial as any substance (other than drug) or combination of substances, synthetic or natural in origin, which can be used for any period of time, as a whole or as a part of a system which treats, augments or replaces any tissue, organ or function in the body [2]. Polymeric biomaterials in medicine include surgical sutures, drug delivery vectors, orthopedic devices and implants, and scaffolds for tissue engineering.

After decades of research many polymeric biomaterials have been developed from synthetic or natural origin. All the polymeric biomaterials have to be evaluated in terms of their biocompatibility, mechanical properties and biodegradation to determine if they are suitable for specific medical applications.

Biocompatibility refers to several characteristics of the biomaterial which leads to the acceptance of the material in the body, such as being non toxic, non carcinogenic, non allergenic, non immunogenic. The materials for in vivo use have to be exposed to hemocompatibility, citotoxicity, mutagenicity, and pyrogenicity test [1].

Mechanical properties like elastic modulus, compression modulus, fatigue, and viscoelasticity are important characteristics to determined their use in the body, for example for bone implants and prosthesis; however micrometric or nanometric characterization is also important in the case of biomaterials for tissue replacement and cell scaffolds, since micro and nano-characteristics are important to manipulate cell proliferation, differentiation, and function to mimic the tissue to be replaced [3].

Biodegradation refers to the rate of breakdown mediated by biological activity, and is an important property for biomaterials used as non permanent scaffolds, implants, drug delivery vectors, and sutures [1].

The most commercial and earliest developed polymers for biomedical applications were the synthetic polymers developed from linear aliphatic polyesters. Synthetic polymers are made by linking small molecules (mers) through primary covalent binding in the main molecular chain backbone, and have a close resemble with natural occurring tissue components like proteins, polysaccharides and deoxyribonucleic acids. Besides synthetic polymers, natural occurring polymers are also used as biomaterials. Many of the natural polymers are synthesized by condensation reactions and the condensing molecule is always the water [4]. Natural polymers and their chemical and mechanical properties specifically provide functions to each of them in the organisms, for example collagen in the dermis, fibrin in the clot, chitin in the exoskeleton of insects and crustaceans [5, 6].

Either synthetic and natural polymers, or bonds among both has been studied and used as materials for medical application; below we shortly describe some examples of the main natural and synthetic polymers and some of their proposed uses in medicine, particularly in the nervous system.

2.1. Natural polymers

Natural polymers are used in clinical applications such as dermal fillers, lubricants, wound sealants and surgical sponges. Other naturally derived polymers have readily available functional groups which facilitate chemical modification.

Agarose is a polysaccharide of D-galactose and 3,6- anhydro-L-galactopyranose derived from the cell walls of red algae. Agarose is biologically inert and is attractive for drug delivery because it has soft, tissue-like mechanical properties, and can form porous gels at low temperatures. Agarose is heated to solubilize the powder in aqueous solutions and then gels through hydrogen bonding upon cooling [7]. Agarose decrease potential immune rejection when inserted into the brain, for example, Brain-derived neurotrophic factor (BDNF) delivered in this way was found to reduce the reactivity of the astrocytes and the production of chondroitin sulfate proteoglycans (CSPGs), and to enhance the number of regenerating fibers that entered the hydrogel into the injured spinal cord in rats [7, 8].

Fibrin is a promising material because of its natural role in wound healing and its current application as a tissue sealant. Obtained from pooled human plasma, fibrin presents an advantage since it could be an autologous source avoiding risk of immune rejection. The most used gel is the fibrin glue and consist of fibrinogen and thrombin enzymatically crosslinked; however it has also used in conjugation with other polymers such as hyaluronic acid [1, 9]. This polymer has the advantage to be injected and polymerized in situ and has been tested for controlled delivery of Nerve growth factor (NGF), Neurotrophin -3 (NT-3) and BDNF in the CNS [7].

Collagen is the main component of connective tissue and is the most abundant protein in mammals, there are at least 19 different types of collagen, for example type I collagen is a fibril forming collagen and is present in the skin and fibrocartilage, type II collagen is found in articular cartilage[10, 11]. Collagen can be isolated from tissue like skin, bone or tendon. Collagen gels alone are quite weak, and are often crosslinked to improve durability. While many applications use unmodified collagen, chemical crosslinkers can be used to inhibit in vivo absorption in applications which require slow degrading constructs, such as drug delivery [7, 12]. Although collagen is abundant in many tissues, is not the main component in the CNS extracellular matrix, therefore some concern is present about their use as CNS cells scaffolds [12], however their use for stably releasing of growth factors like ciliary neurotrophic factor (CNTF), has shown to improve the survival, growth and proliferation of neural stem/progenitor cells (NSPCs) [1].

Alginate is a linear block copolymer of D-mannuronic acid (M) and L-guluronic acid (G) residues. Commercially available, alginate is extracted from brown seaweed algae. Alginate has a high biocompatibility since their hydrophobic nature; however, cannot be enzimatically broken down and has poorly regulated degradation. Partial oxidation of alginate with sodium periodate makes the chains more susceptible to be degraded by hydrolysis [13]. Mammalian cells cannot adhere to alginate unless it is modified with cellular adhesion molecules like laminin, fibronectin, collagen, and RGD sequences, which allow more specific interactions [9]. Covalently modified gels of alginate containing different ration of RGD peptides have been used to encapsulate cells and to induce their differentiation [14].

Hyaluronic acid (HA) is a glycosamine glycane made of residues of N-acetylglucosamine and D-glucuronic. HA is normally presented at high levels in the extracellular matrix of connective, epithelial and neural tissues, and is known to play roles in cellular processes like cell proliferation, morphogenesis, inflammation, and wound repair. However, HA alone does not gel and is rapidly degraded through the action of the enzyme hyaluronidase into smaller oligosacharides, HA can also be degraded by reactive oxygen species at the site of inflammation [15], and is readily cleared due to its high solubility [1]. HA is fabricated into hydrogels using chemical crosslinkers such as glutaraldehyde or carbodiimide, and has also been widely derivatized to form photocrosslinkable and injectable hydrogels. Its polyanionic and hydrophilic characteristics made it highly biocompatible and suitable for applications with minimal cell invasion [9].

Poly(β-1,4-D-glucosamine) or Chitosan is a natural polymer that can be prepared by de-N-acetylation of chitin from crustacean shells, the degree of chitosan deacetylation affects the charge density of the polysaccharide, more deacetylation increase the positive-charge character of the chitosan chains. It can form gels by covalent crosslinking with aldehydes such as glutaraldehyde or ionic crosslinking by polyanions such as sodium citrate or sodium tripolyphosphate [6]. The limited solubility of chitosan in neutral pH provides a unique opportunity to form nanoparticulate drug/gene delivery platforms, but it is also an obstacle if one intends to apply chitosan as a solution in the physiological condition [5]. Chitosan can be easily conjugated with organic materials as well as biomolecules, a number of studies have reported controlled drug delivery using chitosan nanoparticles that incorporate biologically active ingredients such as DNA, proteins, anticancer drugs, and insulin [16]. Chitosan has been extensively investigated as a potential biomaterial in a variety of applications, including drug carriers, wound-healing agents, and in tissue engineering. Chitosan scaffolds have been used to transplant viable peripheral nerve grafts, neural stem cells, and neural progenitor cells into rat spinal cords, resulting in increased axonal regeneration [17].

Methylcellulose (MC) is a chemically modified cellulose derivative in which there is a partial substitution of OH groups with methoxy moieties forming a non-toxic material. MC is a water-soluble polymer at low temperature with thermo reversible gelation at a particular temperature. Thermoreversible characteristics are related to the association of MC hydrophobic groups, and the gelation temperature can be manipulated by salts or ions [18]. MC is widely accepted as a highly compatible material and has been used in traumatic CNS lesion like a scaffold for tissue regeneration [19].

2.2. Synthetic polymers

Synthetic polymers offer exceptional control over polymer composition, architecture, and physical properties not fully accessible with natural polymers. After many years of research in this field numerous polymers have developed like polyesters, polyurethanes, polyanhydrides, polyacrylates, polyphosphoesters, and polydiaxanone. One of the first and now most common uses of polymers in medicine is for resorbable sutures, pins and screws. An extensive review of synthetic polymers is out of this chapter and we will only mention some examples and their main characteristics.

2.2.1. Polyesters

Polyglycolide or polyglycolic acid (PGA) is polymerized from glycolic acid and many of the most important polymer for biomedical use are derived from PGA either through copolymerization or modified glicolide monomers. PGA is one of the most successful and commercially available polymers and is widely used as biodegradable biomaterial in surgery [20].

Poly(lactide-co-glycolide) acid (PLGA) is a PGA derived polymer, is a polyesters obtainable by linear polycondensation of hydroxyacids, or by ring opening of the corresponding lactones. It is the most commonly used biomaterial in medicine. This polymeric agent has been implanted into the brain and has shown good biocompatibility and sustainable drug delivery [8]. In normal untreated animals, polymer microspheres implanted into the brain did not produce gross behavioral or neurological symptoms, and it has been approved for the FDA for repair of human peripheral nerves [5]. Various drugs, especially therapeutic proteins like neurotrophic factors have been encapsulated in this type of brain delivery system, however, the in vivo hydrolysis of PLGA produce and acidic environment that result in a transient pH decline that can compromise the proteins action and stability, and consequently the process of encapsulation and release from biodegradable microspheres must be carefully monitored [5, 21].

Polyethylene glycol (PEG)- based polymers are hydrophilic and water-swellable cross-linked polymers with a high level of elasticity, making them ideal candidates for tissue engineering; more importantly, the degradation rate of the implant can be controlled by simply altering the chemistry of the cross-links within the polymer network [9]. PEG has a variety of applications in drug delivery and tissue engineering, their hydrophilic and non ionic characteristics made it relatively resistant to protein adsorption and highly biocompatible. *In vitro*, they can support the survival of PC12 cells, fetal ventral mesencephalic neural cells, and human neural progenitor cells. Furthermore, in culture, neural cells encapsulated into PEG-based hydrogels survive, maintain phenotype, and extend processes indicating that the hydrogels are not themselves cytotoxic [22, 23]. A recent study in primates found that PEG was completely degraded and the neuroimmune response was less than that found in sham penetrated brains [22, 23].

2.2.2. Polyacrylates

Poly(2-hydroxethyl methacrylate) PHEMA, is one of the earliest polymer used as implantable material. Polymerized from 2-hydroxyethyl methacrylate using free radical precipitation, PHEMA forms a hydrogel biologically inert. One of the main concerns about PHEMA is its low biodegradability, their biodegradation however can be manipulated by modifying the porosity by photopatterning [24]. One of the earliest uses of PHEMA was as an artificial cornea, or keratoprosthesis [25]. Methacrylic-acid- and acrylic-acid-based hydrogels have a high affinity for calcium and other alkaline earth metals, making them more prone to calcification, thereby some calcification episodes has been found after in vivo implantation [9].

Poly(N-isopropylacrylamide) (PNIPAAm) has been widely studied as a temperature responsive drug delivery system. It has the particular ability to undergo a thermally induced phase transition at 32 °C that induces swelling in the polymer network. The phase transition temperature can be tuned via copolymerization of more hydrophilic or hydrophobic co-monomers to achieve desired transitions in relevant *in vivo* environments. At physiological temperatures, PNIPAAm homopolymer gels hold little water and show poor elastic recovery [7, 26].

2.2.3. Poly(ω-hydroxyl acids)

The poly(ω-caprolactone) contains five (CH)2 units in the repeating unit, making the chains much more flexible than PGA, which has one. Therefore, thermal and mechanical properties decreased considerably compared to PGA. However, the rate of biodegradation is slow, making it better suited to slow drug release applications such as one-year implantable contraceptives, biodegradable wound closure staples, etc [27].

2.2.4. Poly(ortho esters)

Poly(ortho esters) (POE) undergo surface degradation, making them ideal as a drug-delivery vehicle. Erosion process is confined predominantly to the surface layers; therefore controlled drug release is possible as well as maintenance of an essentially neutral pH in the interior of the matrix because acidic hydrolysis products diffuse away from the device. The rate of degradation can be controlled by incorporating acidic or basic excipients into the polymer matrix since the orthoester link is less stable in an acid than in a base [27]. The polymer is stable at room temperature when stored under anhydrous conditions. Either solid or injectable materials can be fabricated into different shapes such as wafers, strands, or microspheres that allow drug incorporation by a simple mixing at room temperature and without the use of solvents [28].

2.2.5. Poly(ester-amides)

Poly(ester-amides) (PEAs) combine the high degradability of polyesters with high thermal stability and high modulus and tensil strength of polyamides [29], are non-toxic building blocks and had excellent film forming properties. These polymers were mostly amorphous materials, combine the well-known absorbability and biocompatibility of linear aliphatic polyesters with the high performance and the flexibility of potential chemical reactive sites of amide of polyamides [20]. PEAs can be functionalized to conjugated different drug, peptides or molecules for cells signaling and had been used for microspheres and hydrogels formation [29].

2.2.6. Others

Poly(vinyl alcohol) PVA is prepared from the partial hydrolysis of poly(vinyl acetate). It can be crosslinked into a gel either physically or chemically. In recent studies, PVA was photocured to produce hydrogels as an alternative to chemical crosslinking [9]. PVA is similar to PHEMA in having available pendant alcohol groups that act as attachment sites for biological molecules. In addition to having multiple attachment sites, PVA is also elastic and thus can induce cell orientation or matrix synthesis by enhancing the transmission of mechanical stimuli to seeded cells [30].

3. Biodegradation of polymeric biomaterials

Biodegradation process could be driven by chemical, physical, and biological interactions.

Biodegradation rate within an organism is related to the polymer characteristics and the place in the body where will be exposed. Chemical degradation is influenced by composition and molecular structure, polydispersity, crystallinity, surface area, hydrophilic or hydrophobic characteristics. In general chemical degradation causes the deterioration of the main polymer chains by random cleavage of covalent bounds, depolymerization or crosslinking of linear polymers, interfering with regularly order chain and with cristallinity, decreasing the mechanical properties [1]. Degradation can be by surface degradation or bulk degradation. In the case of bulk degradation, water uptake by hydrophilic polymers is faster than the rate of conversion of polymer into water-soluble materials, bulk degradation causes the collapse of all the material since the degradation process occurs in throughout their volume. Surface degradation appears in hydrophobic polymers, leaving the inner structure intact, these polymers offers a better control of degradation rates [4].

Biodegradation in a biological environment may be defined as a gradual breakdown of a material mediated by a specific biological activity; when materials are exposed to the body fluids may undergo changes in their physicochemical properties as a result of chemical, physical, mechanical, and biological interactions between the material and the surrounding environment. A very important factor in biodegradation is the interaction with the immune system and their specialized cells.

Polymeric materials can be degraded inside the body by at least three general mechanisms, oxidation, hydrolytic, and enzymatic mechanism.

3.1. Hydrolytic mechanism

Hydrolytic degradation of polymers may be defined as the scission of chemical bonds in the polymer backbone by the attack of water to form oligomers and finally monomers. This kind of hydrolysis could not require of specific biological compounds as proteases, although many of the biodegradation process by enzymatic mechanisms mentioned further are hydrolysis reactions.

All biodegradable polymers contain hydrolysable bonds like glycosides, esters, orthoesters, anhydrides, carbonates, amides, urethanes, ureas; while materials with strong covalent bonds in the backbone or non hydrolyzable groups have less biodegradable rates [31].

In the first step of hydrolytic mechanism, water contacts the water-labile bond, by either direct access to the polymer surface or by imbibitions into the polymer matrix followed by bond hydrolysis. Hydrolysis reactions may be catalyzed by acids, bases, or salts. After implantation, the biomaterial absorbs water and swells, and degradation will progress from the exterior of material towards its interior. In general, the first degradation reaction, even after contact with water molecules, is the hydrolytic scission of the polymer chains leadings to a decrease in the molecular weight. While degradation progress, the molecular weight of degradation products is reduced by further hydrolysis which allows them to diffuse from the bulk material to the surface and then to the solution, causing significant weight loss [31]. Rate of hydrolytic degradation is modulated by hydrophilic characteristics of the polymers as mentioned before, therefore materials such as PEG derived hydrogels have a high biodegradation rate [9]. In the case of hydrolysis of aliphatic polyesters such as PLGA, the acid products accelerate biodegradation due to autocatalysis [32].

3.2. Oxidation mechanism

Polymeric biomaterials could be degraded by oxidation when they are exposed to the body fluids and tissues. It is well known that during inflammatory response to foreign materials, inflammatory cells, particularly leukocytes and macrophages are able to produce highly reactive oxygen species such as superoxide (O_2^-), hydrogen peroxide (H_2O_2), nitric oxide (NO), and hypochlorous acid (HOCl). The oxidative effect of these species may cause polymer chain scission and contribute to their degradation. For example superoxide could accelerate the degradation of aliphatic polyesters by the cleavage of ester bonds via nucleophilic attack of O_2^- [20, 31]. A resorbable suture of multifilament of poly(α-hydroxyester), commercially available as Vicryl, for example, exhibited many irregular surface cracks after incubation for 7 and 14 days in an aqueous free radical solution prepared from H_2O_2 and $FeSO_4$, while the same suture in control solution did not have surface cracks, suggesting a role of free radicals in the observed degradation [33]. It has also reported that polyurethanes are attacked initially by neutrophils which secretes reactive oxygen species (ROS) and HOCl, one of the most oxidative compounds [34].

3.3. Enzymatic mechanism

Enzymes are biological catalysts; they accelerate reaction rates in living organisms without undergoing themselves any permanent change. Hydrolysis reactions may be catalyzed by enzymes known as hydrolases, which include proteases, esterases, glycosidases, and phosphatases, among others. For example, it has been shown that the degree of biodegradation of polyurethanes, in the presence of cholesterol esterase enzyme, is about 10 times higher than in the presence of buffer alone [35].

Enzymatic surface degradation occurs when enzymes cannot penetrate the interior of the polymer, due to high cross-link density or limited access to cleavage points, forcing the surface or exterior bonds to cleave first. The mode of interaction between the enzymes and the polymeric chains involves typically four steps:

• Diffusion of the enzyme from the bulk solution to the solid surface.

• Adsorption of the enzyme on the substrate, resulting in the formation of the enzyme–substrate complex.

• Catalysis of the hydrolysis reaction

• Diffusion of the soluble degradation products from the solid substrate to the solution [20, 31].

Enzymatic degradation of natural origin polymer is held by the action of specific enzymes, for example degradation of hyaluronic acid in mammals is carried out by the concerted action of three enzymes: hyaluronidase, b-D-glucuronidase, and b-N -acetyl- D hexosaminidase; in the case of chitin derivatives, lysozyme is the enzyme involved in their degradation inside the body [31].

In vivo degradation rates of a polymers could be faster than *in vitro*; the higher *in vivo* degradation rate have been explained by the effects caused by cellular and enzymatic activities found in the body [36]. *In vitro* degradation tests of polymers in simple aging media are normally conducted to predict the performance of such polymers in the clinical situation; however, taking into account the complexity of the body fluids, it is common to find different results when the same materials are studied both *in vitro* and *in vivo*, mainly because of several oxidation y enzymatic factors are absent in *in vitro* used medias.

4. Immune response and biodegradation

Immune system in the living organism is devoted to continuously surveillance the body to detect self and non-self patterns. Immune system has cellular and molecular entities capable to recognize and induce a response to eliminate the potential dangerous entity or non self elements. There are two kinds of responses, innate immunity and adaptive immunity, the former is a non specific and fast first reaction against a pathogen or a foreign body, while the last is a slow, specific response to a first exposure pathogen or foreign body [37]. The two systems of immunity are related, innate immunity system provides information to the adaptive system by inflammatory mediators and cells such as macrophages and dendritic cells, which process the antigens and present them to T cells of the adaptive immune system [38].

The interaction of inflammatory cells with biomaterials surface-adsorbed proteins constitutes the major immune recognition system for biomaterials; therefore, the study of materials surface properties to avoid the absorption of certain class of proteins has been determinant to understand their compatibility and their degradation process [39-41].

Foreign materials exposed to blood or plasma are immediately covered by proteins commonly called opsonins, the most common opsonins are protein forming the complement system, a group of about 30 proteins that assemble to form a lityc complex and a recognizable system for phagocytic cells [37]. The recognition through specific receptors in the phagocytic cells is determinant to induce phagocytic and inflammatory response. Phagocytosis is one of the main processes undertaken by innate immunity. Different kind of polymorphonuclear leukocytes like neutrophils, and mononuclear leucocytes like monocytes, macrophages, and dendritic cells, engulf the opsonized foreign material. Once inside the cell enzymes and oxidative processes destroy them [37, 42].

The recognition of opsonized material by specific receptors in monocytes like Fc receptors, mannose receptors or complement receptor triggers the rapid induction of proinflamatory cytokines and recruitment of inflammatory cells [42]; while recognition of non opsinized materials by scavenger receptors leads to non inflammatory phagocytic mechanism [43]. The binding of other proteins like fibrin, collagen, albumim, fibronectin and vitronectin present in the blood, plasma and surrounding tissue are also important to mediate an inflammatory process and the recruitment of inflammatory cells [42].

Many biomaterials has to be implanted in the body through chirurgic procedures, the wound healing process after the implantation induce the recruitment of inflammatory cells and an acute inflammatory response, exposing the biomaterials to the immune system.

The first cells present in the injury site are neutrophils, and within a day or two monocytes, macrophages and later lymphocytes arrive to the implantation site forming a chronic inflammatory process (Figure 1).

If the inflammatory process continues and the biomaterial persist, a foreign body reaction can appears, where multinucleated cell, resulted from the fusion of macrophages, invade the implant site and collagen producing cells like fibroblasts arrives to form a fibrotic tissue around the material forming a capsule [44].

In the case of biomaterials directly exposed to the blood, circulating monocytes, platelets, leucocytes and dendritic cells recognize the materials inducing a similar oposinization and inflammatory process, besides, the activation of platelets and blood coagulation system can induce thrombotic occlusion and serious non desired effects in medical devices like artificial valves, hearts or cardiac stents [44, 45].

Neutrophils and macrophages normally phagocyte foreign materials smaller than 10µm like micro and nanoparticles, and degrade them after engulf into the phagolysosomes, a vesicles containing numerous hydrolases within an acidic environment. When biomaterial particles are large, among 10 to 100µm, multinucleated body giant cells can engulf and digest them, however; when biomaterials are larger enough to avoid phagocytosis, frustrated phagocyte response leads to secretion of numerous proteases, toxic oxygen derived metabolites or oxygen radicals contained in phagocytic and phagolysosome degradatory vesicles [46-48]. The amount of enzymes and radicals released depends on the size of the polymer particles, larger particles induce greater amount of enzyme release. The phagocytable form of the biomaterial, particles or powder, can also exert a differential response as compared for example with a less phagocytable material like a film [44]. A recent study however, shows that the induction of pro-

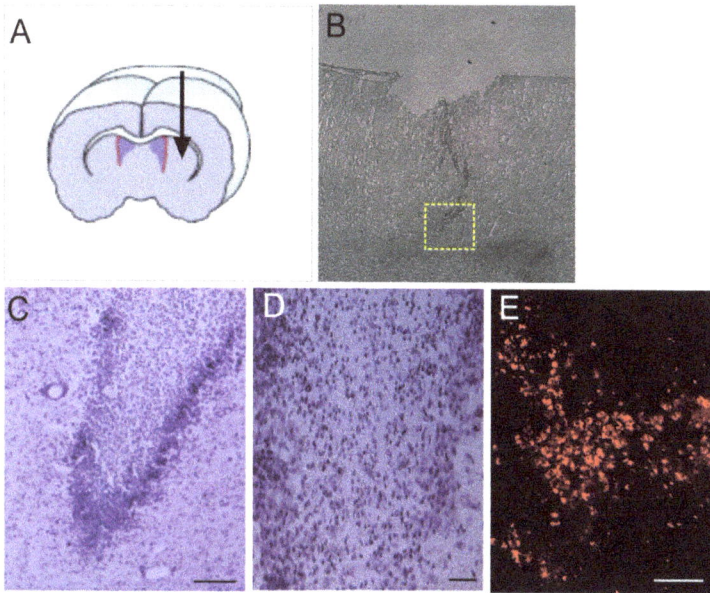

Figure 1. Inflammatory response after 3 days post injection of PEG-Silica gel into rat striatum. (A) Arrow indicates the area of polymer implantation, (B) Phase contrast image of the implanted area, (C) Cresyl Violet staining of the area marked in B, showing cell infiltration at the deepest injection area, (D) Higher magnification showing cell infiltration, (E) Confocal image of a section immunostained to detect activated macrophages at the implanted area. Scales bars: C= 50μm, D and E=100μm.

inflammatory cytokines in macrophages, is held after biomaterials contact with macrophages cell membrane independently of the particles size [49].

Interestingly, fusion of macrophages into giant cells is dependent on the presence of interleukins like IL4, but also of the biomaterials surface properties, for example biomaterials with hydrophilic and anionic surfaces induce the apoptosis or programmed cell dead of macrophages, as compared with hydrophobic and cationic surfaces [50]. The formation of giant cells by macrophages fusion is also modulated by the adsorption of plasma proteins like fibronectin and vitronectin on the materials surface [40, 51].

It is important to mention that there is heterogeneity in the macrophages differentiated from the arrived monocytes into the inflammation site; macrophages known as M1 are the "classical" activated macrophages, stimulated by the TH1 cells or natural killers and microbial products like lipopolysacharides. M1 produce cytotoxic products like reactive nitrogen and oxygen species, and lead to chronic inflammation and formation of foreign body giant cells [52, 53]. M2 macrophages are commonly termed "anti-inflammatory" macrophages. M2 are induced by Interleukine IL4 and IL13 and release high amount of inflammatory suppressors like interleukins IL10, M2 also suppress the nitric oxide release, thereby M2 macrophages are related to immunosuppressive and protective activities like wound healing and fibrosis [53,

54]. Interestingly biomaterials are able to modulate monocytes differentiation to anti-inflamatory, pro regenerative M2 macrophages, or to avoid their differentiation into macrophages, or the fusion of macrophages into multinucleated foreign body giant cells, diminishing the undesirable inflammatory effects and the deleterious biodegradation [52].

Cholesterol esterase has been identified as an enzyme involved in the degradation of polyester(urethane)s and polyether- and polycarbonate-poly(urethane), this enzyme is also used as a marker of monocyte derived macrophages [55]; some materials like degradable polar hydrophobic ionic polyurethane (D-PHI) reduce the expression of esterase activity as compared with cell cultures under polystyrene tissue culture plate, decrease pro-inflammatory interleukins and cytokines like TNFbeta and increase the anti-inflammatory interleukins expression like IL10, suggesting that monocytes can differentiate to an "anti-inflammatory" or M2 phenotypes depending of the biomaterial and therefore influence the regenerative process after wound healing [56].

Enzymes release by activated phagocytes influence the biomaterials degradation by different rates depending upon many characteristic like chemical composition, homogeneity and porosity. Activated macrophages and neutrophils secrete high amount of enzymes like myeloperoxidases, lyzosime and metalloproteinases that contribute to the destruction of invaders and to their migration into the injury or inflammation site; however in the case of chronic inflammatory states enzymes secretion can be deleterious generating an extend and irreversible damage to the tissue [57-59].

As mentioned before surface characteristic of polymeric materials are very important to induce the recruitment, attachment and release of enzymes from neutrophils and macrophages; hydrophilic surfaces induce the interaction with blood containing proteins like fibronectin, fibrin, and albumin, inducing neutrophils adhesion to the proteins and stimulating the degranulation behavior or delivery of primary and secondary granules containing high amounts of enzymes.

For example macrophages cultured on polymers like poly(carbonate-) and poly(ether-) urethanes secrete cholesterol esterase, carboxyl esterase, and serine protease mediating the hydrolytic degradation of the polymers [34].

Hydrolases work under the low pH medium present in the phagolysosme, however when the hydrolases are released outside the cell the extracellular buffer solutions could impair their action; however a highly close interaction of macrophages with the materials forms a tight seal with the substrate, thereby protecting the secreted substances from potential inhibitory environment and therefore a transient and local decrease of pH at the site of phagocytosis improves the hydrolases enzymatic activity [47, 60].

In vitro assays to analyze the hydrolytic degradation of biomaterials are not always very accurate since paradoxically, the *in vitro* rates of degradation could be faster than in vivo rate due to encapsulation of a polymer by fibrotic tissue after a foreign body reaction [36], therefore it is important to perform *in vitro* plus *in vivo* assays in order to characterize the biodegradation rates of polymeric material.

The defense system against pathogens and foreign material includes the release of potent oxidizing agents by activated neutrophils and macrophages. Many of the oxidative products are released to the phagosome after their fusion with the lysosome once the leucocytes has engulf the foreign invader; however, as mentioned before, these potent oxidants could be also released to the extracellular space. After cells activation, oxidant are rapidly formed inside the cells by a process known as oxidative burst that includes an elevated consume of oxygen and the assembly of an enzymatic complex which catalyze the formation of superoxide anion O_2^-. Superoxide anions are further dismutated to hydrogen peroxide H_2O_2, which in the presence of myeloperoxidase generates very potent oxidizing agents like hypochlorous acid and chloroamines. The induction of nitric oxide synthase (iNOS) and the production of reactive nitrogen species and other oxidative species such as hydroxyl radical are also involved in the oxidative mechanisms [61, 62]. In the case of human macrophages however non iNOS induction was detected and therefore some differences as compared with mouse macrophages are reported [62].

Besides their original antibacterial function therefore, the activation of macrophages after contact with materials and the subsequent events like phagocytosis, release of cytokines and foreign body reaction are determinant event in biodegradation of materials. In most of the cases the delivery of high amount of reactive nitrogen and oxygen species and hydrolytic enzymes in response to foreign material, are deleterious to biomaterials but also to the surrounding tissues compromising the compatibility of biomaterials; hence fast biodegradation rates can be beneficial to attenuate undesired inflammatory effects.

4.1. The particular case of the brain

Central nervous system (CNS) is the most highly protected organ in the body; it is located behind the blood-brain barrier (BBB), a specialized structure formed by endothelial cells tightly joint together, a basal membrane, perycites, microglial cells and astrocytes endfeets forming the glia limitants area [63]. Although the BBB provides an isolated environment for the brain and represent a way to isolate and difficult the crosses of blood molecules and cells, brain is not devoid of immune system monitoring and of interchange of proteins and cells, hence the mentioned characteristics of BBB do not impair but delay the immune response [63].

Other characteristic of CNS immune response are the lack of classical dendritic cells in the parenchyma of the CNS, and non conventional lymphoid drainage which impede the migration and interaction of antigen presenting cell with the naïve memory T cell in the lymphoid nodes; therefore, innate immune response in the brain cannot directly initiate the adaptive immune response [64-66].

Immunological surveillance in the brain is held by several cell types, among them resident microglia is considered the immune cells of the CNS. Microglia is generated independent of the bone-marrow cells in adult organism, and is normally present in the brain in a non activated form; however, although different from tissue resident macrophages, is also suitable of activation and cell shape transformation from a ramified neuronal like morphology into a rounded phagocytic macrophage type, able not only to engulf foreign material but to release inflammatory mediators [67].

Astrocytes are neural derived cells that participate in the homeostasis of the tissue, but after an injury or foreign material presence, activated astrocytes deliver inflammatory mediators that can modify the permeability of BBB allowing the entrance of blood circulating leukocytes. Astrocytes can also phagocyte and form a glial scar or fibrous capsules to isolate the foreign material [65, 66, 68].

Although the presence of conventional dendritic cells in the brain parenchyma has been a controversial issue, recently it has reported the presence of them in the meninges and in the endothelial cell layer that filters liquid from the blood and form the cerebrospinal fluid named the choroid plexus, these cells are able to present antigens to induce adaptive immune response [69].

Perivascular cells are bone-marrow derived cells, monocytes and macrophages, located in the space between blood vessels and glia limitants, and can act as antigen presenting cells. The perivascular cells are the first and early arrival periphery cells entering to the brain parenchyma after an injury of inflammatory process [65, 70].

The particular immunological characteristics of CNS are complemented with the secretion, by astrocytes and microglia, of some cytokines like transforming growth factor beta (TGFβ) which suppress immune reaction, creating a particular immunosuppressive environment in the brain parenchyma that attenuates inflammation and protect the brain from deleterious effects of inflammation in neuronal integrity [71].

Polymeric biomaterials for CNS use are in expansion due to their properties as a drug delivery vectors or as cells and nerves scaffolds. BBB is an impediment for parental drug administration; therefore the use of biodegradable polymer for in situ delivery is an alternative to target the CNS. Biodegradable polymers offers a high advantage to control the temporal and spatial delivery of drugs and cells, besides biodegradable polymers circumvents the need of a second surgery to retire the vector as compared with non degradable devices like mini pumps [5, 72, 73].

Some approaches using biodegradable polymers for neurodegenerative disorders like Parkinson's disease, Alzheimer and Huntington disease are being studied in different animal models. For example the use of continuous releasing of neurotrophic factors like Glial Derived Neurotrophic Factor (GDNF) in a rat model of Parkinson's disease has shown to protect neurons from degenerative process [74]. Some other approaches combining the encapsulation of cells and neurotrophic factor releasing microspheres have been recently assayed, obtaining an amelioration of the asymmetric motor activity symptoms [75]. In the case of the treatment of Alzheimer disease (AD), implantation of PLGA microspheres secreting the Neuronal Growth Factor (NGF) promotes the survival of cholinergic neurons in a rodent model that consist in the transaction of septo-hippocampal pathway to induces similar symptoms to AD [76].

The disruption of BBB by injury in *in situ* delivery approaches can induce an immunological reaction that not only implies the particular CNS immune characteristics, but the interaction with periphery immune cells producing an extended inflammatory reaction; therefore, the development of materials for low invasive delivery, like low viscosity injectable hydrogels, has been one of the approaches [5]. Another recent and in expansion approach is the use of functionalized polymeric micro and nanoparticles capable to cross the BBB, offering an advantage to other invasive procedure; however some immunotoxicologycal non desired aspects has to be considered, like the induction of inflammatory response after phagocytosis by microglia and astrocytes, and the subsequent activation of innate inflammatory processes that could rend to uncontrol and extensive damage to the brain and neurodegeneration [77-80].

Different polymers application requires different biodegradation rates, some drug delivery vectors for example depends in the biodegradation rates to deliver the desired drug concentrations, while in the case of cell scaffolds, the time of polymeric degradation could be longer in order to maintain an appropriate environment for cell survival or for isolate them from immune surveillance.

Delivery of growth factors that improves neuronal regeneration and projection using biodegradable polymer such as poly(glycolic acid) (PGA), poly(lactic acid) (PLA), poly(lactic-co-glycolic acid) (PLGA) and poly(e-caprolactone) (PCL) has been extensively studied. These polymers degrade by non enzymatic hydrolytic cleavage of the backbone ester bond to alcohol and carboxylic acid, the latter of which catalyzes further degradation and the biodegradation rate can be controlled by the composition, molecular weight or size of the particles [81, 82]. The blend of polymers at different ratios modifies the biodegradation rates and hence their delivery profiles, for example, PGA derived microspheres with different blend ratio to delivery NGF showed different *in vitro* degradation rates: PLGA 50/50 degrade completely after 84 days of *in vitro* incubation with phosphate buffered saline (PBS) at 37°C, PLGA 85/15 degrade by 80%, and PCL by less than 30% under the same conditions; this profile of degradation coincides with the accumulative release of proteins, with the greatest accumulative release for PLGA 50/50; although the protein release profile is also influenced by the protein loading and the water soluble products exposed to surrounding medium in *in vitro* assays [81].

Besides biodegradation, it is important to consider the biodegradation products generated, the biodegradation of PGA derived polymers leads to the release of carboxylic acid groups that in vitro can be buffered by PBS; however, after implantation in the brain, acidification of local environment by biodegradation products can induce inflammatory response [81]. Inflammatory response induced by biodegradation products has been also studied by analyzing the microglia and astrocytes response of triblocks of lactic acid-b-PEG-b-lactic acid copolymers with two different degradable rates; it has shown an increase in reactive astrocytes in fast degrading hydrogels as compared with low degrading rate gels. This result could be due to the presence of more lactid acid as a product of degradation and a lactid acidosis in the implanted area of fast degrading gel; however, in the case of activated microglia, there were found a higher amount in slow and non degrade hydrogels [23]. In other report of the same research group, long lasting studies shown less microglia and astrocytes activation when the polymer has completely degraded, indicating that long term inflammatory response decrease with fast degrading polymer [22].

The tissue damage induced by biodegradation products could have severe deleterious effects in the CNS as reported for poly (methylidene malonate 2.1.2) (PMM 2.1.2); implanted microspheres of PMM 2.1.2 in rats striatum did not exert an important inflammatory response 1 or

2 months after implantation, and only a transient and mild characteristic foreign body reaction was observed. After 6 months post implantation however, when more degraded microspheres were present, an important microglia and iNOS positive cells infiltrate where observed in the implanted site together with extensive tissue damage. The extensive brain damage could be due to the inflammatory stimulus exerted by acidic polyanions generated during degradation process of PMM 2.1.2, and by the recruitment of activated macrophages and T cells with the subsequent oxidative and enzyme release that contributes to the damage [83].

A delayed inflammatory response after biodegradation of PEG-Silica nanocomposite gel implanted in the rat striatum was also observed [84]. PEG-Silica nanocomposite is suitable for *in vitro* delivery of proteins involve in the stimulation of dopaminergic neurons, without cell toxicity and no inflammatory response at short term in vivo implantation periods; however after 30 days after implantation an increase in activated macrophages and gliosis could be observed. *In vitro* assays using scattered Raman Spectroscopy after PEG-Silica incubation in simulated body fluid, a solution with ions similar to extracellular body fluids, showed an increase in PEG degradation and no modification of silica particles. PEG reduce protein absorption on the surface of silica particles and therefore reduces their immunogenicity [85]; however after PEG degradation, silica nanoparticles could be exposed to the immune system cells inducing an accumulation of activated macrophages and a glial reaction around the implantation site [84].

Biocompatibility and biodegradation studies of the polymeric materials in the CNS requires hence special considerations, the high vulnerability of neurons to long lasting chronic inflammatory reactions, oxidative environments, and biodegradation products, make the use of biodegradable polymeric material in CNS a challenge to avoid important deleterious effects in this tissue [86, 87]; further, delayed immune response in the brain could rend to late inflammation and therefore long term screening of CNS implanted polymers must be performed.

5. On Designed polymers for control biodegradation rates in biomedical applications

The design of specific characteristic to modulate degradation has many biological applications. The degradation of materials by non enzymatic hydrolysis is hard to regulate and only polymers blends and copolymerization can modulate the biodegradation rates. Other approaches to modulate biodegradation are modifications of chemical properties like crystallinity, hydrophobicity, and of surface characteristics like porosity [31].

The design of biomaterials for tissue engineering like cell scaffolds has to consider the mechanical properties and an appropriate architecture to allow the growth, proliferation or differentiation of cells; the modulation of biodegradation rate also contributes to the gradual incorporation of extracellular matrix (ECM) components and cellular in growth [88]. It the other hand, the design of cell responding delivery vectors for sequential delivery of different growth factors are interesting approaches to exert an specific tissue activity, that could lead to a more biological control of drug delivery [89]. A more integrative approach hence is proposed to manipulate biodegradation rates by cells interactions.

On designed polymers can modulate the degradation rate by enzymatic and oxidative processes, and a vast numbers of reports can be found in the literature about this topic; in this chapter we would mention some of the relevant strategies of the on purpose biodegradation polymers and their future on biomedical applications.

The design of enzymatic biodegradable polymeric biomaterials has been explored in the last years and is mainly intended to approach to the biological mechanisms that regulate the functions of ECM. ECM is composed for many different types of proteins and has an instructive role for cells; their biochemical and mechanical properties changes during development and regeneration, and is specific for each tissue. Inspired in the structural and biochemical characteristics of ECM, and by their mechanism of degradation during biological process, it has been developed several polymers that mimic their dynamic characteristics. In vivo release of growth factors (GF) is linked to ECM degradation, since usually GF are interacting with ECM components as a way to control their activity, as well as for inducing a site specific action; for example, by protecting them from enzymatic degradation, or by optimizing their biological activity improving their interaction with their receptors [90, 91]. The control of polymer biodegradation by enzymatic cleavage has been suggested for release of drugs and growth factors in sequence and on demand by the cells of the polymer implantation site, or by the cells introduced inside it. Among the most common enzymes used for on design biodegradable polymer are the matrix metalloproteinases (MMP) enzyme family. MMP are zinc-dependent proteins involved in degradation of ECM during several cell activities like migration, proliferation, adhesion, apoptosis and host defense; and involved in different tissue process like immune reaction, wound healing, morphogenesis, tumor progression and angiogenesis [92].

Among the most common MMP used in biodegradable polymer design are elastase, plasmin and collagenase.

The formation of new blood vessels requires of several growth factors like Vascular Endothelial Growth Factor (VEGF), Fibroblast Growth Factor (FGF), and Transforming Growth Factor (TGF). Blood vessels formation is an important process during wound repair, cell transplantation therapies, and for tissue engineering. Passive delivery of VEGF could lead to a low efficacy because of a rapid clear from the site, and an over or insufficient dosage. Hubbell and co workers pioneered the development of a metalloproteinase (MMP) sensitive PEG hydrogel to deliver VEGF in response to cells, showing that cell demanded release of VEGF supports the endothelial cells growth and the formation of completely remodeled vascularized tissue at the site of the hydrogel implantation [93].

The enhancing of MMP proteolytic degradation of polymeric materials can also improve cell differentiation and function; therefore their use as cell scaffold seems to be a promising approach. Conjugating peptides containing multiple sequences of MMP proteolytic cleavage to poly (ethylene glycol) diacrylate (PEGDA) hydrogels can control degradation rate and the amount of endothelial cells invasion; hydrogels with peptides containing three different proteolytic cleavage sites degraded faster and improved the amount and profundity of vessel invasion of human umbilical vein endothelial cells and human umbilical artery smooth muscle cells, as compared with hydrogels conjugated to peptides with only one proteolytic cleavage site [94].

The release of proteins combined with controlled biodegradable hydrogels can influence the differentiation of stem cells to specific cell types, for example, using MMP sensitive hyaluronic acid hydrogels containing bone morphogenetic protein (BMP) and mesenchymal stem cells, induce almost a completely coverage of rat calvarian bone defect as compared with MMP insensitive gel, showing the importance of cell responding scaffold to stimulate tissue regeneration [95]. The use of PEG hydrogels linked to MMP-1 or collagenase sensitive peptides has also shown a better differentiation of pre-adypocites cells, enhancing the triglyceride accumulation and the formation of adipose tissue like structures in this hydrogel intended for soft tissue augmentation uses [96].

An important aspect during CNS regeneration is the axonal outgrowth of regenerating neurons. After an injury the formation of glial scar and the presence of inhibitory proteins in the ECM impair the projection of neurons to regenerate the damaged connections. MMP play an important role in neuronal outgrowth during morphogenesis and regeneration by modulating the release of guiding factor or by degrading the inhibitory ECM proteins in the scar [97, 98]. In PEGylated fibrinogen hydrogels for example the presence of MMP3 inhibitor impairs the axon outgrowth of dorsal root ganglion neurons; therefore MMP secretion is an important step in the elongation of neurons inside the biomaterial [99]. The projection of neurons can be improved by tuning the degradation rates of PEG-PLA hydrogels [100].

Improve of degradation by oxidative process is another interesting approach. The design of biomaterials responsive to high oxidative environments can be used to modulate release of drug for immunomodulation in pathological conditions like artheriosclerosis, implant rejection sites or vaccines. For example the design of an ABA block copolymeric amphiphiles with PEG as A block hydrophilic polymer and a B block hydrophobic poly(propylene sulfide)-(PPS), exhibited hydrophobic-to hydrophilic changes when is oxidized in the presence of H_2O_2, a behavior that can be used for controlled release applications like vaccine nanoparticles [101, 102]. In a similar approach an oligo(proline)- crosslinked to a PEG, poly(ε-caprolactone), and poly(carboxyl-ε-caprolactone) terpolymer system to form a polymeric scaffold were synthesized, showing and increased biodegradation rate after expose to H_2O_2 or to activated macrophages [103].

6. Conclusions

Manipulation of biodegradation process is fundamental not only to modulate the duration of a material inside the body but to modulate biocompatibility, drug release, and cell invasion. Mimic the dynamic and remarkably important extracellular environment by biomaterials is another characteristic that could improved their functionality and biocompatibility. The interaction among biomaterials and proteins or cells is relevant for functions like drug delivery or cell proliferation and differentiation, and is one of the ongoing challenge that promise the development of specific and bio-responsive materials. Design of synthetic polymers that fully integrate the knowledge accumulated from chemistry, material engineering and biological disciplines like cell biology, biochemistry, and biophysics are the last paradigm in biomaterials. In the particular case of biodegradation of polymeric biomaterials, the profound knowledge about the chemical, physical and the biological mechanisms will render a more comprehensive on purpose biomaterials design.

Acknowledgements

E. Tamariz acknowledges the financial support of Mexican National Council for Science and Technology, CONACYT (82482), and the Public Education Council, PROMEP (UV-PTC-631). A. Rios-Ramirez was supported by CONACYT (82482) scholarship.

Author details

Elisa Tamariz[1] and Ariadna Rios-Ramírez[2]

*Address all correspondence to: etamariz@uv.mx; elisatammx@gmail.com

Biomedical Department, Health Science Institute, Veracruzana University, Veracruz, México

2 Neurobiology Institute, National University of México, Querétaro, México

References

[1] Park, J. B, & Lakes, R. S. Biomaterials An Introduction. New York: Springer; (2007).

[2] NIHClinical applications of biomaterials: NIH Consens Statement; (1982).

[3] Huebsch, N, & Mooney, D. J. Inspiration and application in the evolution of biomaterials. Nature (2009). Nov 26;, 462(7272), 426-432.

[4] Park, J. B, & Lakes, R. S. Polymeric implants materials. In: Park JB, Lakes RS, editors. Biomaterials An Introduction. New York: Springer; (2007). , 173-205.

[5] Straley, K. S, Foo, C. W, & Heilshorn, S. C. Biomaterial design strategies for the treatment of spinal cord injuries. J Neurotrauma (2010). Jan;, 27(1), 1-19.

[6] Hirano, S. Chitin biotechnology applications. Biotechnol Annu Rev (1996). , 2237-258.

[7]   Pakulska, M. M, Ballios, B. G, & Shoichet, M. S. Injectable hydrogels for central nerv-
      ous system therapy. Biomed Mater (2012). Apr;7(2)024101.

[8]   Aurand, E. R, Lampe, K. J, & Bjugstad, K. B. Defining and designing polymers and
      hydrogels for neural tissue engineering. Neurosci Res (2012). Mar;, 72(3), 199-213.

[9]   Slaughter, B. V, Khurshid, S. S, Fisher, O. Z, Khademhosseini, A, & Peppas, N. A.
      Hydrogels in regenerative medicine. Adv Mater (2009). Sep 4;21(32-33)3307-3329.

[10] Harkness, R. D. Biological functions of collagen. Biol Rev Camb Philos Soc (1961).
      Nov;, 36399-463.

[11] Lee, C. H, Singla, A, & Lee, Y. Biomedical applications of collagen. Int J Pharm
      (2001). Jun 19;221(1-2)1-22.

[12] Khaing, Z. Z, & Schmidt, C. E. Advances in natural biomaterials for nerve tissue re-
      pair. Neurosci Lett (2012). Jun 25;, 519(2), 103-114.

[13] Bouhadir, K. H, Lee, K. Y, Alsberg, E, Damm, K. L, Anderson, K. W, & Mooney, D. J.
      Degradation of partially oxidized alginate and its potential application for tissue en-
      gineering. Biotechnol Prog (2001). Sep-Oct;, 17(5), 945-950.

[14] Rowley, J. A, & Mooney, D. J. Alginate type and RGD density control myoblast phe-
      notype. J Biomed Mater Res (2002). May;, 60(2), 217-223.

[15] Noble, P. W. Hyaluronan and its catabolic products in tissue injury and repair. Ma-
      trix Biol (2002). Jan;, 21(1), 25-29.

[16] Bajaj, G, Van Alstine, W. G, & Yeo, Y. Zwitterionic chitosan derivative, a new bio-
      compatible pharmaceutical excipient, prevents endotoxin-mediated cytokine release.
      PLoS One (2012). e30899.

[17] Nomura, H, Tator, C. H, & Shoichet, M. S. Bioengineered strategies for spinal cord
      repair. J Neurotrauma (2006). Mar-Apr;23(3-4)496-507.

[18] Bain, M. K, Bhowmick, B, Maity, D, Mondal, D, Mollick, M. M, Rana, D, et al. Syner-
      gistic effect of salt mixture on the gelation temperature and morphology of methyl-
      cellulose hydrogel. Int J Biol Macromol (2012). Dec;, 51(5), 831-836.

[19] Tate, M. C, Shear, D. A, Hoffman, S. W, & Stein, D. G. LaPlaca MC. Biocompatibility
      of methylcellulose-based constructs designed for intracerebral gelation following ex-
      perimental traumatic brain injury. Biomaterials (2001). May;, 22(10), 1113-1123.

[20] Chu, C. editor. Biodegradable Polymeric Biomaterials: An Updated Overview. The
      Biomedical Engineering Handbook(2000).

[21] Aubert-pouessel, A, Venier-julienne, M. C, Clavreul, A, Sergent, M, Jollivet, C, Mon-
      tero-menei, C. N, et al. In vitro study of GDNF release from biodegradable PLGA mi-
      crospheres. J Control Release (2004). Mar 24;, 95(3), 463-475.

[22] Bjugstad, K. B, Lampe, K, Kern, D. S, & Mahoney, M. Biocompatibility of poly(ethylene glycol)-based hydrogels in the brain: an analysis of the glial response across space and time. J Biomed Mater Res A (2010). Oct;, 95(1), 79-91.

[23] Bjugstad, K. B, & Redmond, D. E. Jr., Lampe KJ, Kern DS, Sladek JR, Jr., Mahoney MJ. Biocompatibility of PEG-based hydrogels in primate brain. Cell Transplant (2008). , 17(4), 409-415.

[24] Bryant, S. J, Cuy, J. L, Hauch, K. D, & Ratner, B. D. Photo-patterning of porous hydrogels for tissue engineering. Biomaterials (2007). Jul;, 28(19), 2978-2986.

[25] Chirila, T. V. An overview of the development of artificial corneas with porous skirts and the use of PHEMA for such an application. Biomaterials (2001). Dec;, 22(24), 3311-3317.

[26] Patenaude, M, & Hoare, T. Injectable, mixed natural-synthetic polymer hydrogels with modular properties. Biomacromolecules (2012). Feb 13;, 13(2), 369-378.

[27] Park, J. B, & Lakes, R. S. Tissue Engineering materials and regeneration. In: Park JB, Lakes RS, editors. Biomaterials An Introduction New York: Springer; (2007). , 485-515.

[28] Heller, J, Barr, J, Ng, S. Y, Abdellauoi, K. S, & Gurny, R. Poly(ortho esters): synthesis, characterization, properties and uses. Adv Drug Deliv Rev (2002). Oct 16;, 54(7), 1015-1039.

[29] Rodriguez-galan, A, Franco, L, & Puiggali, J. Biodegradable Poly(ester amide) s: Synthesis and Application. In: Felton GP, editor. Biodegradable Polymers: Processing, Degradation and Application Nova Science Publishers, Inc.; (2011). , 207-272.

[30] Schmedlen, R. H, Masters, K. S, & West, J. L. Photocrosslinkable polyvinyl alcohol hydrogels that can be modified with cell adhesion peptides for use in tissue engineering. Biomaterials (2002). Nov;, 23(22), 4325-4332.

[31] Azevedo, H. S, & Reis, R. L. Undestanding the enzymatic degradation of biodegradable polymers and strategies to control their degradation rate. In: Reis R, San Roman J, editors. biodegradable systems in tissue engineering and regenerative medicine. Boca Ratón Florida: CRC Press; (2004).

[32] Yu, L, Zhang, Z, Zhang, H, & Ding, J. Biodegradability and biocompatibility of thermoreversible hydrogels formed from mixing a sol and a precipitate of block copolymers in water. Biomacromolecules (2010). Aug 9;, 11(8), 2169-2178.

[33] Zhong, S, & Williams, D. Are free radicals involved in biodegradation of polymeric medical devices in situ Advanced Materials (1991).

[34] Santerre, J. P, Woodhouse, K, Laroche, G, & Labow, R. S. Understanding the biodegradation of polyurethanes: from classical implants to tissue engineering materials. Biomaterials (2005). Dec;, 26(35), 7457-7470.

[35] Santerre, J. P, Labow, R. S, Duguay, D. G, Erfle, D, & Adams, G. A. Biodegradation evaluation of polyether and polyester-urethanes with oxidative and hydrolytic enzymes. J Biomed Mater Res (1994). Oct;, 28(10), 1187-1199.

[36] Jiang, H. L, Tang, G. P, Weng, L. H, & Zhu, K. J. In vivo degradation and biocompatibility of a new class of alternate poly(ester-anhydrides) based on aliphatic and aromatic diacids. J Biomater Sci Polym Ed (2001). , 12(12), 1281-1292.

[37] Paul, P. W. editor. Fundamental Immunology. Philadelphia, PA, U. S. A: Lippincott Williams & Wilkins; (2008).

[38] Iwasaki, A, & Medzhitov, R. Regulation of adaptive immunity by the innate immune system. Science (2010). Jan 15;, 327(5963), 291-295.

[39] Jenney, C. R, & Anderson, J. M. Adsorbed serum proteins responsible for surface dependent human macrophage behavior. J Biomed Mater Res (2000). Mar 15;, 49(4), 435-447.

[40] Keselowsky, B. G, Bridges, A. W, Burns, K. L, Tate, C. C, & Babensee, J. E. LaPlaca MC, et al. Role of plasma fibronectin in the foreign body response to biomaterials. Biomaterials (2007). Sep;, 28(25), 3626-3631.

[41] Wilson, C. J, Clegg, R. E, Leavesley, D. I, & Pearcy, M. J. Mediation of biomaterial-cell interactions by adsorbed proteins: a review. Tissue Eng (2005). Jan-Feb;11(1-2)1-18.

[42] Aderem, A, & Underhill, D. M. Mechanisms of phagocytosis in macrophages. Annu Rev Immunol (1999). , 17593-623.

[43] Plourde, N. M, Kortagere, S, Welsh, W, & Moghe, P. V. Structure-activity relations of nanolipoblockers with the atherogenic domain of human macrophage scavenger receptor A. Biomacromolecules (2009). Jun 8;, 10(6), 1381-1391.

[44] Anderson, J. M, Rodriguez, A, & Chang, D. T. Foreign body reaction to biomaterials. Semin Immunol (2008). Apr;, 20(2), 86-100.

[45] Salacinski, H. J, Tiwari, A, Hamilton, G, & Seifalian, A. M. Cellular engineering of vascular bypass grafts: role of chemical coatings for enhancing endothelial cell attachment. Med Biol Eng Comput (2001). Nov;, 39(6), 609-618.

[46] Henson, P. M. The immunologic release of constituents from neutrophil leukocytes. II. Mechanisms of release during phagocytosis, and adherence to nonphagocytosable surfaces. J Immunol (1971). Dec;, 107(6), 1547-1557.

[47] Henson, P. M. Mechanisms of exocytosis in phagocytic inflammatory cells. Parke-Davis Award Lecture. Am J Pathol (1980). Dec;, 101(3), 494-511.

[48] Kou, P. M, & Babensee, J. E. Macrophage and dendritic cell phenotypic diversity in the context of biomaterials. J Biomed Mater Res A (2011). Jan;, 96(1), 239-260.

[49] Malik, A. F, Hoque, R, Ouyang, X, Ghani, A, Hong, E, Khan, K, et al. Inflammasome components Asc and caspase-1 mediate biomaterial-induced inflammation and foreign body response. Proc Natl Acad Sci U S A (2011). Dec 13;, 108(50), 20095-20100.

[50] Brodbeck, W. G, Shive, M. S, Colton, E, Nakayama, Y, Matsuda, T, & Anderson, J. M. Influence of biomaterial surface chemistry on the apoptosis of adherent cells. J Biomed Mater Res (2001). Jun 15;, 55(4), 661-668.

[51] Mcnally, A. K, Jones, J. A, Macewan, S. R, Colton, E, & Anderson, J. M. Vitronectin is a critical protein adhesion substrate for IL-4-induced foreign body giant cell formation. J Biomed Mater Res A (2008). Aug;, 86(2), 535-543.

[52] Bryers, J. D, Giachelli, C. M, & Ratner, B. D. Engineering biomaterials to integrate and heal: the biocompatibility paradigm shifts. Biotechnol Bioeng (2012). Aug;, 109(8), 1898-1911.

[53] Gordon, S. Alternative activation of macrophages. Nat Rev Immunol (2003). Jan;, 3(1), 23-35.

[54] Mosser, D. M, & Edwards, J. P. Exploring the full spectrum of macrophage activation. Nat Rev Immunol (2008). Dec;, 8(12), 958-969.

[55] Labow, R. S, Meek, E, & Santerre, J. P. Synthesis of cholesterol esterase by monocyte-derived macrophages: a potential role in the biodegradation of poly(urethane)s. J Biomater Appl (1999). Jan;, 13(3), 187-205.

[56] Mcbane, J. E, Matheson, L. A, Sharifpoor, S, Santerre, J. P, & Labow, R. S. Effect of polyurethane chemistry and protein coating on monocyte differentiation towards a wound healing phenotype macrophage. Biomaterials (2009). Oct;, 30(29), 5497-5504.

[57] Bretz, U, & Baggiolini, M. Biochemical and morphological characterization of azurophil and specific granules of human neutrophilic polymorphonuclear leukocytes. J Cell Biol (1974). Oct;, 63(1), 251-269.

[58] Swirski, F. K, & Nahrendorf, M. Leukocyte behavior in atherosclerosis, myocardial infarction, and heart failure. Science (2013). Jan 11;, 339(6116), 161-166.

[59] Xu, X, & Hakansson, L. Degranulation of primary and secondary granules in adherent human neutrophils. Scand J Immunol (2002). Feb;, 55(2), 178-188.

[60] Heiple, J. M, Wright, S. D, Allen, N. S, & Silverstein, S. C. Macrophages form circular zones of very close apposition to IgG-coated surfaces. Cell Motil Cytoskeleton (1990). , 15(4), 260-270.

[61] Labro, M. T. Interference of antibacterial agents with phagocyte functions: immunomodulation or "immuno-fairy tales"? Clin Microbiol Rev (2000). Oct;, 13(4), 615-650.

[62] Murray, P. J, & Wynn, T. A. Protective and pathogenic functions of macrophage subsets. Nat Rev Immunol (2011). Nov;, 11(11), 723-737.

[63] Galea, I, Bechmann, I, & Perry, V. H. What is immune privilege (not)? Trends Immunol (2007). Jan;, 28(1), 12-18.

[64] Weller, R. O, Engelhardt, B, & Phillips, M. J. Lymphocyte targeting of the central nervous system: a review of afferent and efferent CNS-immune pathways. Brain Pathol (1996). Jul;, 6(3), 275-288.

[65] Wraith, D. C, & Nicholson, L. B. The adaptive immune system in diseases of the central nervous system. J Clin Invest (2012). Apr 2;, 122(4), 1172-1179.

[66] Fournier, E, Passirani, C, Montero-menei, C. N, & Benoit, J. P. Biocompatibility of implantable synthetic polymeric drug carriers: focus on brain biocompatibility. Biomaterials (2003). Aug;, 24(19), 3311-3331.

[67] Hanisch, U. K, & Kettenmann, H. Microglia: active sensor and versatile effector cells in the normal and pathologic brain. Nat Neurosci (2007). Nov;, 10(11), 1387-1394.

[68] Ransohoff, R. M, & Brown, M. A. Innate immunity in the central nervous system. J Clin Invest (2012). Apr 2;, 122(4), 1164-1171.

[69] Agostino, D, Gottfried-blackmore, P. M, Anandasabapathy, A, & Bulloch, N. K. Brain dendritic cells: biology and pathology. Acta Neuropathol (2012). Nov;, 124(5), 599-614.

[70] Guillemin, G. J, & Brew, B. J. Microglia, macrophages, perivascular macrophages, and pericytes: a review of function and identification. J Leukoc Biol (2004). Mar;, 75(3), 388-397.

[71] Pratt, B. M, & Mcpherson, J. M. TGF-beta in the central nervous system: potential roles in ischemic injury and neurodegenerative diseases. Cytokine Growth Factor Rev (1997). Dec;, 8(4), 267-292.

[72] Popovic, N, & Brundin, P. Therapeutic potential of controlled drug delivery systems in neurodegenerative diseases. Int J Pharm (2006). May 18;, 314(2), 120-126.

[73] Zhong, Y, & Bellamkonda, R. V. Biomaterials for the central nervous system. J R Soc Interface (2008). Sep 6;, 5(26), 957-975.

[74] Gouhier, C, Chalon, S, Aubert-pouessel, A, Venier-julienne, M. C, Jollivet, C, Benoit, J. P, et al. Protection of dopaminergic nigrostriatal afferents by GDNF delivered by microspheres in a rodent model of Parkinson's disease. Synapse (2002). Jun 1;, 44(3), 124-131.

[75] Delcroix, G. J, Garbayo, E, Sindji, L, Thomas, O, Vanpouille-box, C, Schiller, P. C, et al. The therapeutic potential of human multipotent mesenchymal stromal cells combined with pharmacologically active microcarriers transplanted in hemi-parkinsonian rats. Biomaterials (2011). Feb;, 32(6), 1560-1573.

[76] Pean, J. M, Menei, P, Morel, O, Montero-menei, C. N, & Benoit, J. P. Intraseptal implantation of NGF-releasing microspheres promote the survival of axotomized cholinergic neurons. Biomaterials (2000). Oct;, 21(20), 2097-2101.

[77] Gagliardi, M, Bardi, G, & Bifone, A. Polymeric nanocarriers for controlled and enhanced delivery of therapeutic agents to the CNS. Ther Deliv (2012). Jul;, 3(7),
875-887.

[78] Graber, D. J, Snyder-keller, A, Lawrence, D. A, & Turner, J. N. Neurodegeneration by
activated microglia across a nanofiltration membrane. J Biochem Mol Toxicol (2012).
Feb;, 26(2), 45-53.

[79] Minami, S. S, Sun, B, Popat, K, Kauppinen, T, Pleiss, M, Zhou, Y, et al. Selective targeting of microglia by quantum dots. J Neuroinflammation (2012).

[80] Sharma, H. S, & Sharma, A. Nanoparticles aggravate heat stress induced cognitive
deficits, blood-brain barrier disruption, edema formation and brain pathology. Prog
Brain Res (2007). , 162245-273.

[81] Cao, X, & Schoichet, M. S. Delivering neuroactive molecules from biodegradable microspheres for application in central nervous system disorders. Biomaterials (1999).
Feb;, 20(4), 329-339.

[82] Porjazoska, A, Goracinova, K, Mladenovska, K, Glavas, M, Simonovska, M, Janjevic,
E. I, et al. Poly(lactide-co-glycolide) microparticles as systems for controlled release
of proteins-- preparation and characterization. Acta Pharm (2004). Sep;, 54(3),
215-229.

[83] Fournier, E, Passirani, C, Colin, N, Sagodira, S, Menei, P, Benoit, J. P, et al. The brain
tissue response to biodegradable poly(methylidene malonate 2.1.2)-based microspheres in the rat. Biomaterials (2006). Oct;, 27(28), 4963-4974.

[84] Tamariz, E, Wan, A. C, Pek, Y. S, Giordano, M, Hernandez-padron, G, Varela-echa-
varria, A, et al. Delivery of chemotropic proteins and improvement of dopaminergic
neuron outgrowth through a thixotropic hybrid nano-gel. J Mater Sci Mater Med
(2011). Sep;, 22(9), 2097-2109.

[85] Xu, H, Yan, F, Monson, E. E, & Kopelman, R. Room-temperature preparation and
characterization of poly (ethylene glycol)-coated silica nanoparticles for biomedical
applications. J Biomed Mater Res A (2003). Sep 15;, 66(4), 870-879.

[86] Leach, J. B, Achyuta, A. K, & Murthy, S. K. Bridging the Divide between Neuropros-
thetic Design, Tissue Engineering and Neurobiology. Front Neuroeng (2010).

[87] Mcconnell, G. C, Rees, H. D, Levey, A. I, Gutekunst, C. A, Gross, R. E, & Bellamkonda, R. V. Implanted neural electrodes cause chronic, local inflammation that is correlated with local neurodegeneration. J Neural Eng (2009). Oct;6(5)056003.

[88] Lutolf, M. P. Biomaterials: Spotlight on hydrogels. Nat Mater (2009). Jun;, 8(6), 451-453.

[89] Lee, K, Silva, E. A, & Mooney, D. J. Growth factor delivery-based tissue engineering: general approaches and a review of recent developments. J R Soc Interface (2011). Feb 6;, 8(55), 153-170.

[90] Ramirez, F, & Rifkin, D. B. Cell signaling events: a view from the matrix. Matrix Biol (2003). Apr;, 22(2), 101-107.

[91] Schultz, G. S, & Wysocki, A. Interactions between extracellular matrix and growth factors in wound healing. Wound Repair Regen (2009). Mar-Apr;, 17(2), 153-162.

[92] Vu, T. H, & Werb, Z. Matrix metalloproteinases: effectors of development and normal physiology. Genes Dev (2000). Sep 1;, 14(17), 2123-2133.

[93] Zisch, A. H, Lutolf, M. P, Ehrbar, M, Raeber, G. P, Rizzi, S. C, Davies, N, et al. Cell-demanded release of VEGF from synthetic, biointeractive cell ingrowth matrices for vascularized tissue growth. FASEB J (2003). Dec;, 17(15), 2260-2262.

[94] Sokic, S, & Papavasiliou, G. Controlled proteolytic cleavage site presentation in biomimetic PEGDA hydrogels enhances neovascularization in vitro. Tissue Eng Part A (2012). Dec;18(23-24)2477-2486.

[95] Kim, J, Kim, I. S, Cho, T. H, Kim, H. C, Yoon, S. J, Choi, J, et al. In vivo evaluation of MMP sensitive high-molecular weight HA-based hydrogels for bone tissue engineering. J Biomed Mater Res A (2010). Dec 1;, 95(3), 673-681.

[96] Brandl, F. P, Seitz, A. K, Tessmar, J. K, Blunk, T, & Gopferich, A. M. Enzymatically degradable poly(ethylene glycol) based hydrogels for adipose tissue engineering. Biomaterials (2010). May;, 31(14), 3957-3966.

[97] Bai, G, & Pfaff, S. L. Protease regulation: the Yin and Yang of neural development and disease. Neuron (2011). Oct 6;, 72(1), 9-21.

[98] Zuo, J, Ferguson, T. A, Hernandez, Y. J, Stetler-stevenson, W. G, & Muir, D. Neuronal matrix metalloproteinase-2 degrades and inactivates a neurite-inhibiting chondroitin sulfate proteoglycan. J Neurosci (1998). Jul 15;, 18(14), 5203-5211.

[99] Sarig-nadir, O, & Seliktar, D. The role of matrix metalloproteinases in regulating neuronal and nonneuronal cell invasion into PEGylated fibrinogen hydrogels. Biomaterials (2010). Sep;, 31(25), 6411-6416.

[100] Mahoney, M. J, & Anseth, K. S. Three-dimensional growth and function of neural tissue in degradable polyethylene glycol hydrogels. Biomaterials (2006). Apr;, 27(10), 2265-2274.

[101] Napoli, A, Valentini, M, Tirelli, N, Muller, M, & Hubbell, J. A. Oxidation-responsive polymeric vesicles. Nat Mater (2004). Mar;, 3(3), 183-189.

[102] Thomas, S. N, Van Der Vlies, A. J, Neil, O, Reddy, C. P, Yu, S. T, & Giorgio, S. S. TD, et al. Engineering complement activation on polypropylene sulfide vaccine nanoparticles. Biomaterials (2011). Mar;, 32(8), 2194-2203.

[103] Yu, S. S, Koblin, R. L, Zachman, A. L, Perrien, D. S, Hofmeister, L. H, Giorgio, T. D, et al. Physiologically relevant oxidative degradation of oligo(proline) cross-linked polymeric scaffolds. Biomacromolecules (2011). Dec 12;, 12(12), 4357-4366.

Chapter 2

Aerobic Biodegradation of Surfactants

Encarnación Jurado, Mercedes Fernández-Serrano,
Francisco Ríos and Manuela Lechuga

Additional information is available at the end of the chapter

1. Introduction

Surfactants are a wide group of chemical compounds which have a large number of applications due to their solubility properties, detergency, endurance of water hardness, as well as emulsifying, dispersing, and wetting properties. Surfactants have a characteristic structure, with one or several hydrocarbon chains that form the lipophilic part of the molecule (or the hydrophobic part of the molecule) and one or several polar groups that form the hydrophilic part. These compounds, also called surface-active agents, can have different lengths and degrees of unsaturation in the hydrocarbon chains, as well as in the polar groups, giving rise to a wide variety of surfactants with different properties.

Surfactants can be classified as ionic or non-ionic, depending on the nature of the hydrophilic group. The ionic surfactants are disassociated in water, forming ions. Notable within this group are organic acids, and their salts are anionic surfactants, while bases—amines of different degrees of replacement— and their salts are cationic surfactants. Some surfactants contain both acid and basic groups. These surfactants may be anionic or cationic and are therefore called amphoteric, or ampholytic.

Surfactants constitute a group of substances in which the main characteristic is their accumulation in the interfaces, solid-liquid or liquid-gas, weakening the surface tension of the liquid. This property enables the formation of foams and the penetration of solids as a wetting agent, leading to wide and varied applications of these compounds [1].

These substances are widely used in household cleaning detergents, personal-care products, textiles, paints, polymers, pesticides, pharmaceuticals, mining, oil recovery, and the pulp and paper industries. Detergents and cosmetics involve the mayor use of these compounds. After use, residual surfactants and their degradation products are discharged to sewage-treatment plants or directly to surface waters. Several of these compounds are not biologically degradable and, depending on their concentration, may be harmful to fauna and flora in surface waters. Surfactants can also produce waste which can react with some water components and generate toxic products harmful to human health. For example, endometriosis or decreased sperm quality appear to be consequences (though unconfirmed) associated with the presence of surfactants in the environment.

Due to the enormous economic importance of surfactants and their contribution to the deterioration of the environment when these persist in nature, it is necessary to establish the structural characteristics that govern the susceptibility of these molecules to be degraded. The massive worldwide use of surfactants requires them to be as innocuous as possible for the environment, i.e.: low toxicity and easily biodegradability [2].

Balson and Felix [3] described biodegradation as the destruction of a chemical by the metabolic activity of microorganisms. Degradation of surfactants through microbial activity is the primary transformation occurring in the environment and an important process to treat surfactants in raw waste in sewage-treatment plants. During biodegradation, microorganisms can either utilize surfactants as substrates for energy and nutrients or co-metabolize surfactants by microbial metabolic reactions [4].

The biodegradation process of organic compounds is affected by many factors, the most important of which are the physiochemical characteristics of the compounds (solubility, concentration, structure, etc.), the physiochemical conditions of the environmental media (dissolved oxygen, temperature, pH, light, nutrient concentration, etc.) and the microorganisms present in the aquatic environment. Most surfactants can be degraded by microbes in the environment, although some surfactants may be persistent under anaerobic conditions [5]. Different types of surfactants have different degradation behaviour in the environment.

Biodegradation tests can be used to evaluate the primary and ultimate biodegradability of anionic and non-ionic surfactants. The comparison of different types of surfactants by various biodegradability tests will identify the least damaging to the aquatic media and will determine the influence of the surfactant structure on the biodegradation process. On this basis, simple methods can be chosen to evaluate the biodegradability because results depend on the biodegradation test used. Tests can also determine the choice for including them in detergent formulas, also taking into account their effectiveness in the wash.

A wide variety of surfactants are used in detergents formulas. A mix of several surfactants is selected to find the formulation more appropriate for each kind of soiling. This chapter examines the biodegradation of some anionic and non-ionic surfactants which have noteworthy properties for use in the detergent formulas, and which represent one of the major families of surfactants used today.

Non-ionic surfactants: *Fatty alcohol ethoxylates* (FAEs) represent the economically most important group of non-ionic surfactants. Commercial FAEs generally consist of a mixture of several homologues differing in alkyl chain length and degree of ethoxylation. FAEs are widely used in domestic and commercial detergents, household cleaners, and personal care products. Thus, the major route of disposal of FAEs is down the drain, through sewage systems, and into municipal sewage-treatment plants (WWTP) [6]. *Nonylphenol polyethoxylate* (NPEO), as a result of its field of application, its resistance to biodegradation at low temperatures, and the generation during the degradation process of some persistent metabolites which are much more toxic than the original compound [7], the use of NPEOs has been banned in domestic formulations in some countries of the European Union (Germany, Spain, and the United Kingdom, [8] as well as Switzerland and Canada [9]. The *alkylpolyglucosides* (APGs) belong to non-ionic surfactants of growing use. Because of their good foaming properties, as well as synergy with other surfactants, they have found application in dishwashing and laundry detergents, and in other cleaning products [10]. Also, their good skin tolerance makes them suitable for mild personal-care products [11]. They are prepared on the basis of renewable raw materials, namely (starch/sugar) and fatty alcohols (vegetable oils). As these chemicals belong to a new type of surfactant, few studies have addressed their environmental properties [12]. *Amine-oxide-based surfactants* constitute a particular type of non-ionic surfactants, they are classified as nitrogen non-ionic surfactants, exhibit cationic behaviour in acid solution, and can be ionized depending on the pH of the test medium. They show good foaming properties and are skin compatible [13]. These compounds, the consumption of which only in Westeren Europe is estimated at 14 ktonnes/year [14] are widely used in detergents, toiletries, and antistatic preparations, usually together with other surfactants. They are compatible with anionic surfactants and can be used to give synergistic advantage to formulations [15] and [16].

Anionic surfactants: *Linear-chain alkylbenzenesulfonate* types are the most popularly used synthetic anionic surfactants. They have been extensively used for over 30 years with an estimated global use of 2.8 million tonnes in 1998 [17]. There has been an emphasis over the past few years on the development of surfactants and builders with improved biodegradability and also non-polluting characteristics [18]. This growing concern has led to the development and use of other surfactants, such as *ether carboxylic derivative surfactants*. These anionic surfactants improve the foaming quality of the detergent, reducing the irritation level, and therefore they are used as co-surfactants in detergents which have to be in contact with the skin [19] [20]. These surfactants are marketed in concentrated acid form. For these surfactants, aerobic biodegradation has been studied employing standarized methods which use micro-organism to degrade the surfactant. The results enable us to analyse the behaviour of the surfactant in the environment or in the sewage-treatment plants, and then evaluate their biodegradability to evaluate the suitability of including them in detergent formulas.

2. Materials and methods

2.1. Non-ionic surfactants

The non-ionic surfactants used were: Fatty-alcohol ethoxylates (FAEs) and amine-oxide-based surfactants supplied by Kao Corporation S.A. (Tokyo, Japan), nonylphenol polyethoxylate (degree of ethoxylation 9.5) supplied by Tokyo Chemical Industry (Tokyo, Japan) and the alkylpolyglucosides from Henkel-Cognis (Dusseldorf, Germany) supplied by Sigma.

The FAEs used in this study were: FAE-$R_{10}E_3$, FAE-$R_{10}E_6$, FAE-$R_{12-14}E_4$, FAE-$R_{12-14}E_{11}$, FAE-$R_{16-18}E_6$, and FAE-$R_{16-18}E_{11}$. The alkylpolyglucosides were Glucopone 650 EC (APG-$R_{8-14}DP_{1.35}$),

Glucopone 600 CS UP (APG-$R_{12-14}DP_{1.59}$), and Glucopone 215 CS UP (APG-$R_{8-10}DP_{1.42}$). The amine-oxide-based surfactants used in this study were AO-R_{14}, AO-R_{12}, and AO-Cocoamido.

2.2. Anionic surfactants

The anionic surfactants tested were: linear alkyl benzene sulphonate and ether carboxylic derivative surfactants EC-R_8E_8, EC-$R_{12-14}E_3$, EC-$R_{12-14}E_{10}$, EC-R_8E_5, EC-$R_{6-8}E_{3-8}$, EC-$R_{4-8}E_{1-8}$, supplied by Kao Corporation S.A. (Tokyo, Japan).

Table 1 shows the structure of the surfactants and the abbreviations used in this study.

Family	Surfactant	Structure	Abbreviation
Non-ionic	Fatty-alcohol ethoxylate	$R(-O-CH_2-CH_2)_E-OH$	FAE-R_VE_Z
	Nonylphenol polyethoxylate		NPEO
	Alkylpolyglucoside		APG
	Amine oxide	 (AO-Cocoamido (AO-R_{14} & AO-R_{12}))	AO-R_V
Anionic	Linear alkyl benzene sulphonate	$CH_3(CH_2)_5CH(CH_2)_4CH_3SO_3 \cdot Na^+$	LAS
	Ether carboxylic derivative	$R-O(CH_2-CH_2O)_E-CH_2-COO-X$	EC-R_VE_Z

R: alkyl chain length. E: degree of ethoxylation. X= H^+ or Na. DP: average number of glucose units per alkyl radical

Table 1. Chemical structure and abbreviation of the surfactants used in the tests

2.3. Biodegradation tests

Screening test: The test was conducted according to the OECD 301 E test for ready biodegradability [21]. A solution of the surfactant, representing the sole carbon source for the microorganisms, was tested in a mineral medium, inoculated and incubated under aerobic conditions in the dark at 25ºC for 21 days. The procedure consists of placing 1.2 liters of surfactant solution in a 2-liter Erlenmeyer flask and inoculating the solution with 0.5 mL of water from a secondary treatment of a sewage-treatment plant (STP) that operates with active sludge. The Erlenmeyer flask was plugged with a cotton stopper and left in darkness in a thermostatically controlled chamber at 25ºC. The constant rotary speed of the orbital shaker (125 sweep/min) provided the necessary aeration. The surfactant solution was prepared by dissolving the desired quantity of surfactant in the nutrient solution.

The primary biodegradation was monitored by means of the residual-surfactant concentration over time using colorimetric methods in which the absorbance is directly proportional to the surfactant concentration. For the absorbance measurements, a double-beam spectrophotom-eter, VARIAN Cary 100 Bio, was used. The fatty-alcohol ethoxylates and the nonylphenol polyethoxylate were determined by the iodine-iodide colorimetric method [22]: 0.25 mL of iodine-iodide reagent was added on 10 mL of the test sample, after stirring and maintaining for 5 minutes at room temperature, the absorbance was measured against air at 500 nm in the spectrophotometer. The alkylpolyglucosides were quantified by a modification of the anthrone method proposed by Buschmann and Wodarczak [23]: 5 ml of solution of 0.8 % (w/w) anthrone in concentrated sulfuric acid was dropped into 2 mL of degradation liquor. The mixture was hydrolyzed for 5 min in boiling water and then quickly cooled in cold water for 10 min. The absorbance of this mixture at 622 nm was determined by spectrophotometer. The linear alkyl benzene sulphonate was quantified by a simplified spectrophotometric method for determin-ing anionic surfactants, based on the formation of the ionic-pair anionic surfactant-methylene blue [24]: 5 mL of sample in 10-mL glass vials were made alkaline to pH 10.0 by adding of 200 µL 50 mM sodium tetraborate, pH 10.5, and then 100 µL of methylene blue 1 g/L stabilized were added. Finally, 4 mL of chloroform was added and, after stirring and 5-min wait, the absorbance at 650nm was measured against air or against a blank with chloroform. The biodegradation was calculated according the following equation:

$$Biodegradation\ \% = \frac{[s]_i - [s]_t}{[s]_i} \cdot 100 \tag{1}$$

Where $[S]_i$ is the initial surfactant concentration and $[S]_t$ is the surfactant concentration at each time.

The biodegradation process for the amine oxides and ether carboxylic derivates was monitored by measuring the residual-surfactant concentration over time (21 days) by dissolved organic carbon (DOC) measurements. In the TOC-analyser used, the organic compounds were first oxidized to carbon dioxide, and then the CO_2 released was measured quantitatively by an IR-detector. The oxidation method was high-temperature catalytic oxidation. The Shimazdu VCSH/CSH TOC analyser equipped with an auto-sampler was used. Samples were filtered through a 0.45-µm polyvinylidene fluoride filter (Millipore S.A.) prior to TOC analysis. The biodegradation was calculated according the following equation:

$$Biodegradation\ \% = \frac{(DOC_i - DOC_t)}{DOC_i} \cdot 100 \tag{2}$$

Where DOC_i is the initial DOC concentration and DOC_t is the DOC concentration measured at each time.

Confirmatory Test: The test was performed according to the OECD 301 E test for ready biodegradability [21]. This test is used for surfactants which have failed in the screening test to confirm or reject the results. It consists of inoculating a small amount of microorganisms, from a secondary effluent-treatment plant which works preferably with domestic wastewater. The biodegradation process was performed in a small activated sludge plant at laboratory scale, where synthetic wastewater was used with a surfactant concentration of 10 mg/L at flow rate of 1 L/h. The test was run at room temperature (18-25°C).

Chemical oxygen demand (COD), and dissolved organic carbon (DOC) were measured daily to determine the biodegradation efficiency.

$$COD \ reduction \ \% = \frac{(COD_i - COD_t)}{COD_i} \cdot 100 \tag{3}$$

Where COD_i is the initial COD and COD_t is the COD measured at each time.

$$Mineralization \ \% = \frac{(DOC_i - DOC_t)}{DOC_i} \cdot 100 \tag{4}$$

Where DOC_i is the initial DOC concentration and DOC_t is the DOC concentration measured at each time.

Respirometry Test: The test was made using the system Oxitop Control® (WTW, Weilheim, Germany), which determines the manometric changes that occur when oxygen is consumed to transform the surfactant into CO_2 by the microorganisms inoculated (from a mixed population and aerated) in a mixture formed by the nutrient solution and the surfactant. The Oxitop system offers an individual number of reactors consisting of glass bottles (510 nominal volume) with a carbon dioxide trap (sodium hydroxide) in the headspace. The volume of the test mixture is usually 164 mL. The bottles were furnished with a magnetic stirrer and sealed with a cap containing an electronic pressure indicator. An incubator box was used to maintain the respirometer units at constant temperature (25°C) during a test run. The decrease in headspace pressure in the closed test vessel was continuously recorded and the biochemical oxygen demand (BOD) was calculated according the following equation:

$$DBO = \frac{M(O_2)}{R \cdot T} \cdot \left(\frac{V_{Total} - V_{Liquid}}{V_{Total}} + \alpha \cdot \frac{T_{25}}{T_0} \right) \cdot \Delta p \left(O_2 \right) \tag{5}$$

Where $M(O_2)$ is the molecular weight of oxygen (32g/mol), R is the gas constant (83.144 mbar/(molK)), T_0 is the temperature at 0 °C (273.15 K), T_{25} is the incubation temperature, 25°C (298.15 K), V_{Total} is the total volume in the test vessel, V_{Liquid} is the volume of the test mixture, α is the Bunsen absorption coefficient (0.03103) and $\Delta p(O_2)$ is the difference of the partial pressure of oxygen (mbar). Biodegradation of the test compound was calculated from the measured DBO as a percentage of its theoretical oxygen demand (ThOD).

Pseudomonas putida **biodegradation test:** A monoculture strain *P. putida* CECT 324, provided by the Spanish Type Culture Collection (Valencia, Spain), was used in the biodegradation test. Erlenmeyer flasks were filled with the surfactant solution, enriched with an inorganic medium and with a trace mineral solution [25] [26]. The pH was adjusted to 7.0 and the flasks were inoculated with bacterial stock of *P. putida*. Flasks were incubated at 30°C on a rotary platform shaker for 72 h. At the beginning and after 72 h, a sample of each flask was filtered and used to determine the dissolved organic carbon (DOC). Biodegradation efficiency (E_f) was evaluated as a percentage by:

$$E_f\% = \frac{\left[DOC_i - (DOC_f - DOC_m)\right]}{DOC_i} \cdot 100 \tag{6}$$

Where DOC_i is the initial DOC concentration, DOC_f is the DOC concentration measured at the end of the incubation (72 h) and DOC_m is the minimum concentration that cannot be metabolized by the bacteria [26].

3. Results and discussions

3.1. Screening test

The screening test was applied to the amine-oxide-based surfactants AO-R_{14}, AO-R_{12}, AO-Cocoamido, to the ether carboxylic derivative surfactants EC-R_8E_5, EC-$R_{6-8}E_{3-8}$, EC-R_8E_8, EC-$R_{12-14}E_3$, EC-$R_{12-14}E_{10}$, to the fatty-alcohol ethoxylates FAE-$R_{10}E_3$, FAE-$R_{10}E_6$, FAE-$R_{12-14}E_4$, FAE-$R_{12-14}E_{11}$, FAE-$R_{16-18}E_6$, FAE- $R_{16-18}E_{11}$, to the alkylpolyglucosides APG-$R_{8-14}DP_{1.35}$, APG-$R_{12-14}DP_{1.59}$, APG-$R_{8-10}DP_{1.42}$, to the nonylphenol polyethoxylate and linear alkyl benzene sulphonate.

Results for ether carboxylic derivative, fatty-alcohol ethoxylates, alkylpolyglucosides, nonylphenol polyethoxylate and linear alkyl benzene sulphonate, show that the biodegradability is influenced by the initial concentration of surfactant; that is, the degree of biodegradation achieved is higher when the initial concentration of surfactant is lower. Lower concentrations, 15mg/L and 25 mg/L, result in a percentage of biodegradation close to or above 90%. Current legislation requires a minimum level of biodegradation of over 80% for surfactants to be considered biodegradable, when the OECD test is applied. For the amine-oxide-based surfactants the effect of the concentration is the opposite, the biodegradation is higher when the initial concentration is higher.

Figure 1 shows the influence of the concentration for one example of each family of surfactants tested.

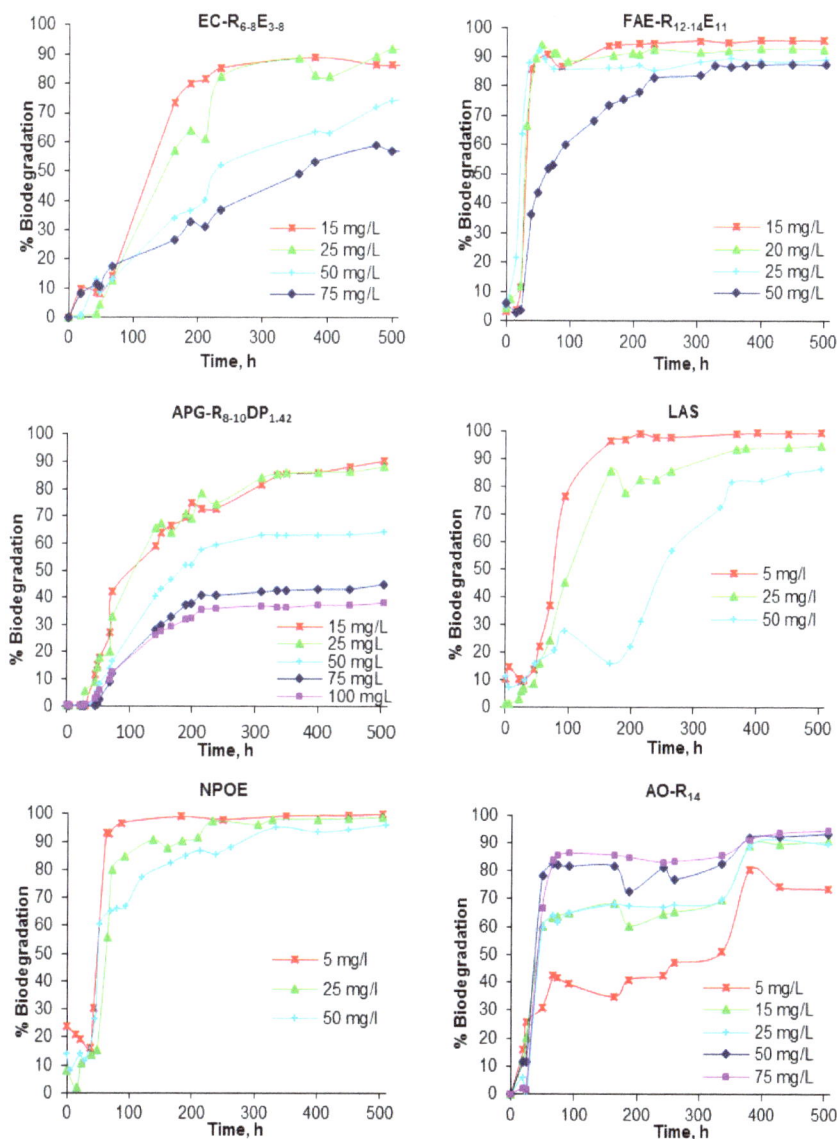

Figure 1. Screening-test results for the surfactants. Influence of the initial concentration.

The degrees of biodegradation achieved for linear alkyl benzene sulphonate and nonylphenol polyethoxylate are among the highest (Figure 1), but NPOE reportedly produces toxic byproducts [27] which can be harmful to human health. This surfactant has been withdrawn in most European countries and North America [28].

An analysis of the screening-test results indicates that for all tested concentrations of fatty-alcohol ethoxylates, there was a preferential surfactant biodegradation of the surfactant with longer alkyl chain and higher degree of ethoxylation. Figure 2 shows the comparison of three fatty-alcohol ethoxylates with different alkyl length and degree of ethoxylation. The results for ether carboxylic derivate surfactants show that the biodegradability was higher for the surfactants with shorter alkyl chains. For the surfactants with the same chain length, biodegradability is higher for those with higher degrees of ethoxylation (Figure 2). For amine-oxide-based surfactants, the results indicate that AO-Cocoamido is less biodegradable than AO-R_{12} and AO-R_{14}, the AO-R_{14} (with the longest alkyl chain) being the most biodegradable amine oxide tested (Figure 2). These surfactants can be considered readily biodegradable, according to García et al. [13], because amine-oxide-based surfactants are rapidly and easily converted into carbon dioxide, water, and biomass under aerobic conditions.

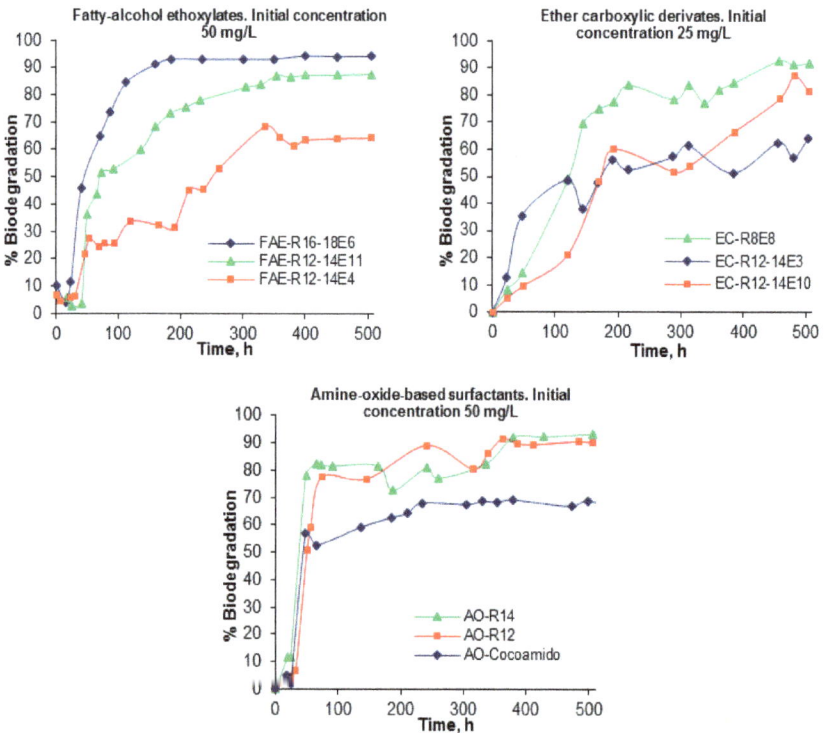

Figure 2. Screening-test results for the surfactants. Influence of the alkyl chain length and the degree of ethoxylation.

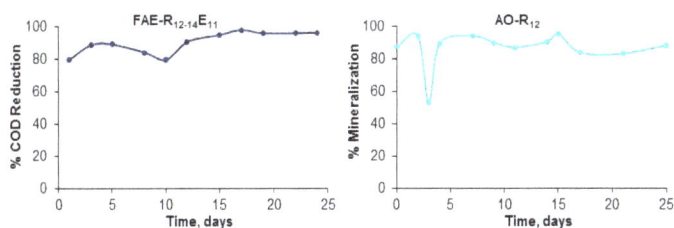

Figure 3. Confirmatory test results.

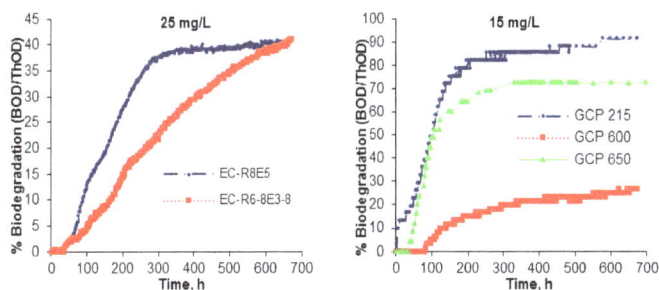

Figure 4. Respirometry test results. Influence of the alkyl chain length, degree of ethoxylation, and average number of glucose units.

3.2. Confirmatory test

After the screening test, a confirmatory test was performed over a period of 21 days for the amine-oxide-based surfactants and two fatty-alcohol ethoxylates. The results show that the surfactants tested can be considered easily biodegradable because, after a few days from the start, biodegradation exceeded 90% and remained steady for 21 days. Figure 3 shows one example for each family of surfactants tested. In case of fatty-alcohol ethoxylates, FAE-$R_{12-14}E_{11}$, the evolution is shown for the COD reduction (Eq. 3) between the synthetic wastewater in the feedtank and the treated water at the outlet. For the amine oxide, AO-R_{12}, Figure 3 shows the evolution of the mineralization achieved at the outlet, calculated on the basis of the DOC, (Eq. 4).

3.3. Respirometry test

The respirometry test was applied for the ether carboxylic derivative surfactants, for the alkylpolyglucosides and for the fatty-alcohol ethoxylates.

In this test, unlike the screening, the biodegradation of ether carboxylic derivatives was not higher for the surfactant with a shorter alkyl chain (Figure 4). However, the surfactant with the highest degree of ethoxylation was the most biodegradable, as in the screening test.

In the case of the alkylpolyglucosides, the comparison of the biodegradability between the three surfactants tested depended on the initial concentration. Thus, for low concentrations, 15 mg/L, 25 mg/L, and 50 mg/L, the most biodegradable was the APG-$R_{8-10}DP_{1.42}$, with the shorter alkyl chain and a middle number of glucose units. However, instead, for higher concentrations, 75 mg/L and 100 mg/L, the most biodegradable alkylpolyglucoside was the APG-$R_{8-14}DP_{1.35,}$ which had the lowest number of glucose units and with a medium-length alkyl chain (data not shown). Notably, for all the concentrations tested, the biodegradability proved lowest for the surfactant with the longest alkyl chain and greatest number of glucose units (APG-$R_{12-14}DP_{1.59}$).

Analysing the influence of the initial concentration of surfactant, (Figure 5), for the ether carboxylic derivate surfactants, as in the screening method, the results show that the biodegradability was higher when the initial concentration was lower. For the fatty-alcohol ethoxylates tested, FAE-$R_{16-18}E_6$, the biodegradation was lower at the highest initial concentration, but the lowest initial concentration did not give the highest percentage of biodegradation, as was expected.

The influence of the initial concentration on the biodegradation process was the same for the three alkylpolyglucosides tested. The biodegradability was higher when the initial concentration of alkylpolyglucoside was lower (Figure 5).

3.4. *Pseudomonas putida* biodegradation test

The *P. putida* biodegradation test was applied for the amine-oxide-based surfactants at different initial surfactant concentrations. This test did not provide a comparable biodegradation value with the other biodegradation tests, but the results can be used to compare the surfactants in order to make decisions concerning their use in the surfactant formulations. The results show that the surfactant AO-Cocoamido was the most biodegradable amine-oxide-based surfactant tested; this surfactant is different from the others because incorporates an amino group in the alkyl chain, which probably increases the hydrophilic character of the surfactant [13]. For the amine oxides with the same structure but different alkyl-chain lengths, AO-R_{12} and AO-R_{14}, the biodegradability was similar, although the AO-R_{12}, with a shorter alkyl chain, was slightly more biodegradable. Figure 6 shows the results for 30 mg/L of initial concentration.

The biodegradation process is influenced by the initial concentration, with the biodegradation efficiency (Eq. 6) being higher when the initial concentration is lower. This trend was found for the three amine-oxide-based surfactants. Figure 7 shows the biodegradation efficiency at different values of initial concentration for the AO-$R_{12.}$

3.5. Biodegradation parameters

Biodegradation profiles resulting from the screening test as well as from the respirometry test allowed us to determine the kinetics of the biodegradation process, this being to evaluate the persistence of surfactants and to assess the risks of exposure to humans, animals, and plants.

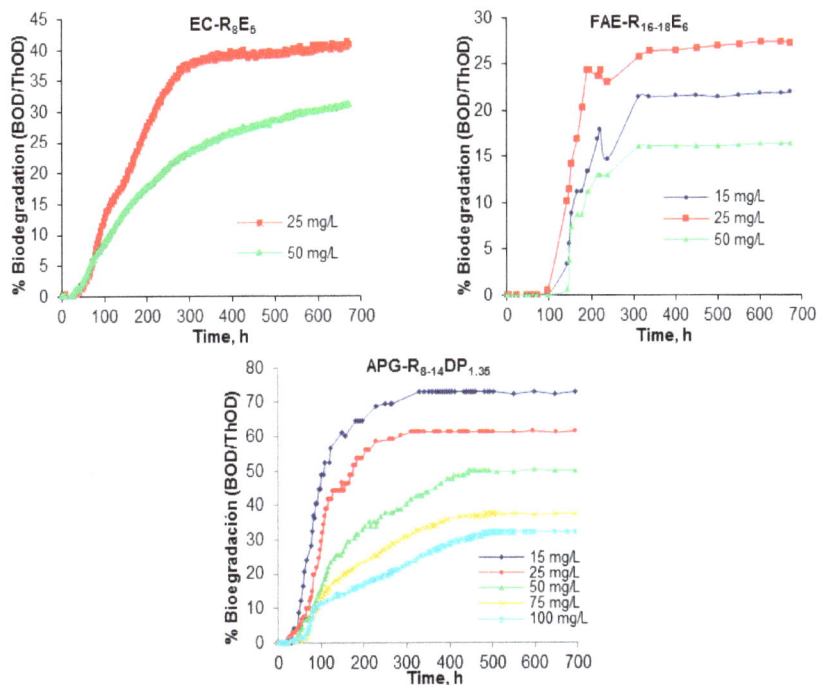

Figure 5. Respirometry test results. Influence of the initial concentration.

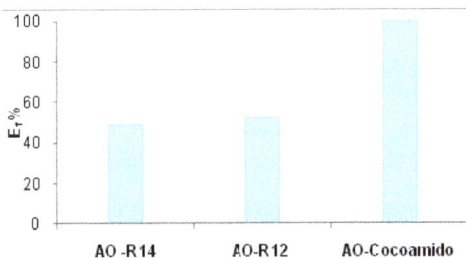

Figure 6. *P. putida* biodegradation test at 30 mg/L to amine-oxide-based surfactants

This is also useful for the design of industrial plants and equipment needed to eliminate these products.

Using the profiles of the biodegradation process, we can define and evaluate some character-istic parameters for the comparison and quantification of the biodegradation assays [29]. In this study two were selected:

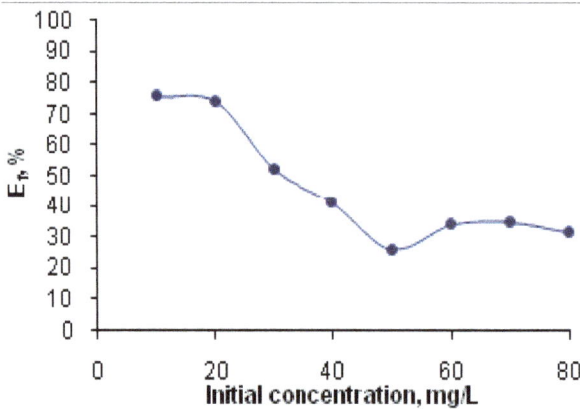

Figure 7. *P. putida* biodegradation test for the amine-oxide-based surfactant AO-R$_{12}$. Effect of the concentration.

- *Latency time* (t_L) is the time needed for the non-adapted microorganisms to acclimatize themselves to the new substrate. The latency or acclimation period prior to the biodegradation process of organic compounds in the aquatic environment can have several causes, such as a lack of nutrients, enzymatic induction, predation by protozoa, mutation of species, growth of a microbial population capable of metabolising the substrate, or simply the adaptation to the presence of toxic agents. This time corresponds to the period during which a mild change occurs in the residual concentration. For the screening test, it was calculated by drawing two tangents to the adaptation and biodegradation stages. The latency term was the cut-off point of both straight lines. For the respirometry test, it was calculated as the time necessary to achieve 10% biodegradation.

- *Mean biodegradation rate* (V_M) has been defined as the mean velocity of biodegradation reached until achieving 50% biodegradation of the surfactant and it has been calculated as the quotient between the percentage of biodegradation reached and the time needed to reach this biodegradation value. This parameter provides the speed of the biodegradation process.

Figure 8 shows the latency times obtained for the surfactants tested at the initial concentration of 25 mg/L, for the screening test and the respirometry test.

The latency times obtained show that the behaviour of the microorganism varies considerably depending on the surfactant tested, as well as the test used. According to the result, the non-adapted microorganisms need more time to acclimatize when the surfactant tested is the APG-R$_{12-14}$DP$_{1.59}$ in case of the screening test, and the EC-R$_{6-8}$E$_{3-8}$ when the respirometry test is employed.

The mean biodegradation rate was also evaluated in cases where possible, Figure 9 shows the mean biodegradation rate for the surfactants tested at the initial concentration of 25 mg/L, in the screening test and the respirometry test.

The mean biodegradation rate V_M, enabled us to compare the biodegradation processes of the surfactants. According to the results (Figure 9), the mean biodegradation rate was higher for the fatty-alcohol ethoxylates while the carboxylic derivative surfactants showed the slowest mean biodegradation rate in case of the screening test. In case of the respirometry test, the carboxylic derivative surfactants registered the best values and the alkylpoliglucosides the worst.

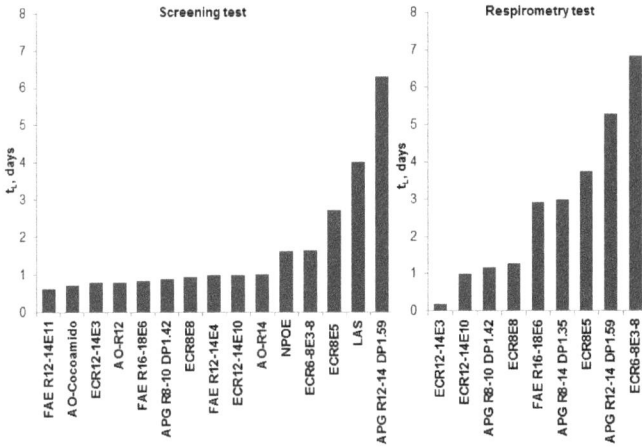

Figure 8. Latency time at the initial concentration of 25 mg/L.

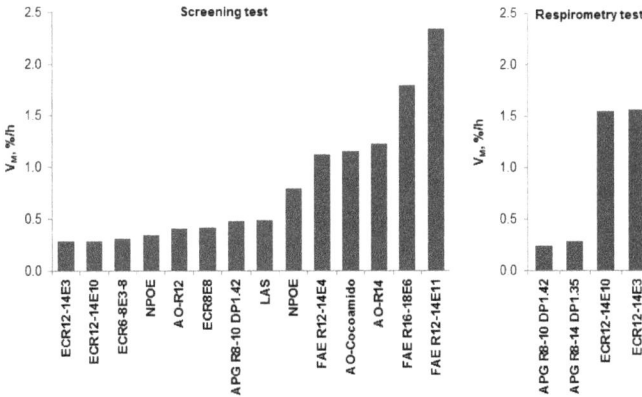

Figure 9. Mean biodegradation rate at the initial concentration of 25 mg/L.

3.6. Comparison of the biodegradation test used

The results of biodegradation and the biodegradation parameters for each family of surfactants tested may vary significantly depending on the biodegradation test used. It is therefore useful to determine the method which is the best to perform the aerobic biodegradation test of surfactants. For this, it is important to ascertain which method most accurately represents the actual conditions in the environment where the surfactants are dumped and the advantages and disadvantages when the method is applied.

Table 2 shows the advantages and disadvantages of the different biodegradation tests used in this study.

Biodegradation Test	Advantages	Disadvantages
Screening	Easy to prepare and to carry out Easy analysis of the results	Problems in the analysis of low surfactant concentrations Each inoculums differs Duration: 21 days, representing many measures.
Confirmatory	Similar conditions to a	Bulking sludge. This causes solid loss by flotation Each sludge for each test is different Duration: more than one month because of the adaptation phase of the sludge This requires much time and experimentation
Respirometry	Easy to prepare and to carry out Automatic tracking	
P. putida	Fast, 72h Bacteria *P. putida* are commonly present in activated sludge. Reproducibility Involve a definite living system	Complications to measure the DOC for low concentrations, but high concentrations cannot be tested because they can cause inhibition. The bacterial stock provides DOC that interferes mainly at low concentrations. Possibility of contamination of the strain. Need to work in sterile conditions

Table 2. Advantages and disadvantages of the aerobic biodegradation tests.

Based on these considerations, the screening test and the respirometry test are the most reproducible and the easiest to perform, and they supply more information.

4. Conclusions

According to the analysis, the biodegradation results depend on the biodegradation test used, the microorganisms used in the test, and the family of the surfactants tested. An important aspect is the adaptation of the microorganisms to the type of surfactant used as a sole carbon source.

Taking into account the screening test results, we can demonstrate the influence of the surfactant structure on the biodegradability. Regarding the length alkyl chain, the effect depends on the family of surfactant: for the fatty-alcohol ethoxylates and amine-oxide-based surfactants the biodegrability is higher when the alkyl chain is longer, while, for the carboxylic derivative surfactants and alkylpolyglucosides, the opposite occurs.

With respect to the influence of the initial surfactant concentration, the importance that this parameter has on the biodegradability has been evidenced. For all the surfactants tested, the greater the initial concentration is, the lower the biodegradability is, except for the amine oxides for which the effect is otherwise.

For the surfactants analyzed in this study, the fatty-alcohol ethoxylates and especially the FAE-$R_{12-14}E_{11}$ can be considered the most biodegradable but the carboxylic derivative surfactants the least biodegradable, according to the mean biodegradation rate.

Nomenclature

AO: amine-oxide-based surfactant

APG: alkylpolyglucoside

BOD: biological oxygen demand

COD: chemical oxygen demand

COD_i: initial chemical oxygen demand

COD_t: chemical oxygen demand at time t

DOC: dissolved organic carbon

DOC_f: dissolved organic carbon at the end of the incubation

DOC_i: initial dissolved organic carbon

DOC_m: minimum dissolved organic carbon that cannot be metabolized

DOC_t: dissolved organic carbon at time t

DP: average number of glucose units per alkyl radical

E: degree of ethoxylation

EC: ether carboxylic derivate

E_f: biodegradation efficiency

FAE: fatty-alcohol ethoxylate

LAS: linear alkyl benzene sulphonate

NPEO: nonylphenol polyethoxylate

R: alkyl chain length

$[S]_i$= initial surfactant concentration

$[S]_t$= surfactant concentration at time t

ThOD: theoretical oxygen demand

t_L: latency time

TOC: total organic carbon

V_M: mean biodegradation rate

Author details

Encarnación Jurado, Mercedes Fernández-Serrano, Francisco Ríos and Manuela Lechuga

*Address all correspondence to: mferse@ugr.es

Department of Chemical Engineering, University of Granada, Granada, Spain

References

[1] Holmberg K. Handbook of Applied Surface and Colloid Chemistry. Chichester: Wiley; 2001.

[2] Jurado E, Fernández-Serrano M, Núñez-Olea J, Luzón G, Lechuga M. Acute toxicity and relationship between metabolites and ecotoxicity during the biodegradation process of non-ionic surfactants: Fatty-alcohol ethoxylates, nonylphenol polyethoxylate and alkylpolyglucosides. Water Sci Technol. 2009;59(12):2351-2358.

[3] Balson T, Felix MSB. The biodegradability of non-ionic surfactants. Karsa DR, Porter MR (eds.) Biodegradability of Surfactants. Blackie Academic and Professional; 1995. p204-230.

[4] Ying GG. Fate, behavior and effects of surfactants and their degradation products in the environment. Environ Int. 2006;32(3):417-431.

[5] Scott MJ, Jones MN. The biodegradation of surfactants in the environment. Biochim Biophys Acta. 2000:1508(1-2):235-251.

[6] Wind T, Stephenson RJ, Eadsforth CV, Sherren A, Toy R. Determination of the fate of alcohol ethoxylate homologues in a laboratory continuous activated-sludge unit study. Ecotoxicol Environ Saf. 2005;64(1):42-60.

[7] Maguire R. Review of the persistence of nonylphenol and nonylphenol ethoxylates in aquatic environments. Water Qual Res J Can. 1999;34(1):37-78.

[8] European Commission Directive 2003/53/EC. Restrictions on the marketing and use of certain dangerous substances and preparations (nonylphenol, nonylphenol ethoxylate and cement) Official Journal L 178, 17/07/2003 P. 0024-0027.

[9] Soares A, Murto M, Guieysse B. Biodegradation of nonylphenol in a continuous bioreactor at low temperatures and effects on the microbial population. Appl Microbiol Biot.2006;69(5): 597-606.

[10] Jurado E, Fernández-Serrano M, Núñez-Olea J, Lechuga M, Jiménez JL, Ríos F. Acute toxicity of alkylpolyglucosides to vibrio fischeri, daphnia magna and microalgae: A comparative study. Bull Environ Contam Toxicol. 2012;88(2):290-295.

[11] Kuhn AV, Neubert RHH. Characterization of mixtures of alkyl polyglycosides (Plantacare) by liquid chromatography-electrospray ionization quadrupole time-of-flight mass spectrometry. Pharm Res. 2004;21(12):2347-2352.

[12] Jurado E, Fernández-Serrano M, Núñez-Olea J, Lechuga MM., Jimenez-Pérez J, Rios-Ruiz F. Effect of concentration on the primary and ultimate biodegradation of alkylpolyglucosides in aerobic biodegradation tests. Water Environ Res. 2010;83(2):1-10.

[13] García MT, Campos E, Ribosa I. Biodegradability and ecotoxicity of amine oxide based surfactants. Chemosphere. 2007;69(10):1574-1578.

[14] Merrettig-Bruns U, Jelen E. Study on the anaerobic biodegradation of detergent Surfactants. Final Report. Fraunhofer Institute for Environmental, Safety and Energy Technology UMSICHT, Oberhausen. 2003.

[15] Cross J. Cationic Surfactants. In: Cross J, Singer J. (eds.). Analytical and Biological Evaluation, Surfactant Science Series. New York: Marcel Dekker; 1994. (53) p140-175.

[16] Domingo A. A guide to the surfactants World. Barcelona: Proa; 1995

[17] Ying GG. Fate, behavior and effects of surfactants and their degradation products in the environment. Environ Int. 2006;32(3):417-431.

[18] Yu Y, Zhao J, Bayly A. Development of surfactants and builders in detergent formulations. Chinese J Chem Eng. 2008;16(4):517-527.

[19] Jurado E, Fernández-Serrano M, Lechuga M, Ríos F. Environmental impact of ether carboxylic derivative surfactants. J Surfactants Deterg. 2012;15(1):1-7.

[20] Jurado E, Fernández-Serrano M, Núñez-Olea J, Lechuga M, Ríos F. Ecotoxicity of anionic surfactants AKYPO®. WIT Transactions on Ecology and the Environment 2011;144:497-505.

[21] OECD Organization for Economic Cooperation and Development. OECD Guidelines for the Testing of Chemicals, Volume 1, Section 3: Degradation and Accumulation, OECD, Paris. France. 1993.

[22] Jurado E, Fernández-Serrano M, Núñez-Olea J, Luzón G, Lechuga M. Comparison and use of methods for the determination of non-ionic surfactants in biodegradation processes.Tenside Surfact Det. 2002;39(5):154–159.

[23] Buschmann N, Kruse A, Wodarczak S. Analytical methods for alkylpolyglucosides. I: Colorimetric Determination. Tenside Surfact Det.1995;32(4): 336-339.

[24] Jurado E, Fernández-Serrano M, Núñez-Olea J, Luzón G, Lechuga M. (2006) Simplified spectrophotometric method using methylene blue for determining anionic surfactants: Applications to the study of primary biodegradation in aerobic screening tests. Chemosphere. 2006; 65(2): 278-285.

[25] Shim H, Shin E, Yang ST. A continuous fibrous-bed bioreactor for BTEX biodegradation by acoculture of Pseudomonas putida and Pseudomonas fluorescens. Adv Environ Res. 2002;7(1):203–216.

[26] Ballesteros Martín MM, Casas López JL, Oller I, Malato S, Sánchez Pérez JA. A comparative study of different tests for biodegradability enhancement determination during AOP treatment of recalcitrant toxic aqueous solutions. Ecotoxicol Environ Saf. 2010;73(6):1189–1195.

[27] Sherrard KB, Marriott PJ, McCormick MJ, Cotton R, Smith G. Electrospray mass spectrometric analysis and photocatalytic degradation of polyethoxylate surfactants used in wool scouring. Anal Chem. 1994;66(20):3394-3399.

[28] Directive 2003/53/EC of the european parliament and of the council of 18 June 2003 amending for the 26th time Council Directive 76/769/EEC relating to restrictions on the marketing and use of certain dangerous substances and preparations (nonylphenol, nonylphenol ethoxylate and cement).

[29] Jurado E, Fernández-Serrano M, Nuñez-Olea J, Luzón G, Lechuga M. Primary biodegradation of commercial fatty-alcohol ethoxylate surfactants: characteristic parameters. J. Surfactants Deterg. 2007;10(3):145-153.

Chapter 3

Biosurfactants: Production and Applications

R.S. Reis, G.J. Pacheco, A.G. Pereira and D.M.G. Freire

Additional information is available at the end of the chapter

1. Introduction

Great emphasis has recently been given to the environmental impacts caused by chemical surfactants due to their toxicity and difficulty in being degraded in the environment [1]. Increasing environmental concerns, the advance in biotechnology and the emergence of more stringent laws have led to biosurfactants being a potential alternative to the chemical surfactants available on the market [2, 3]. Although biosurfactants have promising use in bioremediation processes, their industrial scale production is currently difficult due to high raw-material costs, high processing costs and low manufacturing output [3]. As a result, the current research challenges are to increase the yield and to reduce the cost of raw materials [4].

The number of publications and patents involving biosurfactants has recently increased considerably [5]. Although many biosurfactants and their manufacturing processes have been patented, only some of them have been commercialized. EC-601 (EcoChem Organics Company), a dispersive agent of water-insoluble hydrocarbons containing rhamnolipids, and PD5 (Pendragon Holdings Ltd), an additive for fuels based on a mixture of rhamnolipid biosurfactants and enzymes, are examples of biosurfactant-based products commercially available [6]. Several studies have aimed to optimize the biosurfactant production process by changing the variables that influence the type and amount of biosurfactant produced by a microorganism. Important variables are carbon and nitrogen sources [7], potential nutrient limitations and other physical and chemical parameters such as oxygen [6], temperature and pH [4]. Recent studies have also focused on *in situ* production from renewable substrates, resulting in the so-called new generation of biosurfactants production [3], as well as metabolic engineering strategies and strain improvements to enhance the metabolic fluxes towards the product [4].

Among the biological surfactants, rhamnolipids reportedly have a good chance of being adopted by the industry as a new class of renewable resource-based surfactants [5]. Strain-engineering may be a promising strategy to improve manufacturing output, and the production by recombinant and non-pathogenic strains has been shown as possible and favorable [8]. Rhamnolipids have been the focus of many studies and are the better characterized biosurfactants in terms of production, metabolic pathways and gene regulation. Several bacterial species have been reported to produce the glycolipidic biosurfactants rhamnolipids [9]. In *Pseudomonas aeruginosa*, these biosurfactants are the products of the convergence of two metabolic pathways; the biosynthesis of dTDP-L-rhamnose – formation of lipopolysaccharide (LPS) – and the diversion of the β-hydroxydecanoyl-ACP intermediate from the FASII cycle by RhlA. The enzymes rhamnosyltransferases RhlB and RhlC catalyse the transfer of dTDP-L-rhamnose to either HAA, or to a previously generated mono-rhamnolipid, respectively [10]. High carbon to nitrogen ratio, exhaustion of nitrogen source, stress conditions and high cell densities are among the conditions that favor higher levels of production [11]. Rhamnolipids production in *P. aeruginosa* is tightly controlled by mulitple layers of gene regulation that respond to a wide variety of environmental and physiologic signals, and are capable of combining different signals to generate unique and specific responses [12].

Biosurfactants can potentially replace virtually any synthetic surfactant and, moreover, introduce some unique physico-chemical properties. Currently, the main application is for enhancement of oil recovery and hydrocarbon bioremediation due to their biodegradability and low critical micelle concentration (CMC) [13]. The use of biosurfactants has also been proposed for various industrial applications, such as in food additives [14], cosmetics, detergent formulations and in combinations with enzymes for wastewater treatment [13, 15].

In this chapter we intend to present an introduction of biosurfactants and their various applications with emphasis on bioremediation. Due to the relevance of rhamnolipids compared to other biosurfactants, their metabolic pathways and genetic regulation in *P. aeruginosa* will be revised. We additionally will discuss the critical factors and parameters for improved production of rhamnolipids.

2. Application of biosurfactants

Biosurfactants are potentially replacements for synthetic surfactants in several industrial processes, such as lubrication, wetting, softening, fixing dyes, making emulsions, stabilizing dispersions, foaming, preventing foaming, as well as in food, biomedical and pharmaceutical industry, and bioremediation of organic- or inorganic-contaminated sites. Glycolipids and lipopeptides are the most important biosurfactants (BS) for commercial purpose (Table 1). Shete et al. (2006) [16] mapped the patents on biosurfactants and bioemulsifiers (255 patents issued worldwide) showing high number of patents in the petroleum industry (33%), cosmetics (15%), antimicrobial agent and medicine (12%) and bioremediation (11%). Sophorolipids (24%), surfactin (13%) and rhamnolipids (12%) represent a large portion of the patents, however, this may be underestimated since many patents do not specify the producer organism restricting to the specific use of the BS only.

Biosurfactant class		Microorganism	Application
Glycolipids	Rhamnolipids	*P. aeruginosa and P. putida*	Bioremediation
		P. chlororaphis	Biocontrol agent
		Bacillus subtilis	Antifungal agent
		Renibacterium salmoninarum	Bioremediation
	Sophorolipids	*Candida bombicola and C. apicola*	Emulsifier, MEOR, alkane dissimilation
	Trehalose lipids	*Rhodococcus spp.*	Bioremediation
		Tsukamurella sp. and Arthrobacter sp.	Antimicrobial agent
	Mannosylerythritol lipids	*Candida antartica*	Neuroreceptor antagonist, antimicrobial agent
		Kurtzmanomyces sp	Biomedical application
Lipopeptides	Surfactin	*Bacillus subtilis*	Antimicrobial agent, biomedical application
	Lichenysin	*B. licheniformis*	Hemolytic and chelating agent

Table 1. Major types of biosurfactants.

Improvement of detection methods together with increased concerns with environmental issues are pushing researchers and policymakers towards more environmentally friendly solutions for waste management and replacements for non-biodegradable substances. Organic aqueous wastes (e.g., pesticides), organic liquids, oils (e.g., petroleum-based) and organic sludges or solids (e.g., paint-derived) are common environmental organic chemical hazards and are source of soil and aquatic contaminations that are normally difficult to be removed. Another commonly found environmental hazard are the heavy metals, such as lead, mercury, chromium, iron, cadmium and copper, which are also linked to activities of our modern society. The remediation of contaminated sites is usually performed via soil washing or in situ flushing, in case of soil contamination, and bioremediation or use of dispersants, in case of aquatic areas. Soil washing/flushing is heavily dependent on the solubility of the contaminants, which can be very challenging when dealing with poorly soluble hazards. Hydrophobic contaminants usually require use of detergents or dispersants, both in soil or aquatic environment, and the process is often followed by their biodegradation. Heavy metal, however, cannot be biodegraded and are converted to less toxic forms instead. Hence, the commonly found combination of inorganic and organic contamination demands a complex remediation process. High hydrophobicity and solid-water distribution ratios of some pollutants result in their interaction with non-aqueous phases and soil organic matter. Those interactions reduce dramatically the availability for microbial degradation, since bacteria preferentially degrade chemicals that are dissolved in water [17].

Bioremediation is a process that aims the detoxification and degradation of toxic pollutants through microbial assimilation or enzymatic transformation to less toxic compounds [18]. The success of this process relies on the availability of microbes, accessibility of contaminants and conduciveness of environment. A typical bioremediation process consists of application of nutrients (containing nitrogen and phosphorous), under controlled pH and water content, together with an emulsifier and surface-active agents. Biostimulation is the bioremediation based on the stimulation of naturally indigenous microbes by addition of nutrients directly to the impacted site, whereas bioaugmentation is based on addition of specific microbes and nutrients to the impacted site. Bioaugmentation has been subject of several reports including use of genetically engineered microorganisms (reviewed in Gentry et al., 2004 [19]). Biostimulation success relies on microorganism targeting the pollutant as a primarily food source, which is supported by available electron donors/acceptors and nutrients (reviewed in Smets & Pritchard, 2003 [20]).

The bioavailability of a chemical in general is governed by physical-chemical processes such as sorption and desorption, diffusion and dissolution. Microorganisms improve bioavailability of potential biodegradable nutrients by production of biosurfactants [21], and the success of microbes in colonize a nutrient-restricted environment is often related to their capacity of producing polymers with surfactant activity.

The best-studied biosurfactant are the glycolipids, which contain mono- or disaccharides linked to long-chain aliphatic acids or hydroxyaliphatic acids. Rhamnolipids are better known glycolipid class, which are normally produced as a mixture of congeners that varies in composition according to the bacterium strain and medium components, which provides specific properties to rhamnolipids derived from different isolates and production processes [7]. This class of biosurfactant has been implied in several potential applications such as in bioremediation, food industry, cosmetics and as an antimicrobial agent. Several reports have been shown rhamnolipids to be efficient in chelating and remove/wash heavy metals, perhaps due to the interaction between the polar glycosidic group with the metal ions. Whereas their interaction with organic compounds increases their bioavailability or aids their mobilization and removing in a washing treatment. Rhamnolipids have been shown to be effective in reducing oil concentration in contaminated sandy soil [22] and their addition at relatively low concentration (80 mg/L) to diesel/water system substantially increased biomass growth and diesel degradation [23]. Interestingly, rhamnolipids combined with a pool of enzyme produced by Penicillium simplicissimum enhanced the biodegradation of effluent with high fat content from poultry processing plant, suggesting a synergistic interaction between biosurfactant and enzymes in waste treatment [15].

2.1. Surfactants in removal of hydrophobic pollutants

Highly hydrophobic contaminants can bind very tightly to soil, therefore inaccessible to biodegradation. Surfactants have the potential to promote desorption of the contaminants from soil. Usually, 1-2% (w/w) of surfactant is used for washing contaminant soil, whereas in aqueous solution the concentration of surfactant can be as low as 10 times less than in soil. Rhamnolipids were effective in removing polycyclic aromatic hydrocarbons (PAHs) and pentachlorophenol from soil with 60-80% removal efficiency, which varied with contact time and biosurfactant concentration [24, 25]. Addition of rhamnolipid biosurfactants to phenanthrene-degrading bacteria (Pseudomonas strain R and isolate P5-2) increased phenanthrene mineralization in Fallsington sandy loam with high phenanthrene-sorption capacity, and

addition of this biosurfactant at concentrations above the CMC resulted in enhanced phenanthrene release from soil [26]. Interesting, phenanthrene pseudosolubilization was increased in
the presence of less hydrophobic PAHs [27]. The explanation for this cooperative effect could
be that less hydrophobic compounds were accommodated at the interfacial region of a
hydrophobic core, consequently, the interfacial tension between the core and water was
reduced, and the reduced interfacial tension may support a larger core volume for the same
interfacial energy [17]. Moreover, rhamnolipids were shown to enhance partitioning rate of
PAHs such as fluorene, phenanthrene and pyrene [24].

Mixture of different surfactants often presents better properties than the individual surfactants, due to synergistic effect. An improved strategy to surfactant-enhanced remediation
(SER) is to apply mixture of surfactants at reduced concentration of individual surfactants,
which reduces the cost while maintaining the efficiency of remediation [17]. Solubilization of
hydrophobic contaminants is improved by using mixtures of anionic and nonionic surfactants,
which was shown to exhibit synergistic interaction, suggesting that appropriate combinations
of surfactants have the potential to enhance the efficiency of soil washing and flushing and to
facilitate the bioavailability of pollutants [28, 29].

Trehalose lipids produced by *Rhodococcus erythropolis* were shown to have good solubilization
capacity for hydrophobic compounds such as phenanthrene, and great potential for applications in bioremediation of sites contaminated with PAHs [30]. PAHs sorption/desorption ratio
was reduced with combined treatment with chemical surfactant (Tween 80) and biosurfactants
under thermophilic conditions, and substantial amount of PAHs were desorbed from soil into
the aqueous phase when surfactant concentration was above CMC [31]. Combining Tween 80,
Triton X100 and biosurfactants from P. aeruginosa strains resulted in effective enhanced
solubility of phenanthrene at 50 ^{0}C compared with addition of the individual surfactants [32].
Interesting, degradation of phenanthrene was completely inhibited for all the surfactants
tested at concentration higher that their CMC, suggesting that the combination of surfactant
and biosurfactant has potential in bioremediation however it requires a research in a case-by-
case basis. The phenanthrene-degrading strain B-UM lacks the ability to produce surfactants
for dissolution of phenanthrene, and the direct adhesion of cells to phenanthrene surface might
be the major pathway for B-UM to take up this PAH [33]. As expected, addition of surfactants
inhibited the phenanthrene degradation by B-UM, and similar effect were found with addition
of Triton X100 to *Arthrobacter* sp. growing on n-hexadecane [34], suggesting that surfactant can
inhibit biodegradation of hydrocarbons by de-adhesion of cells from the liquid/solid-water
surfaces preventing cells to contacting the surface of phenanthrene, therefore, the degradation
is impaired. Kumar et al. (2006) [35] isolated and characterized a *P. putida* strain named IR1. It
was shown that this strain is capable of utilizing up to four-ring PAHs but not hexadecane and
octadecane as a sole carbon and energy source, and the authors identified the presence of both
tension-active and emulsifying activities, suggesting that IR1 produces biosurfactants on both
water miscible and immiscible substrates. Rhamnolipids together with anthracene-degrading
bacteria had a dramatic increase in the solubility of anthracene by the bacterium strains and,
interesting, it was observed the metabolism of biosurfactant by one of the strains which appears
to be an important event on this process [36]. Addition of Tween 60, nonionic surfactant, to a

Rhodococcus rhodochrous strain in liquid media was also shown to improve biodegradation of fluorene by being an additional carbon source to the bacterial cells [37]. Moreover, the combination of biostimulation and bioaugmentation was shown to result in significant removal of phenanthrene under Antarctic environmental conditions. The authors combined a complex organic source of nutrients (fish meal), a surfactant (Brij 700) and a psychrotolerant PAH-degrading bacterial consortium and it was shown that the combined treatment is more efficient than the biostimulation or bioaugmentation isolated [38].

2.2. Surfactants in petroleum industry

Indigenous or injected biosurfactant-producing microorganisms are exploited in oil recovery in oil-producing wells. Microbial enhanced oil recovery (MEOR) is often implemented by direct injection of nutrients with microbes that are able of producing desired products for mobilization of oil, by injection of a consortium or specific microorganisms or by injection of the purified microbial products (e.g., biosurfactants). These processes are followed by reservoir repressurization, interfacial reduction of tension/oil viscosity and selective plugging of the most permeable zones to move the additional oil to the producing wells. Oil recovery was showed to be increased by 30-200 % with injection of biosurfactants, bacteria (e.g., *P. aeruginosa*, *X. campestri*, *B. licheniformis*) and nutrients [18]. However, application of MEOR requires a thoroughly research on a case-by-case basis taking in account the physical-chemical conditions and soil and rock formation characteristics. The characteristics of the oil that has been already recovered from the well will also impact the MEOR application. MEOR is a powerful technique to recover oil, especially from reservoirs with low permeability or crude oil with high viscosity, but the uncertainties on the results and costs are a major barrier to its widespread.

Oilfield emulsions are formed at various stages of petroleum exploration, production and oil recovery and processing, and represent one of the major problems for the petroleum industry [39], which requires a de-emulsification process in order to recover oil from theses emulsions. Traditionally, de-emulsification is obtained by centrifugation, heat treatment, electrical treatment and/or chemicals [39]. Biosurfactants have the potential to replace the use of chemical de-emulsifiers in situ, saving on transport of the oil emulsion and providing a more environmentally-friendly solution. Among the bacteria species, Acinetobacter and Pseudomonas species are the main de-emulsifiers in the mixed cultures [40]. The microorganisms exploit the dual hydrophobic/hydrophilic nature of biosurfactants or hydrophobic cell surfaces to disrupt the emulsions. Glycolipids (e.g., rhamnolipids), glycoproteins, phospholipids and polysaccharides are among the microbial tools to displace the emulsifiers from the oil-water interface [4]. The major advantage of using microorganisms or their products over chemical products is the disposal of the de-emulsifier in the aqueous phase and its removal from the oil phase, since emulsion formation is required in further processing steps. Microbes and biosurfactants are in general readily biodegradable, which allows a cheap and easy removal of the de-emulsifier after this process.

Surfactants have potential application for oil recovery from petroleum tank bottom sludges and facilitating heavy crude transport through pipelines. It was shown that rhamnolipids can be used to remove soaked oil from the used oil sorbents [30]. More than 95% of oil removal was achieved using commercial rhamnolipids (JBR215, Jeneil Biosurfactant Company, USA), and the main factor affecting oil removal were the sorbent pore size and washing time.

2.3. Washing and bioremediation of crude oil-contaminated environments

Petroleum hydrocarbons are an essential raw material in our current society, but they also constitute a major environmental pollutant that is very difficult to be bioremediated. Crude oils have very low water solubility, high adsorption onto soil matrix and present limited rate of mass transfer for biodegradation [2]. Oil-contaminated soil is especially difficult for bioremediation since oil excess forms droplets or films on soil particles, which is a powerful barrier against microbial degradation [41]. Bioavailability of contaminants in soil to the metabolizing organisms is influenced by factors such as desorption, diffusion and dissolution. Biosurfactants are produced to decrease the tension at the hydrocarbon-water interface aiming to pseudosolubilize the hydrocarbons, thus increasing mobility, bioavailability and conse-quent biodegradation [42]. Several biosurfactant are produced by a diversity of microorgan-isms in order to survive in an oil-rich environment, and this adaptation process selected for surfactants with highly adaptable phycal-chemical properties. Biosurfactants are, therefore, very suitable for applications in the oil industry and this is reflected in the market, where the large majority of biosurfactants produced are in petroleum-related applications [21]. The applications are, in general, in oil recovery, oil spill management, MEOR and as oil dispersants and demulsifiers [18].

Purified rhamnolipid biosurfactants were applied in the removal of oil from contaminated sandy soil [22]. The authors optimized the biosurfactant and oil concentrations in the removal of oil applying statistical experimental design tool that generates a surface response. Sandy oil contained predominantly aromatic and paraffinic hydrocarbons (5-10% w/w) mixed with reduced concentration of rhamnolipid (6.3-7.9 g/L) resulted in removal of oil by up to 91 and 78%, respectively. Rhamnolipids, when added above CMC, enhanced the apparent aqueous solubility of hexadecane, the biodegradation of hexadecane, n-paraffins, octadecane, creosotes in soil and promoted biodegradation of petroleum sludges [43, 44]. Rhamnolipids produced by Nocardioides sp. A-8 allows this bacterium to grown on aromatic hydrocarbons or n-paraffin as sole carbon source by lowering the surface tension and emulsifying the aromatic hydrocarbons [45]. The authors found similar results for the strain Pantoea sp. A-13, which also produces rhamnolipid to grow on n-paraffin or kerosene. Both A-8 and A-13 strains were isolated together with other 15 aerobic microbial isolates from oil-contaminated sites in Antarctica and appear to be very promising source for application in accelerated environmen-tal bioremediation at low temperatures.

Urum and Pekdemir (2004) [41] applied different biosurfactants (rhamnolipid, aescin, saponin, lecithin and tannin) in washing oil-contaminated soil and observed significant removal of crude oil with different concentrations of biosurfactant solution. Oil mobilization was the main cause for its removal, which was triggered by the reduction of surface and interfacial tensions, rather than oil solubilization or emulsification. This work was followed by a comparison of a biosurfactant, rhamnolipid, and a chemical surfactant, SDS, in removal of crude oil from soil

[46]. More than 80% of crude oil was removed with application of SDS or rhamnolipid, however, the biosurfactant was effective in a wider range of tested conditions. Urum et al (2006) [47] then compared the efficiency on oil removal from soil using rhamnolipid and saponin biosurfactants and SDS, and the results showed a preferential removal of oil by the surfactants. SDS was more effective for aliphatic than aromatic hydrocarbons, whereas biosurfactants removed more aromatics. These results provide insights on the formulation of surfactant combinations, suggesting that the strategy should consider the degree of aromaticity in the crude oil-contaminated soil. The combination of oil-degrading bacteria and biosurfactant or biosurfactant-producing bacteria has also been tested by research groups. *P. putida*, an oil-degrading bacterium, was co-cultured with a biosurfactant-producing bacterium and improved degradation was observed in both aqueous and soil matrix in comparison with to the individual bacterium cultures [35]. This treatment resulted in increased oil emulsification and also adhesion of hydrocarbon to the bacteria cell surface.

2.4. Other applications

Rhamnolipids have potential microbial activity. It has been shown that these biosurfactants are very efficient bacteriostatic agent against Listeria monocytogenes, an important food-related pathogen, and showed synergistic effect when combined with nisin, a broad-spectrum bacteriocin [48]. Both biosurfactants and surfactin were shown to be able to reduce bacterial adhesion to polystyrene surfaces more efficient than the chemical surfactant sodium dodecyl sulfate. Moreover, purified rhamnolipid inhibited virtually 100% of the growth of strongly adherent L. monocytogenes strain [49]. Moreover, rhamnolipids were shown to significantly reduce the rate of deposition and adhesion, in rinsed chamber with these biosurfactants, of several bacterial and yeast strains isolated from explanted voice prostheses [50]. Rhamnolipids are also known to remove heavy metals. It was shown that these biosurfactant was able to remove niquel and cadmium from soils with efficiencies of 80-100%, under controlled environment, and 20-80% removal was reported in field samples [51].

MEL, a glycolipid, is a potent antimicrobial agent, especially against gram-positive bacteria. MEL has also been shown to induce growth arrest, apoptosis and differentiation of mouse malignant melanoma cell cultures [52, 53]. MEL-A produced by Pseudozyma antartica T-34 has been shown to have cell differentiation activities against human leukemia cells, mouse melanoma and PC12 cells [54]. Delivery of siRNA into the cytoplasm is still a challenge and a barrier for more gene silencing-based individualized drugs. MEL-A-containing cationic liposomes has been shown to directly deliver siRNA into the cytoplasm by the membrane fusion in addition to endocytotic pathway with better results compared with viral vectors in clinical applications [55].

Lipopeptides, such as surfactins, are particularly interesting because of their high surface activities and antibiotic potential. These biosurfactants have been reported as antibiotics, antiviral and antitumor agents, also as immunomodulators and inhibitors of specific toxins and enzymes. Surfactin, a cyclic lipopeptide, is known to be active in several biological activities, such as induction of ion channel formation, antiviral and antitumor, anti-inflammatory agent (reviewed by Rodrigues et al., 2006 [50]). Moreover, pre-coated vinyl urethral catheters with surfactin were shown to reduce biofilm formation by several Salmonella species and other infectious bacteria [56]. Surfactin has been shown to be more efficient than chemical surfactants (sodium dodecyl sulphate or dodecylamine) in improving floatability of metal-laden sorbents under similar conditions [57]. Also, surfactin was shown to contribute to reduce colonization of pathogenic bacteria, such as *L. monocytogenes*, *Enterobacter sakazakii* and *Salmonella enteritidis*, when applied to solid surfaces prior to infection [58, 59].

3. Metabolism and gene regulation of rhamnolipid production

Biosurfactants are considered ideal for environmental application, due to their numerous advantages over their chemical counterparts, such as biodegradability and less toxicity [60]. Among the biosurfactants, rhamnolipids are the better studied and promising candidates for large scale production especially because of their notable tensoative and emulsifying properties [61-63].

3.1. Rhamnolipid production

P. aeruginosa produces two major types of rhamnolipid in liquid cultures: the monorhamno-lipid, rhamnosyl-β-hydroxydecanoyl-β-hydroxydecanoate (Rha-C10-C10) and the dirhamno-lipid, rhamnosyl-rhamnosyl-β-hydroxydecanoyl-β-hydroxydecanoate (Rha-Rha-C10-C10) [64]. Besides these, twenty-five rhamnolipid congeners have been described in *P. aeruginosa*, varying in chain length and/or extent of saturation, showing that the addition of a hydrocarbon chain to dTDP-L-rhamnose is not specific to the carbon chains [65].

In *Pseudomonas aeruginosa*, these biosurfactants are the products of the convergence of two metabolic pathways: the biosynthesis of dTDP-L-rhamnose and the diversion of the β-hydroxydecanoyl-ACP intermediate from the FASII cycle by RhlA to synthesize the fatty acid dimer moiety of rhamnolipids and free 3-[3-hydroxyalkanoyloxy) alkanoic acid (HAA). The rhamnosyltransferases RhlB and RhlC catalyse the transfer of dTDP-L-rhamnose to either HAA, or a previously generated mono-rhamnolipid, respectively [10].

Free HAAs also show surface-tension activities and have been directly related to the promotion of swarming motility [10]. Recent studies suggest that RhlA is responsible for diverting the β-hydroxydecanoyl-ACP intermediate from the FASII cycle by directly competing with FabA and FabI for this intermediate [66]. In addition, RhlA is the only protein required to convert two molecules of β-hydroxyacyl- ACP into an HAA. The diversion of the β-hydroxydecanoyl-ACP intermediate from the FASII cycle providing a substrate for the enzyme RhlAB to produce the hydrocarbon chain in the rhamnolipid molecule is an important step, both biotechnolog-ically and clinically, but is still not fully understood.

3.2. Biosynthesis of dTDP-L-rhamnose

Rhamnose sugar is widely found in bacteria and plants, but not in humans. The activated L-rhamnose is derived from a glucose scaffold in four sequential steps, yielding deoxy-thymidine di-phospho (dTDP)-L-rhamnose. The first enzyme in the dTDP-L-rhamnose pathway is glucose-1-phosphate thymidylyltransferase (RmlA, EC 2.7.7.24) which catalyzes the transfer of a thymidylmonophosphate nucleotide to glucose-1-phosphate. The catalytic activity of RmlA is allosterically regulated by the final product of the pathway, dTDP-L-rhamnose [67], which makes RmlA the regulatory sensor for the downstream pathway. RmlA is a homotetramer with the monomer consisting of three functional domains: one core domain that shares the sequence similarity with nucleotidyltransferases, and two other domains that contain the recognition and binding sites for the nucleotide and sugarphosphate [67]. The second enzyme, dTDP-D-glucose 4,6-dehydratase (RmlB, EC 4.2.1.46), catalyzes an oxidation of the C4 hydroxyl group of the D-glucose residue, followed by dehydration, leading to the formation of dTDP-4-keto-6-deoxy-D-glucose [68]. The third enzyme, dTDP-4-keto-6-deoxy-D-glucose 3, 5-epimerase (RmlC, EC 5.1.3.13), catalyzes a double epimerization reaction at the C3 and C5 positions of the 4-keto-6-deoxy-D-glucose ring [69]. Finally, dTDP-4-keto-6-deoxy-Lmannose reductase (RmlD, EC 1.1.1.133) reduces the C4 keto group of the 4-keto-6-deoxy-L-mannose moiety and leads to the formation of dTDP-L-rhamnose [69]. All four enzyme genes are organized as a single operon in P. aeruginosa, called *rmlBDAC*.

3.3. Regulation of rhamnolipids synthesis

Rhamnolipids production in *P. aeruginosa* is tightly controlled by multiple layers of gene regulation that respond to a wide variety of environmental and physiologic signals, and are capable of combining different signals to generate unique and specific responses [12]. High carbon to nitrogen ratio, exhaustion of nitrogen source, stress conditions and high cell densities are among the conditions that favor higher levels of production [11]. Apparently, the key regulatory targets for rhamnolipid production in *P. aeruginosa* are the regulation of the *rmlBDAC* and *rhlAB* operons. The *rhlAB* operon is transcriptionally and posttranscriptionally regulated by manifold factors, commonly associated to the quorum sensing system, some of which also participates of the regulation of *rmlBDAC* operon.

3.4. Role of quorum sensing systems

The quorum sensing (QS) system is a bacterial communication system characterized by the secretion and detection of signal molecules – autoinducers – within a bacterial population. When it reaches a population "quorum", in which the autoinducers threshold is achieved, the bacterial population coordinates its responses to environmental inputs. QS is a global regulatory system found in most bacterial species controlling several and diverse biological functions, such as virulence, biofilm formation, bioluminescence and bacterial conjugation [70]. The main components of a quorum sensing system are the QS signal synthase, the signal receptor (regulatory protein), and the signal molecule [71]. The complex autoinducer/regulatory protein modulates the activity of the QS-regulated genes. There are two known conventional QS systems in *P. aeruginosa*, las and rhl. The autoinducer synthases LasI and RhlI produce the homoserine lactones 3OC12-HSL and C4-HSL, respectively, which complex with their correspondent transcriptional regulators, LasR and RhlR, to modulate the transcription of 5-10 % of the entire P. aeruginosa genome [72]. A third distinct QS system is formed by the transcriptional factor PqsR (also called MvfR) [73], responsible for activating the gene clusters pqsABCDE and phnAB both required for the production of Pseudomonas quinolone signal (PQS) and 4-hydroxy-2-alkylquinolones (HAQs), respectively, which is known to influence the production of QS-dependent factors, such as elastase, pyocyanin, PA-1L lectin, and rhamnolipids [74].

There has been postulated that sequences conserved in some promoters regions of RhlR and LasR-regulated genes are responsible by these regulation [75]. It can be verified that some QS-regulated genes belong specifically to *rhl* regulon, and some sequences RhlR-specific can be determined in promoters regions of some genes regulated by *rhl* system [76]. Although the *rhl* system has already been considered *las*-dependent [77], it has been shown that the expression of *rhl* system is maintained in a *lasR* mutant [72].

Rhamnolipid production in *P.aeruginosa* was shown to be directed linked to QS system by the transcriptional regulator RhlR. RhlR acts as an activator of *rhlAB* and *rmlBDAC* transcription when complexed to C4-HSL and as a repressor in the absence of the autoinducer [78, 79]. The *rhlR* gene is known to have four different transcription start sites [78]. In rich medium conditions, the expression of RhlR is dependent on LasR; however, under phosphate-limiting conditions, the expression of RhlR is regulated from multiple promoters through different transcriptional activators, including Vfr, RhlR, and the sigma factor $\sigma 54$, and it is also known that low-phosphate condition upregulates rhlR and RhlR-regulated genes, including rhamnolipids, even under low AHL levels [80]. In addition, in the last few years many other factors related to the QS system have been identified and reported to directly or indirectly influence rhamnolipid production.

3.5. Stationary phase and quorum sensing-related factors

Stationary phase is a physiological state frequently related to nutrient scarcity. Rhamnolipids is also associated with physiological roles in the uptake of poorly accessible substrates and is often associated with bacterial response to nutrient-deficient environments. The production is repressed in exponential phase and low cell density while is activated in stationary phase and high cell densities. This environmental regulation is related to several factors that directly or indirectly control transcription and post-transcription levels of expression.

The production in stationary phase is related to factors which contribute to the production in this phase and with other factors that inhibit the production in the exponential phase. The main factor that contributes to the rhamnolipid production in stationary phase appears to be the *rhl* system. As mentioned, the *rhl* system activates the *rmlBDAC* and *rhlAB* operons when complexed to C4-HSL [78, 79]. This QS system is activated in high cell densities, so it represents a direct link between high cell population and rhamnolipid production activation.

The second factor that guarantees the rhamnolipid production on stationary phase is related to a posttranscriptional regulation. It has been shown that most QS-regulated genes are not induced before the stationary growth phase, even when exogenous acyl-HSL signals are present in early growth phases [81]. These findings clearly indicate the other factors are involved in the expression of QS-regulated genes. The posttrancriptional regulator GidA was shown to primarily modulate the expression of RhlR and RhlR-dependent genes in *P. aeruginosa*. In *gidA* mutants, the levels of *rhlR* mRNA are similar to the levels in wild-type, whereas *rhlA* mRNA levels were significantly decreased, suggesting that the rhamnolipid production is controlled via posttranscriptional modulation of RhlR levels by GidA [82].

In relation to factors that inhibit the rhamnolipid production in the exponential phase, two regulators can be identified. The first of them is the QscR factor. It has been demonstrated that the *rhlAB* operon is present in the subset of genes repressed by QscR in the exponential phase [83]. This factor is a luxR-homologue protein that integrates the QS regulatory network and controls a distinct but overlapping regulon with the *las* and *rhl* systems [83]. QscR forms inactive heterodimers with LasR and RhlR in low AHL levels, while in increased AHL concentrations, QscR-AHL complexes are formed, and LasR-3OC12-HSL and RhlR-C4-HSL interactions occur, as a result of the dissociation of the heterodimer [84]. This can explain the effect of the QscR over the *rhlAB* operon. The second regulator is the DksA protein. DksA overexpression was shown to reduce *rhlI* and *rhlAB* expression in *P. aeruginosa*, while *dksA* mutants have higher *rhlI* expression [85]. It was demonstrated that DksA is required for *rhlAB* translation in *E.coli* heterologous expression, but does not affect its transcription [85]. DksA synthesis reaches its maximum during the exponential growth phase and it is posttranscriptionally downregulated during the late exponential and stationary growth phase [86]. This evidence indicates that DksA inhibits rhamnolipid production by repressing C4-HSL production.

Beyond the factors related to the growth phase, an important factor related to the QS systems which influences rhamnolipid production in *P. aeruginosa* is the *Pseudomonas* quinolone signal (PQS). PQS is involved in a complex regulatory network of QS connecting systems *las* and *rhl*. While PQS directly activates the *rhl* system in a *las* independent manner [87, 88], PQS synthesis is driven by *pqsABCDE* operon, which in activated by PqsR (MvfR), a LysR-type regulator. PqsR is activated by *las* QS system and repressed by *rhl* QS system, evidencing a complex network of regulation [89]. PQS was shown to be related to rhamnolipid production in different ways. Firstly, PQS production occurs in the late logarithmic phase and reaches its maximum in the late stationary phase [87, 88], with a similar profile to that of rhamnolipid biosynthesis. It was shown that *pqsR* and *pqsE* mutants have reduced rhamnolipid production, even when exposed to wild-type levels of C4-HSL [78]. This indicates a direct participation of PqsE, PqsR and/or PQS in rhamnolipid synthesis. Also, it has been hypothesized that PqsR is essential for rhamnolipid production, since PQS does not overcome the absence of *pqsR* in the regulation of phz1 operon, also involved in *rhl*-dependent phenotype [74]. As PqsR controls PQS production and rhamnolipid production is abolished in the absence of PQS signaling [87], it is possible that this effect is indirect. Rhamnolipid regulation also requires PqsE, which is involved in bacterial response to PQS and PqsR [74]. An alternative hypothesis is the regulation of rhamnolipid production by the LasR/RhlR ratio, known to modulate pqsR expression.

3.6. Regulatory factors related to stress conditions

An important environmental condition which highly influences the rhamnolipid production in *P. aeruginosa* is related to stress. Conditions such as nutrient deprivation and nitrogen exhaustion, even in a QS-independent manner, contribute to an increase in rhamnolipid production. Thus, several regulatory factors have been identified in last years which connect this condition to specific gene regulation patterns. Relevant ones are the sigma factors of RNA polymerase RpoS (σ^S or σ^{38}) and RpoN (σ^{54}).

The sigma factor of stationary phase, RpoS, plays an important role in the response to different stress conditions in *P. aeruginosa*. *rpoS* mutants have been shown to be more susceptible to carbon starvation, heat shock, high osmolarity, low pH, and hydrogen peroxide [90]. RpoS levels increase at the onset of stationary phase and in response to nutrient deprivation, even at low cell densities [91]. RpoS regulon comprises virtually all genes regulated in stationary phase, and has several overlaps with *las* and *rhl* regulons [92]. RpoS is involved in rhamnolipid production by two different manners, which indicates a genetic link between rhamnolipid production and nutrient deprivation and environment stress adaptation. Firstly, the *rhlAB* operon integrates the RpoS regulon and has been shown to be upregulated and partially RpoS-dependent [93]. Secondly, the *rmlBDAC* operon was shown recently to be activated by RpoS, which interacts in a different promoter region than the RhlR-C4HSL [79], which suggests a regulation related to stress in a QS-independent manner. Furthermore, RpoS was shown to be required for swarming motility [94], a phenotype related to HAAs and rhamnolipids [10].

The sigma factor RpoN (σ^{54}) is known to be involved in nitrogen metabolism in bacteria. The rhamnolipid production is largely dependent on nitrogen exhaustion in *P. aeruginosa* [95], and different nitrogen sources can act as inhibitors (e.g. ammonium, glutamine, asparagine and arginine) or activators (nitrate, glutamate, and aspartate) of rhamnolipid production [95, 96]. Moreover, nitrate has been shown to provide the highest yields of rhamnolipid production [7], related to upregulation of glutamine synthase under nitrogen-limiting conditions by RpoN [97]. The RpoN activity on the expression of catabolic pathways is related to CbrA-CbrB and NtrB-NtrC, two-component regulatory systems, which are regulated by nitrogen availability [98]. On the other hand, the activation of rpoN-dependent promoters seems to be CbrB and NtrC-dependent [99]. The *rhlAB* operon and *rhlR* are controlled indirectly and directly, respectively, by RpoN [93], since under nitrogen-limiting conditions, *P. aeruginosa* seems to improve its nitrogen assimilation through ATP-dependent pathways (e.g. glutamine synthase) and the increase of nitrogen uptake via, e.g., upregulating rhamnolipid biosynthesis.

Another important factor that can correlate rhamnolipid production to stress conditions, but also with quorum sensing, is the PQS. PQS is involved in stress response in *P. aeruginosa* and has been reported to be related to oxidative stress, UV irradiation resistance and antimicrobial agents [100]. Rhamnolipids were shown to act in PQS solubilization/ assimilation, and therefore, in its activity as a transcriptional regulator [101]. This clearly indicates the interconnection between rhamnolipid and PQS synthesis.

3.7. Regulatory factors related to virulence

The great adaptability of *P. aeruginosa* to a wide range of habitats is also related to its ability to produce several virulence factors such as pyocyanin, elastases, proteases and rhamnolipids, which provides antimicrobial activity [102], haemolytic activity in human pathogenesis [103] and promotion of swarming motility [10]. Different regulatory factors that control virulence in *P. aeruginosa* have been shown to activate rhamnolipid production.

Vfr, the global regulator of the virulence in *P. aeruginosa*, is a cAMP binding protein that activates the RhlR expression in a LasR-independent manner [78]. It affects the expression of multiple virulence factors downstream in the QS cascade, including rhamnolipid production. Vfr is also involved in RpoS synthesis [104], which has been demonstrated to influence rhamnolipid production acting on *rmlBDAC* and *rhlAB* operons. Another factor related to virulence is the VqsR protein. This is a LuxR homologue that has been shown to modulate the expression of QS genes and others metabolic process, seeming to be a global regulator in *P. aeruginosa* [105]. Rhamnolipid production is reduced in *vqsR* mutants, which can be related to the influence of this factor in the QS system, since most of the VqsR-regulated genes were previously identified as QS-regulated [99, 105]. PtxR, a LysR-type transcriptional regulator, which modulates the production of virulence factors in *P. aeruginosa* by repressing PQS-genes expression [106], is also involved in the biosurfactant production. The rhamnolipid synthesis increase in *ptxR* mutants, which can be related to its modulation of the QS system, since the *ptxR* mutants showed upregulation of *rhlI* and *lasI* genes, as well as the C4-HSL and 3OC12-HSL autoinducers, when compared to the wild type strain [106].

3.8. Regulatory factors related to biofilm formation

Biofilm formation provides several advantages to the bacteria, providing protection that may enhance bacterial survival under environmental stress conditions. Rhamnolipids play a major role in the architecture of biofilms produced by *P. aeruginosa*. The cell detachment of the biofilm structure and the formation of water channels have been shown to be dependent of the rhamnolipids synthesis. It has been reported that rhamnolipid production is related to biofilms in *P. aeruginosa* through AlgR, a regulatory factor related to biosynthesis of alginate [107]. AlgR was shown to be the main repressor of rhamnolipid production within adherent biofilms and during its development, acting as a repressor of the expression of *rhlI* and *rhlAB* [108]. No such effect was reported on planktonic growth so far, therefore, it was hypothesized that AlgR acts through a contact-dependent or biofilm-specific mode of regulation. BqsS-BqsR, a two component system, has also been reported to be related to biofilm formation in P. aerugionosa [109]. In *P. aeruginosa*, bqsS-bqsR mutant showed reduced rhamnolipids production, which might be an indirect regulation, since the production of C4-HSL, as well as PQS, were reduced in this mutant. Although the environmental stimuli that trigger BqsS-BqsR activity are still unknown, evidences support the existence of opposite effects of AlgR and BqsS-BqsR on rhamnolipid production in the context of biofilm formation [109].

3.9. Regulatory factors related to unknown stimuli

P. aeruginosa is a ubiquitous and highly adaptable bacterium, and rhamnolipids play different roles in its adaptation processes. Besides the known environmental stimuli, others are likely to play important role in rhamnolipid production. Some regulatory factors have been identified as involved on rhamnolipids production, however their environmental stimuli are still to be discovered. GacS-GacA is a well characterized two-component system in *P. aeruginosa* that modulates gene expression through promoting the synthesis of two small RNAs, RsmY and RsmZ. These small RNAs regulate gene transcriptional by modulating the activity of the small RNA-binding protein RsmA [110]. RsmA is a translational regulator and acts by preventing the translation initiation of target RNAs. However, RsmA also activates indirectly gene expression, by acting over repressor factors. GacS-GacA can have opposite effects on rhamnolipids production, evidencing a complex regulation. RsmA represses the QS system, including C4-HSL and 3OC12-HSL production [111], whereas, RsmA acts as an activator of rhamnolipid synthesis, since *rhlB* was downregulated in the *rsmA* mutant [112]. Therefore, it has been hypothesized that RsmA might stabilize *rhlAB* mRNA and/or facilitate its translation initiation. Another global regulator involved in rhamnolipids production is VqsM, an AraC-type transcriptional regulator. VqsM has been shown to modulate the expression of several genes, including QS regulators. *vqsM* mutants show reduced expression of *rhlAB* [113], *rhlR* and *rpoS*, evidencing an indirect regulation upstream of the QS regulatory network. The environmental stimuli for this regulation are also unknown.

4. Economic aspects of biosurfactant production

Several microorganisms are known to produce biosurfactants that can vary in structure and chemical composition. These variations are dependent on the producing microorganism, raw matter used for fermentation and conditions of fermentative process [17]. According to recent data, global biosurfactants market was worth USD 1,7 bi in 2011 and is expected to reach USD 2,2 bi in 2018, based on a growth rate of 3.5% per annum. The global biosurfactants market volume is expected to reach 476,512.2 tons by 2018, due to increasing demand from the Asia, Africa and Latin America, which should account for 21 % of it [114].

The number of publications related to identification, optimization of production process and better understanding of the metabolic pathways has increased in recent years [5]. Many biosurfactants and their production processes have been patented, but only some of them have been commercialized [6]. Some examples of products based on biosurfactants that are available in the market are shown in Table 2.

Besides the research efforts, the cost for biosurfactant production is approximately three to ten times more than the cost to produce a chemical surfactant. Biosurfactants are typically produced by microorganisms growing in hydrocarbons as a carbon source, which are usually expensive increasing the production cost [6]. In addition, the downstream cost, low productivity and intense foaming formation during the biosurfactant production currently is a barrier for an economically viable production of biosurfactant [3, 6]. Therefore, most researches have been focusing on increasing the production yield, reducing raw material cost and developing oxygenation strategies to reduce foaming formation [4, 6, 61]. Several approaches have been applied in order to improve biosurfactant productivity such as optimization of growth conditions (e.g., growth medium, temperature, pH, oxygenation, fermentation phases), genetic modifications (mutation, gene knockout or amplification, altered regulation), com-

Biosurfactant	Origin	Supplier	Application
BioFuture	Bacterial rhamnolipid	BioFuture Ltd, Ireland	Bioremediation of contaminated soil with hydrocarbon
EC-601	Bacterial rhamnolipid	Ecochem Ltd, Canada	Dispersive agent of water-insoluble hydrocarbons
EC-1800	Bacterial consortium	Ecochem Ltd, Canada	Cleans up oil spills in soil, sand and gravel
EC-2100W	Bacterial consortium	Ecochem Ltd, Canada	Degrades hydrocarbon based compounds in waste water treatment plants, lagoons, storage tanks, sumps and other aqueous environments
JBR products	Microbial rhamnolipid	Jeneil Biosurfactant Co., LTC, USA	Used in different industries, such food and agro-industrial markets
Petrosolv	Bacterial unknown	Enzyme Technologies Inc, USA	Oil removal; oil recovery and processing
Surfactin	*Bacillus subtilis*	Sigma-Aldrich Co. LLC, USA	Antifungal, antibacterial and antitumor activities

Table 2. Examples of commercialized products containing biosurfactants

bined omics analysis (genomics, transcriptomics, proteomics, metabolomics, fluxomics) and computational modeling [115].

5. Strategies for improvement of biosurfactant production

5.1. Optimization of medium composition

Rhamnolipids are mainly produced by Pseudomonas species, such as *Pseudomonas aeruginosa*. This bacterium produces rhamnolipids as secondary metabolite and their production coincides with the stationary growth phase [65, 95, 116]. Rhamnolipids can be produced using varies carbon sources, such as vegetable oils (e.g. olive, soy and corn), glucose, glycerol and n-alcanes [7]. The carbon and nitrogen sources are important factors in the production of these biosurfactants and have great influence their production cost and considerable efforts have been done towards the use of agro-industrial byproducts and renewable resources as substrates in the production process [3].

Studies have been shown that use of inexpensive substrates, such as crude or waste materials, dramatically affects the production costs of biosurfactants [3]. Different waste substrates have been used for rhamnolipid production, such as fatty acids from soybean oil refinery wastes [117], glycerin from biodiesel production waste [118] and sunflower-oil refinery waste [119]. Nevertheless, the potential of rhamnolipids production from renewable resources is so far not fully exploited. According to Henkel et al [3], use of waste substrates in the production processes is likely to increase its influence on the field of microorganism-based production, since they are usually cheaper, maximize the utilization efficiency regarding the overall production process and makes the process more environmentally friendly.

Nitrogen source, C/N ratio and mineral salts are also important variables to achieve improved productivity of rhamnolipid [7]. Currently, the main nitrogen source used to promote rhamnolipid production is nitrate [5]. Interestingly, in contrast with nitrate, ammonium has been shown to prevent rhamnolipid production [120-122]. Growth limiting conditions are known to promote rhamnolipids production, as well as other secondary metabolites. High carbon and nitrogen (C/N) ratio [7, 121] and carbon and inorganic phosphorous (C/P) [96] have been shown to increase rhamnolipid productivity. Also, limiting concentration of multivalent ions such as Mg, Ca, K, Na, and trace element salts increase rhamnolipid yield [121]. The growth conditions that influence rhamnolipid overproduction with *P. aeruginosa* have been intensively studied. The mainresearches on this field can be divided into (a) batch cultivations under growth-limiting conditions, (b) agitated flasks and bioreactors cultivation, which include continuous cultivation, fed-batch strategies and controlled aeration systems, (c) downstream processing and (d) biochemical and molecular strategies, aiming to improve strain capability to produce these biosurfactants [5].

5.2. Cultivation strategies and aeration systems

Cultivation strategies described for rhamnolipid and other biossurfactants include shake flask, batch, fed-batch, continuous, and integrated microbial/enzymatic processes [5]. Most of these studies involve the optimization of culture conditions in shake flasks, due to its simplicity, whereas bioreactors provides high control over the relevant parameters, such as oxygen transfer and pH during all the cultivation process, and allows upscaling studies [123, 124]. Bioreactor is often applied in continuous or fed-batch fermentation in biosurfactant produc-tion. Both alternatives have been proposed to be more economically viable than simple batches, due to the suppression of several steps regarding the inoculum preparation, sterilization and finalization of production process by carrying out the production for long periods [124].

One of the main limitations for the biosurfactant production in bioreactor is the intense foaming formation when the solution is aerated and agitated caused by the surfactant [6]. Particularly with *P. aeruginosa*, high foam formation is further increased by the presence of extracellular proteins, which results in increased production cost. Mechanic foam breakers are not very efficient, and chemical antifoams agents can alter the quality of the product and the pollutant potential of the final effluent of the bioreactor [6]. Several strategies have been proposed to overcome this limitation using. For example, non-dispersive devices have been successfully applied to promote the oxygen transfer from the gas phase to the liquid phase without dispersion of these phases (Santa Anna et al, 2004). This process has been described by Gruber et al (1993) and a similar process, used in alcoholic fermentations, has been patented by L'Air Liquide (Cutayar et al, 1990). Although this oxygenation process is promising for the use in bioreactors, the manual control of the oxygenation was shown difficult and even inefficient. Recently, Kronemberger et al [6] developed a non-dispersive oxygenated device controlled by a programmable logic controller (PLC), which allows setting the dissolved oxygen concentration in a process for rhamnolipid production by *P. aeruginosa*. The better control over dissolved oxygen also allowed the demonstration that rhamnolipid production by *P. aeruginosa* is partially dependent of the dissolved oxygen concentration in the medium. Moreover, the relationship between rhamnolipid production and oxidative stress has been recently studied and proteins related to oxidative stress pathways and rhamnolipid production by *P. aeruginosa* was identified [125].

5.3. Downstream processing

Comparable to the selection of cheap raw materials and cultivation strategies, reduction of downstream costs is an essential factor towards the establishment of an economic process. However, the purification costs will depend on the employed strain, the rhamnolipid mixtures produced and the application of the product, since for some industrial applications a high purity grade will not be necessary and thus reducing purification costs [5]. For some industrial applications it is proposed that downstream costs can account for approximately 60% of the total production cost [126].

Conventional methods for recovery of biosurfactants have been used, such as acid precipitation, solvent extraction, crystallization, ammonium sulfate precipitation and centrifugation [126]. More recently, other methods for biosurfactants recovery have been reported, including foam fractionation, ultrafiltration and adsorption-desorption on polystyrene resins and ion exchange chromatography. These procedures take advantage of biosurfactant properties, such as their surface activity or their ability to form micelles or vesicles, and are particularly applicable for large-scale continuous recovery of extracellular biosurfactants from culture broth. In addition, these methods can operate in continuous mode for recovering biosurfactants with high level of purify [4].

Recovery techniques that promote high level of purify often require application of solvents (e.g.,acetone, methanol and chloroform), which are toxic in nature and harmful to the environment [4]. Rhamnolipids are usually precipitated by acidification and extracted using ethyl acetate; extraction of sophorolipids is normally carried out with n-hexane, while for trehalolipids, the preferred solvent is a mixture of chloroform and methanol. Although these techniques are already well established for lab-scale applications, their cost does not allow scaling-up for industrial production of biosurfactants. Thus, the research effort has been directed towards the development of low-cost extraction and purification procedures, avoiding the use of hazardous and costly organic solvents [13].

Rhamnolipid produced by *P. aeruginosa* is often secreted with other virulence factors and studies have been conducted to better understand the molecular mechanisms related to rhamnolipid production. Ideally, rhamnolipids should be produced with minimum presence of other virulence factors (such as pyocianin, elastase and protease), which may allow the use of crude biosurfactant with the suppression of purification stage for environmental applications and bioremediation [125].

5.4. Metabolic and cellular strategies for strain improvement

Rational metabolic and cellular engineering approaches have been successfully applied to improve strain performance in several cases of biotechnological production of small-volume and high-value bioproducts. However, such attempts are limited to the manipulation of only selected group of genes encoding enzymes and regulatory proteins, which are selected using available information and research experience [115]. Recent advances in high-throughput experimental techniques supported by bioinformatics (genomic, transcriptomic, proteomic, metabolomic) have resulted in rapid accumulation of data, providing a basis for in-depth understanding of biological processes. Although our ability to integrate these data is currently limited, the information obtained with these approaches, together with experimental observations *in vivo* and predictions of modeling and simulation *in silico*, can provide solutions to understand the features and functions of biological systems [115].

Despite the advances described, optimization of rhamnolipid production has been performed with non-genetically modified and randomly mutated strains of *P. aeruginosa* [5]. The genomic and proteomic approaches for biotechnology is still very limited, despite the genus Pseudomonas being extensively studied due to its versatility and ability to environmental biodegradation of contaminants [73]. In *P. aeruginosa*, studies are related predominantly with clinical interests, many of them aimed at understanding the *quorum sensing* system in this bacterium. Recently, a comparative proteome analysis of *P. aeruginosa* PA1 was done investigating the differences of the intracellular proteome after the cultivation of rhamnolipid producing and non-producing cells [11]. Rhamnolipid production was either induced or suppressed by variation of media conditions and a total of 21 differentially expressed proteins could be identified by MALDI-TOF/TOF mass spectrometry. Under the rhamnolipid favoring conditions, proteins involved in the response to oxidative stress, secretion pathways and quorum sensing were mainly differentially expressed. Pacheco et al [125] treated *P. aeruginosa* with hydrogen peroxide to trigger its oxidative stress response, and the proteome profile was analyzed. There were identified 14 differentially expressed proteins between samples that were treated and not treated with peroxide. Several of these proteins are potentially involved in the rhamnolipid production/ secretion pathway and oxidative stress. The production of elastase, alkaline proteases and rhamnolipids seems to be regulated in a similar way, which makes it necessary to develop molecular strategies that maximize the production of rhamnolipids over other virulence factors.

A system-wide analysis of growth and rhamnolipid production regulation has the potential to reveal several unknown interactions between genes, proteins and metabolites, thus facilitating rational process engineering strategies [127], either with *P. aeruginosa* strains or heterologous hosts [8]. According to Muller and Hausmann [8], the main goals of such approaches with *P. aeruginosa* would be: (a) metabolic engineering and further strain improvement to enhance the metabolic fluxes towards the product, (b) removal/alteration of autologous genes to avoid byproduct formation and (c) introduction of heterologous genes for the use of alternative substrates and enhanced metabolism. Thus, the integration of engineering and biology is possible and profoundly needed in order to achieve high yields and low costs on biosurfactant production. High-throughput "omics" analysis, predictive computational modeling or simulation and experimental perturbation can be combined to generate new knowledge about the cellular physiology and metabolism, in order to design strategies for metabolic and cellular engineering of organisms.

6. Conclusion

In conclusion, the production of the better known biosurfactant, rhamnolipid, can be linked to several regulatory factors that respond to environmental inputs such as population density, nutrient availability and diverse stresses. Rhamnolipids production is highly controlled by multilayered regulatory systems and better knowledge of the effect of environmental stimuli can greatly improve their commercial production. On the other side biosurfactant application in processes such as washing, biorremadiation and MEOR, requires a thoroughly research on a case-by-case basis taking in account the physical chemical properties of biosurfactants and the environmental conditions.

Author details

R.S. Reis[1], G.J. Pacheco[2,3], A.G. Pereira[4] and D.M.G. Freire[2]

1 University of Sydney, School of Molecular Biology, NSW, Australia

2 Department of Biochemistry, Chemistry Institute, Federal University of Rio de Janeiro, RJ, Brazil

3 Laboratory of Toxinology, IOC, Oswaldo Cruz Foundation, RJ, Brazil

4 Department of Cellular Biology, University of Brasília, DF, Brazil

References

[1] Van Hamme, J. D, Singh, A, & Ward, O. P. Physiological aspects. Part 1 in a series of papers devoted to surfactants in microbiology and biotechnology. Biotechnology Advances. (2006). , 24(6), 604-20.

[2] Banat, I. M, Makkar, R. S, & Cameotra, S. S. Potential commercial applications of microbial surfactants. Appl Microbiol Biotechnol. (2000). , 53(5), 495-508.

[3] Henkel, M, Müller, M. M, Kügler, J. H, Lovaglio, R. B, Contiero, J, Syldatk, C, et al. Rhamnolipids as biosurfactants from renewable resources: Concepts for next-generation rhamnolipid production. Process Biochemistry. (2012). , 47(8), 1207-19.

[4] Mukherjee, S, Das, P, & Sen, R. Towards commercial production of microbial surfac-
     tants. Trends Biotechnol. (2006). , 24(11), 509-15.

[5] Müller, M. M, Kügler, J. H, Henkel, M, Gerlitzki, M, Hörmann, B, Pöhnlein, M, et al.
     Rhamnolipids-Next generation surfactants? Journal of Biotechnology. (2012).

[6] Kronemberger FDASanta Anna LMM, Fernandes ACLB, Menezes RRD, Borges CP,
     Freire DMG. Oxygen-controlled biosurfactant production in a bench scale bioreactor.
     Appl Biochem Biotechnol. (2008).

[7] Santos, A. S. Sampaio APW, Vasquez GS, Santa Anna LM, Pereira Jr N, Freire DMG.
     Evaluation of different carbon and nitrogen sources in production of rhamnolipids
     by a strain of Pseudomonas aeruginosa. Applied Biochemistry and Biotechnology-
     Part A Enzyme Engineering and Biotechnology. (2002).

[8] Müller, M, & Hausmann, R. Regulatory and metabolic network of rhamnolipid bio-
     synthesis: Traditional and advanced engineering towards biotechnological produc-
     tion. Applied Microbiology and Biotechnology. (2011). , 91(2), 251-64.

[9] Abdel-mawgoud, A, Lépine, F, & Déziel, E. Rhamnolipids: diversity of structures,
     microbial origins and roles. Applied Microbiology and Biotechnology. (2010). , 86(5),
     1323-36.

[10] Déziel, E, Lépine, F, Milot, S, & Villemur, R. rhlA is required for the production of a
     novel biosurfactant promoting swarming motility in Pseudomonas aeruginosa: 3-(3-
     hydroxyalkanoyloxy)alkanoic acids (HAAs), the precursors of rhamnolipids. Micro-
     biology. (2003). , 149(8), 2005-13.

[11] Reis, R. S. da Rocha SLG, Chapeaurouge DA, Domont GB, Santa Anna LMM, Freire
     DMG, et al. Effects of carbon and nitrogen sources on the proteome of Pseudomonas
     aeruginosa PA1 during rhamnolipid production. Process Biochemistry. (2010). ,
     45(9), 1504-10.

[12] Reis, R. S, Pereira, A. G, & Neves, B. C. Freire DMG. Gene regulation of rhamnolipid
     production in Pseudomonas aeruginosa- A review. Bioresource Technology. (2011). ,
     102(11), 6377-84.

[13] Banat, I, Franzetti, A, Gandolfi, I, Bestetti, G, Martinotti, M, Fracchia, L, et al. Micro-
     bial biosurfactants production, applications and future potential. Applied Microbiol-
     ogy and Biotechnology. (2010). , 87(2), 427-44.

[14] Denise Maria FL√?via de Aj, Frederico K, M√°rcia N. Biosurfactants as Emerging Ad-
     ditives in Food Processing. Innovation in Food Engineering: CRC Press; (2009). ,
     685-705.

[15] Damasceno FRCCammarota MC, Freire DMG. The combined use of a biosurfactant
     and an enzyme preparation to treat an effluent with a high fat content. Colloids and
     Surfaces B: Biointerfaces. (2012). , 95(0), 241-6.

[16] Shete, A. M, Wadhawa, G, Banat, I. M, & Chopade, B. A. Mapping of patents on bioe-mulsifier and biosurfactant: A review. Journal of Scientific and Industrial Research. (2006). , 65(2), 91-115.

[17] Cameotra, S. S, & Makkar, R. S. Biosurfactant-enhanced bioremediation of hydropho-bic pollutants. Pure Appl Chem. (2010). , 82(1), 97-116.

[18] Singh, S, Kang, S. H, Mulchandani, A, & Chen, W. Bioremediation: environmental clean-up through pathway engineering. Current opinion in biotechnology. (2008). , 19(5), 437-44.

[19] Gentry, T, Rensing, C, & Pepper, I. New approaches for bioaugmentation as a reme-diation technology. Critical Reviews in Environmental Science and Technology. (2004). , 34(5), 447-94.

[20] Smets, B. F, & Pritchard, P. Elucidating the microbial component of natural attenua-tion. Current opinion in biotechnology. (2003). , 14(3), 283-8.

[21] Bognolo, G. Biosurfactants as emulsifying agents for hydrocarbons. Colloids and Sur-faces A: Physicochemical and Engineering Aspects. (1999).

[22] Santa Anna LMSoriano AU, Gomes AC, Menezes EP, Gutarra MLE, Freire DMG, et al. Use of biosurfactant in the removal of oil from contaminated sandy soil. J Chem Technol Biotechnol. (2007). , 82(1), 687-91.

[23] Whang, L. M. Liu PWG, Ma CC, Cheng SS. Application of biosurfactants, rhamnoli-pid, and surfactin, for enhanced biodegradation of diesel-contaminated water and soil. Journal of hazardous materials. (2008). , 151(1), 155-63.

[24] Garcia-junco, M, Gomez-lahoz, C, Niqui-arroyo, J. L, & Ortega-calvo, J. J. Biosurfac-tant-and biodegradation-enhanced partitioning of polycyclic aromatic hydrocarbons from nonaqueous-phase liquids. Environmental science & technology. (2003). , 37(13), 2988-96.

[25] Mulligan, C. N, & Eftekhari, F. Remediation with surfactant foam of PCP-contami-nated soil. Engineering geology. (2003). , 70(3), 269-79.

[26] Dean, S. M, Jin, Y, Cha, D. K, Wilson, S. V, & Radosevich, M. Phenanthrene degrada-tion in soils co-inoculated with phenanthrene-degrading and biosurfactant-produc-ing bacteria. Journal of environmental quality. (2001). , 30(4), 1126-33.

[27] Lan Chun CLee JJ, Park JW. Solubilization of PAH mixtures by three different anion-ic surfactants. Environmental Pollution. (2002). , 118(3), 307-13.

[28] Zhou, W, & Zhu, L. Efficiency of surfactant-enhanced desorption for contaminated soils depending on the component characteristics of soil-surfactant-PAHs system. Environmental Pollution. (2007). , 147(1), 66-73.

[29] Zhou, W, & Zhu, L. Enhanced soil flushing of phenanthrene by anionic-nonionic mixed surfactant. Water research. (2008).

[30] Wei, Y. H, Chou, C. L, & Chang, J. S. Rhamnolipid production by indigenous Pseu-domonas aeruginosa J4 originating from petrochemical wastewater. Biochemical Engineering Journal. (2005). , 27(2), 146-54.

[31] Cheng, K, Zhao, Z, & Wong, J. Solubilization and desorption of PAHs in soilaqueous system by biosurfactants produced from Pseudomonas aeruginosa under thermophilic condition. Environmental technology. (2004). , CG3.

[32] Wong JWCFang M, Zhao Z, Xing B. Effect of surfactants on solubilization and degradation of phenanthrene under thermophilic conditions. Journal of environmental quality. (2004). , 33(6), 2015-25.

[33] Makkar, R. S, & Rockne, K. J. Comparison of synthetic surfactants and biosurfactants in enhancing biodegradation of polycyclic aromatic hydrocarbons. Environmental toxicology and chemistry. (2009). , 22(10), 2280-92.

[34] Efroymson, R. A, & Alexander, M. Biodegradation by an Arthrobacter species of hydrocarbons partitioned into an organic solvent. Applied and Environmental Microbiology. (1991). , 57(5), 1441-7.

[35] Kumar, M, & Leon, V. Materano ADS, Ilzins OA, Galindo-Castro I, Fuenmayor SL. Polycyclic aromatic hydrocarbon degradation by biosurfactant-producing Pseudomonas sp. IR1. Zeitschrift fur Naturforschung C-Journal of Biosciences. (2006).

[36] Cui, C. Z, Wan, X, & Zhang, J. Y. Effect of rhamnolipids on degradation of anthracene by two newly isolated strains, Sphingomonas sp. 12A and Pseudomonas sp. 12B. Journal of microbiology and biotechnology. (2008). , 18(1), 63-6.

[37] Kolomytseva, M. P, Randazzo, D, Baskunov, B. P, Scozzafava, A, Briganti, F, & Golovleva, L. A. Role of surfactants in optimizing fluorene assimilation and intermediate formation by *Rhodococcus rhodochrous* VKM B-2469. Bioresource Technology. (2009). , 100(2), 839-44.

[38] Ruberto LAMVazquez SC, Curtosi A, Mestre MC, Pelletier E, Mac Cormack WP. Phenanthrene biodegradation in soils using an Antarctic bacterial consortium. Bioremediation journal. (2006). , 10(4), 191-201.

[39] Manning, F. S, & Thompson, R. E. Oilfield Processing of Petroleum: Crude Oil: Pennwell Corporation; (1995).

[40] Nadarajah, N, Singh, A, & Ward, O. P. De-emulsification of petroleum oil emulsion by a mixed bacterial culture. Process Biochemistry. (2002). , 37(10), 1135-41.

[41] Urum, K, Pekdemir, T, & Çopur, M. Surfactants treatment of crude oil contaminated soils. Journal of Colloid and Interface Science. (2004). , 276(2), 456-64.

[42] Rosen, R, & Ron, E. Z. Proteome analysis in the study of the bacterial heat-shock response. Mass Spectrom Rev. (2002). , 21(4), 244-65.

[43] Maier, R. M, & Soberon-chavez, G. Pseudomonas aeruginosa rhamnolipids: biosyn-
      thesis and potential applications. Appl Microbiol Biotechnol. (2000). Epub
      2000/12/29., 54(5), 625-33.

[44] Rahman, K. S, Rahman, T. J, Kourkoutas, Y, Petsas, I, Marchant, R, & Banat, I. M. En-
      hanced bioremediation of n-alkane in petroleum sludge using bacterial consortium
      amended with rhamnolipid and micronutrients. Bioresour Technol. (2003). Epub
      2003/08/05., 90(2), 159-68.

[45] Vasileva-tonkova, E, & Gesheva, V. Glycolipids produced by Antarctic *Nocardioides*
      sp. during growth on *n*-paraffin. Process Biochemistry. (2005). , 40(7), 2387-91.

[46] Urum, K, Pekdemir, T, Ross, D, & Grigson, S. Crude oil contaminated soil washing in
      air sparging assisted stirred tank reactor using biosurfactants. Chemosphere. (2005). ,
      60(3), 334-43.

[47] Urum, K, Grigson, S, Pekdemir, T, & Mcmenamy, S. A comparison of the efficiency
      of different surfactants for removal of crude oil from contaminated soils. Chemo-
      sphere. (2006). , 62(9), 1403-10.

[48] Magalhães, L, & Nitschke, M. Antimicrobial activity of rhamnolipids against *Listeria
      monocytogenes* and their synergistic interaction with nisin. Food Control. (2012).

[49] De Araujo, L. V, Abreu, F, & Lins, U. Anna LMMS, Nitschke M, Freire DMG. Rham-
      nolipid and surfactin inhibit *Listeria monocytogenes* adhesion. Food Research Interna-
      tional. (2011). , 44(1), 481-8.

[50] Rodrigues, L. R, Banat, I. M, Van Der Mei, H. C, Teixeira, J. A, & Oliveira, R. Interfer-
      ence in adhesion of bacteria and yeasts isolated from explanted voice prostheses to
      silicone rubber by rhamnolipid biosurfactants. J Appl Microbiol. (2006). , 100(3),
      470-80.

[51] Wang, S, & Mulligan, C. N. Rhamnolipid foam enhanced remediation of cadmium
      and nickel contaminated soil. Water, Air, & Soil Pollution. (2004). , 157(1), 315-30.

[52] Zhao, X, Wakamatsu, Y, Shibahara, M, Nomura, N, Geltinger, C, Nakahara, T, et al.
      Mannosylerythritol lipid is a potent inducer of apoptosis and differentiation of
      mouse melanoma cells in culture. Cancer research. (1999). , 59(2), 482-6.

[53] Zhao, X, Geltinger, C, Kishikawa, S, Ohshima, K, Murata, T, Nomura, N, et al. Treat-
      ment of mouse melanoma cells with phorbol 12-myristate 13-acetate counteracts
      mannosylerythritol lipid-induced growth arrest and apoptosis. Cytotechnology.
      (2000). , 33(1), 123-30.

[54] Wakamatsu, Y, Zhao, X, Jin, C, Day, N, Shibahara, M, Nomura, N, et al. Mannosyler-
      ythritol lipid induces characteristics of neuronal differentiation in PC12 cells through
      an ERK-related signal cascade. European Journal of Biochemistry. (2003). , 268(2),
      374-83.

[55] Inoh, Y, Furuno, T, Hirashima, N, Kitamoto, D, & Nakanishi, M. Rapid delivery of small interfering RNA by biosurfactant MEL-A-containing liposomes. Biochemical and Biophysical Research Communications. (2011).

[56] Mireles, J. R, Toguchi, A, & Harshey, R. M. Salmonella enterica serovar Typhimurium swarming mutants with altered biofilm-forming abilities: surfactin inhibits biofilm formation. Journal of Bacteriology. (2001). , 183(20), 5848-54.

[57] Zouboulis, A, Matis, K, Lazaridis, N, & Golyshin, P. The use of biosurfactants in flotation: application for the removal of metal ions. Minerals Engineering. (2003). , 16(11), 1231-6.

[58] Nitschke, M, Araújo, L, Costa, S, Pires, R, Zeraik, A, Fernandes, A, et al. Surfactin reduces the adhesion of food-borne pathogenic bacteria to solid surfaces. Letters in applied microbiology. (2009). , 49(2), 241-7.

[59] Korenblum, E, De Araujo, L. V, Guimarães, C. R, De Souza, L. M, Sassaki, G, Abreu, F, et al. Purification and characterization of a surfactin-like molecule produced by Bacillus sp. H2O-1 and its antagonistic effect against sulfate reducing bacteria. BMC microbiology. (2012).

[60] Nitschke, M. Costa SGVAO, Haddad R, GonÃ§alves LAG, Eberlin MN, Contiero J. Oil wastes as unconventional substrates for rhamnolipid biosurfactant production by Pseudomonas aeruginosa LBI. Biotechnol Prog. (2005). , 21(5), 1562-6.

[61] Heyd, M, Kohnert, A, Tan, T. H, & Nusser, M. KirschhÃfer F, Brenner-Weiss G, et al. Development and trends of biosurfactant analysis and purification using rhamnolipids as an example. Analytical and Bioanalytical Chemistry. (2008). , 391(5), 1579-90.

[62] Cha, M, Lee, N, Kim, M, Kim, M, & Lee, S. Heterologous production of Pseudomonas aeruginosa EMS1 biosurfactant in Pseudomonas putida. Bioresour Technol. (2008). , 99(7), 2192-9.

[63] Mulligan, C. N. Environmental applications for biosurfactants. Environ Pollut. (2005). , 133(2), 183-98.

[64] Deziel, E, Lepine, F, Milot, S, & Villemur, R. Mass spectrometry monitoring of rhamnolipids from a growing culture of Pseudomonas aeruginosa strain 57RP. Biochim Biophys Acta. (2000). Epub 2000/06/01.

[65] Déziel, E, Lépine, F, Dennie, D, Boismenu, D, Mamer, O. A, & Villemur, R. Liquid chromatography/mass spectrometry analysis of mixtures of rhamnolipids produced by Pseudomonas aeruginosa strain 57RP grown on mannitol or naphthalene. Biochim Biophys Acta. (1999).

[66] Zhu, K, & Rock, C. O. RhlA converts b-hydroxyacyl-acyl carrier protein intermediates in fatty acid synthesis to the b-hydroxydecanoyl-b-hydroxydecanoate component of rhamnolipids in Pseudomonas aeruginosa. J Bacteriol. (2008). , 190(9), 3147-54.

[67]   Blankenfeldt, W, Giraud, M. F, Leonard, G, Rahim, R, Creuzenet, C, Lam, J. S, et al. The purification, crystallization and preliminary structural characterization of glucose-phosphate thymidylyltransferase (RmlA), the first enzyme of the dTDP-L-rhamnose synthesis pathway from *Pseudomonas aeruginosa*. Acta Crystallogr D Biol Crystallogr. (2000). Pt 11):1501-4. Epub 2000/10/29., 1.

[68]   Allard, S. T, Giraud, M. F, Whitfield, C, Graninger, M, Messner, P, & Naismith, J. H. The crystal structure of dTDP-D-Glucose 4,6-dehydratase (RmlB) from *Salmonella enterica* serovar *Typhimurium*, the second enzyme in the dTDP-l-rhamnose pathway. J Mol Biol. (2001). Epub 2001/03/13., 307(1), 283-95.

[69]   Rahim, R, Ochsner, U. A, Olvera, C, Graninger, M, Messner, P, Lam, J. S, et al. Cloning and functional characterization of the Pseudomonas aeruginosa rhlC gene that encodes rhamnosyltransferase 2, an enzyme responsible for di-rhamnolipid biosynthesis. Mol Microbiol. (2001). , 40(3), 708-18.

[70]   Williams, P, & Camara, M. Quorum sensing and environmental adaptation in Pseudomonas aeruginosa: a tale of regulatory networks and multifunctional signal molecules. Curr Opin Microbiol. (2009). Epub 2009/03/03., 12(2), 182-91.

[71]   Williams, P. Quorum sensing, communication and cross-kingdom signalling in the bacterial world. Microbiology. (2007). Pt 12):3923-38. Epub 2007/12/01.

[72]   Dekimpe, V, & Deziel, E. Revisiting the quorum-sensing hierarchy in Pseudomonas aeruginosa: the transcriptional regulator RhlR regulates LasR-specific factors. Microbiology. (2009). Pt 3):712-23. Epub 2009/02/28.

[73]   Loh, K. C, & Cao, B. Paradigm in biodegradation using Pseudomonas putida-A review of proteomics studies. Enzyme Microb Technol. (2008). , 43(1), 1-12.

[74]   Deziel, E, Gopalan, S, Tampakaki, A. P, Lepine, F, Padfield, K. E, Saucier, M, et al. The contribution of MvfR to Pseudomonas aeruginosa pathogenesis and quorum sensing circuitry regulation: multiple quorum sensing-regulated genes are modulated without affecting lasRI, rhlRI or the production of N-acyl-L-homoserine lactones. Mol Microbiol. (2005). Epub 2005/02/03., 55(4), 998-1014.

[75]   Whiteley, M, Lee, K. M, & Greenberg, E. P. Identification of genes controlled by quorum sensing in Pseudomonas aeruginosa. Proc Natl Acad Sci U S A. (1999). Epub 1999/11/26., 96(24), 13904-9.

[76]   Schuster, M, & Greenberg, E. P. Early activation of quorum sensing in Pseudomonas aeruginosa reveals the architecture of a complex regulon. BMC Genomics. (2007). Epub 2007/08/24.

[77]   Argenio, D, Wu, D. A, Hoffman, M, Kulasekara, L. R, Deziel, H. D, & Smith, E. EE, et al. Growth phenotypes of Pseudomonas aeruginosa lasR mutants adapted to the airways of cystic fibrosis patients. Mol Microbiol. (2007). Epub 2007/05/12., 64(2), 512-33.

[78] Medina, G, Juarez, K, Valderrama, B, & Soberon-chavez, G. Mechanism of Pseudomonas aeruginosa RhlR transcriptional regulation of the rhlAB promoter. J Bacteriol. (2003). Epub 2003/10/04., 185(20), 5976-83.

[79] Aguirre-ramirez, M, Medina, G, Gonzalez-valdez, A, Grosso-becerra, V, & Soberon-chavez, G. The Pseudomonas aeruginosa rmlBDAC operon, encoding dTDP-L-rhamnose biosynthetic enzymes, is regulated by the quorum-sensing transcriptional regulator RhlR and the alternative sigma factor sigmaS. Microbiology. (2012). Pt 4): 908-16. Epub 2012/01/21.

[80] Jensen, V, Lons, D, Zaoui, C, Bredenbruch, F, Meissner, A, Dieterich, G, et al. RhlR expression in Pseudomonas aeruginosa is modulated by the Pseudomonas quinolone signal via PhoB-dependent and-independent pathways. J Bacteriol. (2006). Epub 2006/10/10., 188(24), 8601-6.

[81] Schuster, M, Lostroh, C. P, Ogi, T, & Greenberg, E. P. Identification, timing, and signal specificity of Pseudomonas aeruginosa quorum-controlled genes: a transcriptome analysis. J Bacteriol. (2003). Epub 2003/03/20., 185(7), 2066-79.

[82] Gupta, R, Gobble, T. R, & Schuster, M. GidA posttranscriptionally regulates rhl quorum sensing in Pseudomonas aeruginosa. J Bacteriol. (2009). Epub 2009/07/14., 191(18), 5785-92.

[83] Lequette, Y, Lee, J. H, Ledgham, F, Lazdunski, A, & Greenberg, E. P. A distinct QscR regulon in the Pseudomonas aeruginosa quorum-sensing circuit. J Bacteriol. (2006). Epub 2006/04/20., 188(9), 3365-70.

[84] Ledgham, F, Ventre, I, Soscia, C, Foglino, M, Sturgis, J. N, & Lazdunski, A. Interactions of the quorum sensing regulator QscR: interaction with itself and the other regulators of Pseudomonas aeruginosa LasR and RhlR. Mol Microbiol. (2003). Epub 2003/03/27., 48(1), 199-210.

[85] Jude, F, Kohler, T, Branny, P, Perron, K, Mayer, M. P, Comte, R, et al. Posttranscriptional control of quorum-sensing-dependent virulence genes by DksA in Pseudomonas aeruginosa. J Bacteriol. (2003). Epub 2003/05/31., 185(12), 3558-66.

[86] Perron, K, Comte, R, & Van Delden, C. DksA represses ribosomal gene transcription in Pseudomonas aeruginosa by interacting with RNA polymerase on ribosomal promoters. Mol Microbiol. (2005). Epub 2005/04/28., 56(4), 1087-102.

[87] Diggle, S. P, Winzer, K, Chhabra, S. R, Worrall, K. E, Camara, M, & Williams, P. The Pseudomonas aeruginosa quinolone signal molecule overcomes the cell density-dependency of the quorum sensing hierarchy, regulates rhl-dependent genes at the onset of stationary phase and can be produced in the absence of LasR. Mol Microbiol. (2003). Epub 2003/09/26., 50(1), 29-43.

[88] Mcknight, S. L, Iglewski, B. H, & Pesci, E. C. The Pseudomonas quinolone signal reg-
      ulates rhl quorum sensing in Pseudomonas aeruginosa. J Bacteriol. (2000). Epub
      2000/04/27., 182(10), 2702-8.

[89] Wade, D. S, Calfee, M. W, Rocha, E. R, Ling, E. A, Engstrom, E, Coleman, J. P, et al.
      Regulation of Pseudomonas quinolone signal synthesis in Pseudomonas aeruginosa.
      J Bacteriol. (2005). Epub 2005/06/22., 187(13), 4372-80.

[90] Suh, S. J, Silo-suh, L, Woods, D. E, Hassett, D. J, West, S. E, & Ohman, D. E. Effect of
      rpoS mutation on the stress response and expression of virulence factors in Pseudo-
      monas aeruginosa. J Bacteriol. (1999). Epub 1999/06/29., 181(13), 3890-7.

[91] Bertani, I, Sevo, M, Kojic, M, & Venturi, V. Role of GacA, LasI, RhlI, Ppk, PsrA, Vfr
      and ClpXP in the regulation of the stationary-phase sigma factor rpoS/RpoS in Pseu-
      domonas. Arch Microbiol. (2003). Epub 2003/07/25., 180(4), 264-71.

[92] Schuster, M, Hawkins, A. C, Harwood, C. S, & Greenberg, E. P. The Pseudomonas
      aeruginosa RpoS regulon and its relationship to quorum sensing. Mol Microbiol.
      (2004). Epub 2004/02/07., 51(4), 973-85.

[93] Medina, G, Juarez, K, Diaz, R, & Soberon-chavez, G. Transcriptional regulation of
      Pseudomonas aeruginosa rhlR, encoding a quorum-sensing regulatory protein. Mi-
      crobiology. (2003). Pt 11):3073-81. Epub 2003/11/06.

[94] Diggle, S. P, Winzer, K, Lazdunski, A, Williams, P, & Camara, M. Advancing the
      quorum in Pseudomonas aeruginosa: MvaT and the regulation of N-acylhomoserine
      lactone production and virulence gene expression. J Bacteriol. (2002). Epub
      2002/04/27., 184(10), 2576-86.

[95] Venkata-ramana, K, & Karanth, N. G. Factors affecting biosurfactant production us-
      ing Pseudomonas aeruginosa CFTR-6 under submerged conditions. Chem Technol
      Biotechnol (1989). , 45(1), 249-57.

[96] Mulligan, C. N, & Gibbs, B. F. Correlation of nitrogen metabolism with biosurfactant
      production by Pseudomonas aeruginosa. Appl Environ Microbiol. (1989). , 55(11),
      3016-9.

[97] Totten, P. A, Lara, J. C, & Lory, S. The rpoN gene product of Pseudomonas aerugino-
      sa is required for expression of diverse genes, including the flagellin gene. J Bacteriol.
      (1990). Epub 1990/01/01., 172(1), 389-96.

[98] Nishijyo, T, Haas, D, & Itoh, Y. The CbrA-CbrB two-component regulatory system
      controls the utilization of multiple carbon and nitrogen sources in Pseudomonas aer-
      uginosa. Mol Microbiol. (2001). Epub 2001/06/13., 40(4), 917 31.

[99] Li, L. L, Malone, J. E, & Iglewski, B. H. Regulation of the Pseudomonas aeruginosa
      quorum-sensing regulator VqsR. J Bacteriol. (2007). Epub 2007/04/24., 189(12),
      4367-74.

[100] Haussler, S, & Becker, T. The pseudomonas quinolone signal (PQS) balances life and death in Pseudomonas aeruginosa populations. PLoS Pathog. (2008). e1000166. Epub 2008/09/27.

[101] Calfee, M. W, Shelton, J. G, Mccubrey, J. A, & Pesci, E. C. Solubility and bioactivity of the Pseudomonas quinolone signal are increased by a Pseudomonas aeruginosa-produced surfactant. Infect Immun. (2005). Epub 2005/01/25., 73(2), 878-82.

[102] Wang, X, Gong, L, Liang, S, Han, X, Zhu, C, & Li, Y. Algicidal activity of rhamnolipid biosurfactants produced by Pseudomonas aeruginosa. Harmful Algae (2005). , 4(1), 433-43.

[103] Fujita, K, Akino, T, & Yoshioka, H. Characteristics of heat-stable extracellular hemolysin from Pseudomonas aeruginosa. Infect Immun. (1988). Epub 1988/05/01., 56(5), 1385-7.

[104] Beatson, S. A, Whitchurch, C. B, Sargent, J. L, Levesque, R. C, & Mattick, J. S. Differential regulation of twitching motility and elastase production by Vfr in *Pseudomonas aeruginosa*. J Bacteriol. (2002). Epub 2002/06/12., 184(13), 3605-13.

[105] Juhas, M, Wiehlmann, L, Huber, B, Jordan, D, Lauber, J, Salunkhe, P, et al. Global regulation of quorum sensing and virulence by VqsR in Pseudomonas aeruginosa. Microbiology. (2004). Pt 4):831-41. Epub 2004/04/10.

[106] Carty, N. L, Layland, N, Colmer-hamood, J. A, Calfee, M. W, Pesci, E. C, & Hamood, A. N. PtxR modulates the expression of QS-controlled virulence factors in the Pseudomonas aeruginosa strain PAO1. Mol Microbiol. (2006). Epub 2006/06/29., 61(3), 782-94.

[107] Gomez, E, Santos, V. E, Alcon, A, Martin, A. B, & Garcia-ochoa, F. Oxygen-uptake and mass-transfer rates on the growth of Pseudomonas putida CECT5279: Influence on biodesulfurization (BDS) capability. Energy and Fuels. (2006). , 20(4), 1565-71.

[108] Morici, L. A, Carterson, A. J, Wagner, V. E, Frisk, A, & Schurr, J. R. Honer zu Bentrup K, et al. Pseudomonas aeruginosa AlgR represses the Rhl quorum-sensing system in a biofilm-specific manner. J Bacteriol. (2007). Epub 2007/09/04., 189(21), 7752-64.

[109] Dong, Y. H, Zhang, X. F, An, S. W, Xu, J. L, & Zhang, L. H. A novel two-component system BqsS-BqsR modulates quorum sensing-dependent biofilm decay in Pseudomonas aeruginosa. Commun Integr Biol. (2008). Epub 2009/06/11., 1(1), 88-96.

[110] Brencic, A, Mcfarland, K. A, Mcmanus, H. R, Castang, S, Mogno, I, Dove, S. L, et al. The GacS/GacA signal transduction system of Pseudomonas aeruginosa acts exclusively through its control over the transcription of the RsmY and RsmZ regulatory small RNAs. Mol Microbiol. (2009). , 73(3), 434-45.

[111] Pessi, G, & Haas, D. Dual control of hydrogen cyanide biosynthesis by the global activator GacA in Pseudomonas aeruginosa PAO1. FEMS Microbiol Lett. (2001). Epub 2001/06/19., 200(1), 73-8.

[112] Heurlier, K, Williams, F, Heeb, S, Dormond, C, Pessi, G, Singer, D, et al. Positive control of swarming, rhamnolipid synthesis, and lipase production by the posttranscriptional RsmA/RsmZ system in Pseudomonas aeruginosa PAO1. J Bacteriol. (2004). Epub 2004/05/06., 186(10), 2936-45.

[113] Dong, YH, Zhang, XF, Xu, JL, Tan, AT, Zhang, LH, & Vqs, . , a novel AraC-type global regulator of quorum-sensing signalling and virulence in Pseudomonas aeruginosa. Mol Microbiol. 2005;58(2):552-64. Epub 2005/10/01.

[114] (Biosurfactants Market- Global Scenario, Raw Material and Consumption Trends, Industry Analysis, Size, Share and Forecasts, 2011- 2018 [database on the Internet]. [cited 10/11/2012]. Available from: http://www.transparencymarketresearch.com).

[115] Lee, S. Y, Lee, D. Y, & Kim, T. Y. Systems biotechnology for strain improvement. Trends Biotechnol. (2005). , 23(7), 349-58.

[116] Santa Anna LMSebastian GV, Pereira Jr N, Alves TLM, Menezes EP, Freire DMG. Production of biosurfactant from a new and promising strain of Pseudomonas aeruginosa PA1. Applied Biochemistry and Biotechnology- Part A Enzyme Engineering and Biotechnology. (2001).

[117] Abalos, A, Pinazo, A, Infante, MR, Casals, M, & Garc, . and antimicrobial properties of new rhamnolipids produced by Pseudomonas aeruginosa AT10 from soybean oil refinery wastes. Langmuir. 2001;17(5):1367-71.

[118] De Sousa, J. R. Da Costa Correia JA, De Almeida JGL, Rodrigues S, Pessoa ODL, Melo VMM, et al. Evaluation of a co-product of biodiesel production as carbon source in the production of biosurfactant by P. aeruginosa MSIC02. Process Biochemistry. (2011). , 46(9), 1831-9.

[119] Benincasa, M, Contiero, J, Manresa, M. A, & Moraes, I. O. Rhamnolipid production by Pseudomonas aeruginosa LBI growing on soapstock as the sole carbon source. Journal of Food Engineering. (2002). , 54(4), 283-8.

[120] Kohler, T, Curty, L. K, Barja, F, Van Delden, C, & Pechere, J. C. Swarming of Pseudomonas aeruginosa is dependent on cell-to-cell signaling and requires flagella and pili. J Bacteriol. (2000). , 182(21), 5990-6.

[121] Guerra-santos, L. H, Käppeli, O, & Fiechter, A. Dependence of Pseudomonas aeruginosa continous culture biosurfactant production on nutritional and environmental factors. Appl Microbiol Biotechnol. (1986). , 24(6), 443-8.

[122] Manresa, M. A, Bastida, J, Mercade, M. E, Robert, M, De Andres, C, Espuny, M. J, et al. Kinetic studies on surfactant production by pseudomonas aeruginosa 44T1. J Ind Microbiol. (1991). , 0(2), 133-6.

[123] Veglio, F, Beolchini, F, & Toro, L. Kinetic modeling of copper biosorption by immobilized biomass. Industrial and Engineering Chemistry Research. (1998). , 37(3), 1107-11.

[124] Kronemberger FDABorges CP, Freire DMG. Fed-Batch Biosurfactant Production in a Bioreactor. International Review of Chemical Engineering. (2010). , 2(4), 513-8.

[125] Pacheco, G. J, & Reis, R. S. Fernandes ACLB, da Rocha SLG, Pereira MD, Perales J, et al. Rhamnolipid production: effect of oxidative stress on virulence factors and proteome of Pseudomonas aeruginosa PA1. Appl Microbiol Biotechnol. (2012). , 95, 1519-29.

[126] Desai, J. D, & Banat, I. M. Microbial production of surfactants and their commercial potential. Microbiol Mol Biol Rev. (1997). , 61(1), 47-64.

[127] Vemuri, G. N, & Aristidou, A. A. Metabolic engineering in the-omics era: Elucidating and modulating regulatory networks. Microbiol Mol Biol Rev. (2005). , 69(2), 197-216.

Chapter 4

Ecotoxicological Behavior of some Cationic and Amphoteric Surfactants (Biodegradation, Toxicity and Risk Assessment)

Stefania Gheorghe, Irina Lucaciu, Iuliana Paun, Catalina Stoica and Elena Stanescu

Additional information is available at the end of the chapter

1. Introduction

Detergents industry is a competitive industry, with a large opening to innovation and economical development. Although very good for sanitation, the big domestic and industrial detergents consumption has a significant contribution to surfactants concentrations increase in towns' sewage and implicit to surface water and groundwater contamination [1] (Figure 1). The negative effects manifested by the presence of surfactants in surface water are mostly due to superficial – active proprieties – detergents surfactants characteristic, indifferently of class type. In accordance with molecule charge, the surfactants are grouped in four categories: anionic, cationic, nonionic and amphoteric [2].

This chapter is focus on cationic and amphoteric surfactants frequently used in laundry and dishes detergents, fabric softeners, personal care products and biocides. Cationic and amphoteric surfactants control was not required until 2004, when the European Detergents Regulation no. 648 entered into force, especially because there were no standard methods for quantitative determination of these types of surfactants [3]. Also, the biodegradation assessing was not requested and there is no European standard method for this testing. These surfactants are not currently limited by national or international norms relating to waste waters and surface waters quality. Literature references concerning ecotoxicological characteristics and risk assessment of cationic and amphoteric surfactants are relatively reduced.

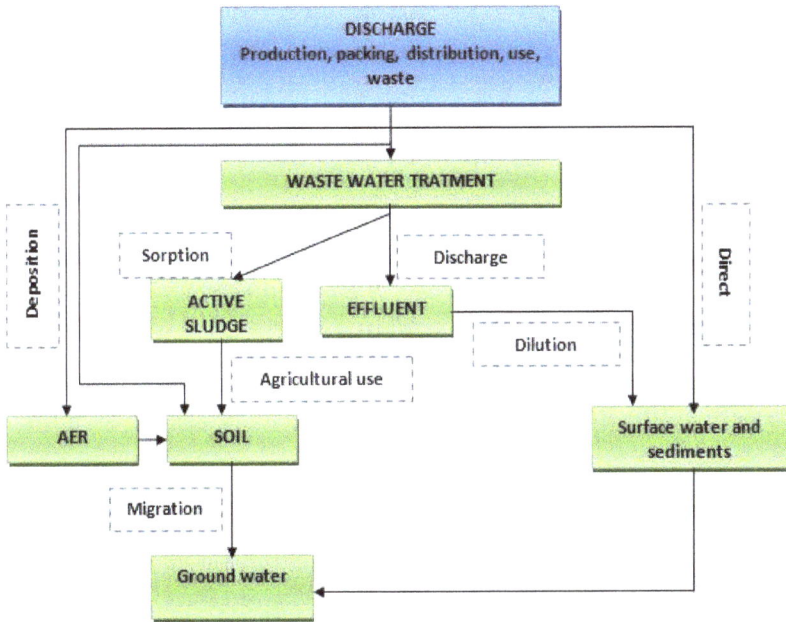

Figure 1. Surfactants environmental contamination [1]

2. Detergents legislative framework

At European level, detergents and cleaning products have a special place in legislative framework of European Community because are manufactured in big quantity and they may affect the environment during both manufacture and using processes. In the last years surfactants biodegradability was the most significant problem.

The chapter broaches a new and important actuality theme at international level, namely *the implementation of the most important European legislative regulations concerning detergents and cleaning products* – Regulation (EC) no. 648/2004 and it amendments. The present Regulation establish strict rules to assure the free circulation of detergents – products for consummators and industrial and institutional products and implicit of surfactants on UE market, so that the human health and environmental protection to be guaranteed at high level. A significant request of Regulation is that each producer/ importer / distributor to attest *ultimate aerobic biodegradability of surfactants used in detergents* [3].

In 2006 have become applicable Regulation (EC) no. 907/2006 – through that is follow the assurance of a environmental higher level protection(impose for detergents (a) *biodegradability* and (b) *conformity with at least one ultimate biodegradability tests* specified in Annex III) and human health (impose requests concerning the information's which must be written on the detergents packages) [4].The last important amendment of Detergents Regulation, is the norm (EU) no. 259/2012 which standardized the use of phosphates and other phosphorous compounds in household laundry detergents and automatic dishwashing for consumers.

Also at European level is applicable the Technical Guidance for stratified approach of Regulation (EC) no. 648/2004, emitted in 2005, which provide that the use of surfactants in detergents is allowable unless that surfactants fulfills the aerobe degradation criteria even if are subject to direct testing as individual substance (mineralization) or through interpolation. For the surfactants which not success to pass one between these mineralization tests, but which respect the primary biodegradability criteria may request a derogation for its utilization in industrial and institutional detergents. These derogations are obtained, in base of environment safety concerning the assessments for the metabolites which may result at the surfactant biodegradation. All assessments will be stratified performed (Figure 2), in accordance with a phased process which will provide all the information's concerning the environmental risks of the recalcitrant metabolites resulted after biodegradation. For passing the complementary risk assessment it is necessary to show that the PEC does not surpass the PNEC of the metabolites [5].

Environmental European legislation showed that only anionic and non-ionic surfactants have set limit values, while the cationic and amphoteric surfactants have not imposed limits in waste waters or surface water, even though they have a frequent use in cleaning products and biocides.

At international level exist some actions to encourage the producers to obtain safe cleaning products, transposed in Regulation (EC) no. 66/2010 concerning UE ecological label. The ecological labeling of products is facultative and promotes the security of detergents on the entire life cycle: from the raw materials, production process, packing, distribution, use, recycling and elimination. Through ecological labeling is trying the reduction of hazardous chemicals use, with effects on water, air and soil and of carcinogenic and allergic risks. The detergents with the European Ecolabel contain no hazardous substances to the aquatic environment; have a increased biodegradability, and an efficient use that does not cause damage to the environment [6].

Beginning with February 2009, the most representative European associations (AISE, CESIO, CEFIC) have informed about their initiative to undertake further researches in order to: establish the surfactants ecotoxicity and assess the potential environmental risk; develop an improved method for measuring the anaerobic biodegradability under sludge digester conditions; and to evaluate the biodegradation of the main organic non-surfactant ingredients from detergents [7, 8].

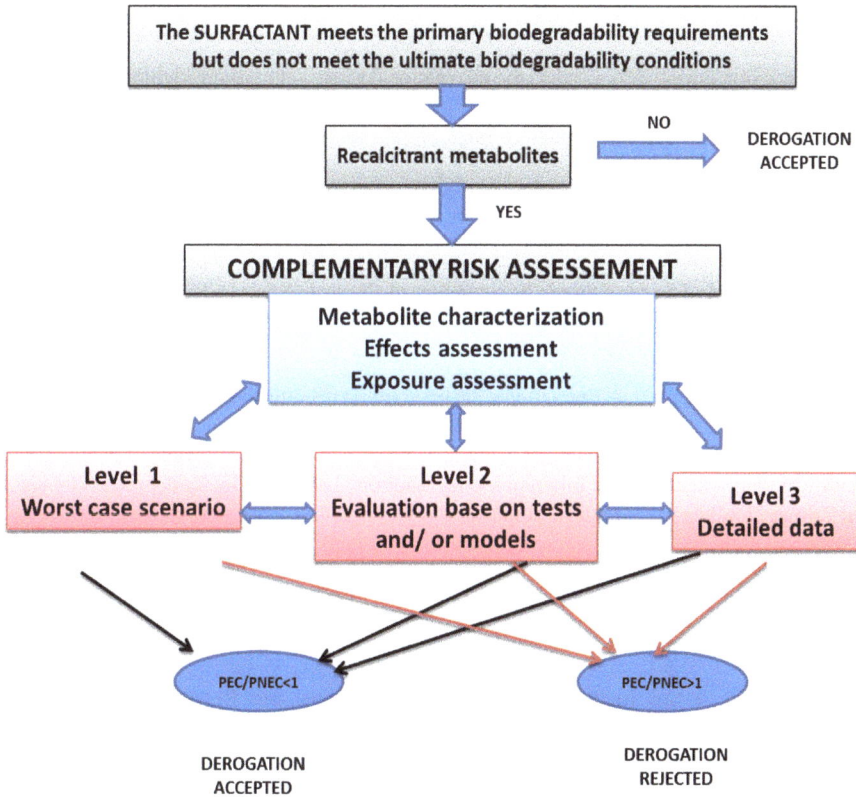

Figure 2. Complementary risk assessment, adapted from CESIO and AISE

3. Surfactants ecotoxicity

In literature there are many studies to evaluate the ecotoxicity of anionic and non-ionic surfactants, and therefore future research should be directed especially to elucidate the toxic effects of cationic and amphoteric surfactants whose ecotoxicological profile is unknown, and their physical and chemical properties can significantly interfere in the results of the toxico-logical studies.

According to CESIO reports, half of detergents consumption has been used in domestic applications and other half in cosmetic industry, metal processing, paper and leather industry. In 2007 the most used surfactants were anionic and non-ionic surfactants covering half of produced surfactants [9]. In 2008 it was estimated that in Eastern Europe were used annually > 4.2 million tons of detergents and 1.2 million tons of softeners, up to 2006 [10].

It was found that during 1990-2010, in the international waste waters were identified following surfactants concentrations: anionic 330 - 9450 μg/L; nonionic 5 - 395 μg/L; cationic 0.1 - 325 μg/L (even 6000 μg/L in hospitals waste waters) [1,10-12]. No data on amphoteric surfactants were identified.

In surface water were estimated following concentrations: anionic < 4 - 81 μg/L; nonionic <0.002 - 31 μg/L; amphoteric < 0.01 to 3.8 μg/L; cationic < 0.1 - 34 μg/L [13, 14].

According to our research studies, in Romania, in the last 10 years, the concentrations of surfactants in waste waters and surface waters were: anionic 0.3-9 mg/L; non ionic 0.05 - 4 mg/L; cationic 0.03 – 0.35 mg/L; amphoteric 0.02 -0.05 mg/L.

According with international regulations, the first criteria in environmental risk assessment of surfactants is to assess their biodegradation. Biological degradation of surfactants could be performed by a several tests which ones decrease in order of stringency as followed: Ultimate / Readily biodegradability tests, Inherent biodegradability tests, Rapidly biodegradability tests and Primary biodegradability tests [15].

The ultimate biodegradability tests are recommended to assess the biodegradation of surfactants, because by using them, we can control whether surfactants are degraded in the presence of microorganisms to the metabolites (non-surfactants), mineral salts, biomass and CO_2 (the measured parameters).

Biodegradability testing methodology is required by Detergents Regulation no. 648/2004 (Annexes no. III and VIII), which provides degradation limits of surfactants used in cleaning products [16]. All data concerning biodegradability, use informations, consumption and current conditions of environmental exposure of the substance, make it possible to PEC (Predict Environment Concentration) of the substance.

Legislation, in force requires a primary biodegradability of cationic and amphoteric surfactants greater than 80%. In terms of ultimate biodegradability (Table 1), these compounds are finally degraded under aerobic (>60%-100%) and anaerobic conditions (64-100%). Some problems are highlighted for quaternary cationic surfactants and amphoteric alkyl betaines in both conditions.

Surfactants products have some negative effects on surface waters as: decreasing of air / water oxygen transfer, water quality damage because of foam, sorption on solid particles preventing the sedimentation, reduction of river self-cleaning capacity, affecting the gases transfer between the microorganism cells and have a great toxicity on the aquatic organisms in trophic level.

Toxicological behavior is the second criterion in environmental risk assessment. Detergents show toxic effects for all aquatic organisms if there are present in sufficient amounts and that include biodegradation products. Most fish die when the detergent concentration in water is about 15 mg/L and also, at concentrations above 5 mg/L cause the death of eggs and affecting the fish reproduction [17]. Another study reported that 0.4 to 40 mg/L of detergents induce toxic effects by damaging the gills, growth delay, alteration of feeding process and the inhibition of the organs chemoreceptors in vertebrates. In case of invertebrates at detergent concentrations below that 0.01 mg/L, the reproduction, growth and development are disturbed [18].

Generally, the toxicity of surfactants is influenced by a range of abiotic and biotic factors. The abiotic factors, eg. physico-chemical properties of water (pH, hardness, other polar substances, dissolved oxygen, suspended matters) lead to a low bioavailability of the compound to aquatic organisms. Also, the physico-chemical properties of the surfactant (the size of aliphatic chain [26, 32], type of surfactant, absorption capacity and concentration) have a great influence of the toxicity level. Biotic factors generally refer to: age of organisms, tested species [33, 34], sensitivity between species [35] and acclimatization at very low concentrations of detergent [10, 18].

As long as, it is practically impossible to perform bioassays on all aquatic food chain, in order to assess the ecotoxicological effects of chemicals (Figure 3), at international level, certain representative aquatic food chain species were established, as follows: microorganisms, algae, crustaceans (benthic and planktonic) and fish. With the REACH Regulation implementation [36], the eco-toxicity tests were diversified by applying of microbiotests [37] as an alternative to conventional methods, in order to reduce or replace animal testing (highlighted in the OECD, ISO and EPA methodology – Table 2).

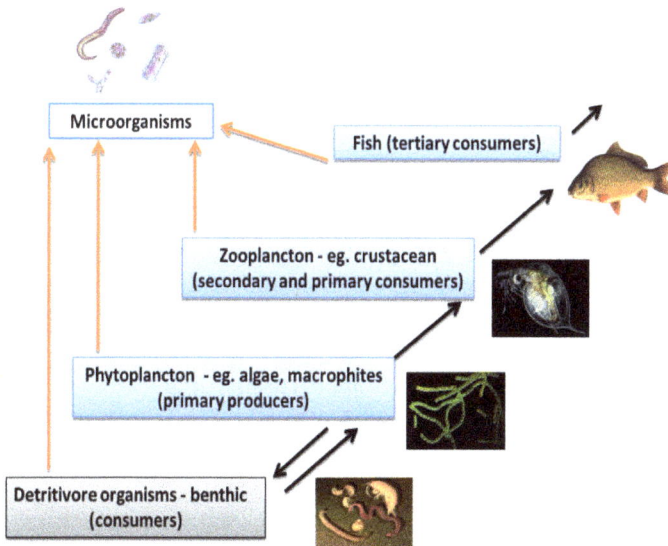

Figure 3. Aquatic food chain

Surfactant type	Biodegradability
CATIONIC SURFACTANTS	
Esterquats (DEEDMAC, HEQ, TEAQ, etc.)	79% (ISO 10708) [1] 80->85% (OECD 301B) [11, 19] 90% (OECD 301F) [20] 75% (OECD 302B) [11] >60% (OECD 301D) [21] 64-100% (ECETOC) [9] 73-100% (ECETOC)) [11, 20]
Diesterquats	92% (OECD 301A) [22] 90% (OECD 301B) [23]
Ammonium quaternary compounds (eg. DSDMAC, DTDMAC, ATMAC)	>5% (OECD 301D) [1] 0 -24%(ECETOC) [9] 40-81% (OECD 301F, 301B, 302A) [24]
Other cationic surfactants (hydrogenated chain)	63% (OECD 301A) [22]
AMPHOTERIC SURFACTANTS	
Alkyl betaines (dimethylaminebetaines / alkyl amidobetaines)	99% (OECD 301A) [22] 0% - >60%(ECETOC) [9] >60%(ThOD) (OECD301D) [25] 60 – 100% (ISO 14593) [26]
Hidroxysulfobetaines	40-47% [25, 27]
Imidazoline derivatives (cocoamphoacetates / alkyl amphoacetates/ alkylamino propionates)	>60% (ECETOC) [9] >60% [25] 80-90% (OECD 301E) [1] 79.8% (ECETOC) [9] 80-100%(ISO 14593) [26] 60-79% (OECD 301D, 301E) [27] 2.5% ThGP (ISO 11734) [27]
Cocamidopropylbetaine / Coco alkyl derivatives	82% (ThOD), 95% (COD) (OECD 301C) [28] 90- 100% [29,30] 97% (OECD 301A) [22] 57 – 84% (ThOD) (OECD 301D) [25] 45 – 75% ThGP (ISO 11734) [27]
Other amphoteric surfactants	97% (DOC) (OECD 302A) [31] 60% (ThCO2), 70% (DOC)(OECD 301B) [31]

Table 1. Cationic and amphoteric surfactants biodegradability

Usually, it is recommended to perform the standardized OECD and ISO methodology for assessing of ecotoxicity. The toxicity level of surfactants is assessed using tests batteries covering all trophic level of aquatic environment, and it is recommended the acute and chronic tests in classical or alternative system. Both cationic and amphoteric surfactants cause high or moderate acute toxicity on fish, crustaceans, algae and bacteria. It is noted that the ranges of toxicity values are very large and diversified, even for the same aquatic organism or test method and for this reason the literature is very permissive (Table no.3).

No.	Toxicity tests	OECD	ISO/EPA	Microbiotests
Vertebrates (tertiary consumers)				
1.	Fish acute toxicity test, (static test/ semi static test/ dynamic test)	203	ISO 7346: 1,2,3 ISO 13216 ISO 10229 ISO/CD 15088-1 EPA 2000.0; 2004.0; 2006.0	(Q)SAR and ECOSAR methods Fish cell cytotoxicity tests Genetic tests Endocrine tests [approved by ICCVAM, ECVAM, OECD, EPA, SETAC, ECETOC in order to reduce / replace the animals used in toxicity tests according to REACH]
2.	Fish, prolonged toxicity test- 14 days study	204	-	
3.	Fish juvenile growth tests	215	-	
4.	Fish, early-life stage toxicity test	210	-	
5.	Fish, Short term toxicity test on embryonic stages	212	ISO 12890 EPA 1000.0	
6.	Bioaccumulation in fish	305	-	
7.	Fish sex development test	234	-	
Zooplankton (primary and secondary consumers)				
8.	*Daphnia magna*, acute immobilization test (static and semi static test)	202	ISO 6341 ISO 14669 EPA 2002.0; 2021.0	Daphtoxkit F, Daphnia IQ Test (*Daphnia magna, Daphnia pulex*) Thamnotoxkit F (*Thamnocephalus platyrus*) Rotoxkit F (*Brachinous calyciflorus*) Rotoxkit M (*Brachinous splicatilis*) Ceriodaphtoxkit K (*Ceriodaphnia dubia*) Ostracodtoxkit (*Heterocypris incongruens*)
9.	*Daphnia magna*, reproduction test	211	ISO 10706 EPA 1002.0	
10.	*Daphnia magna*, chronic toxicity test	-	ISO 10706 ISO/DIS 20665 ISO/WD 20266	
Phytoplankton (primary producers)				
11.	Fresh algal growth inhibition test, *Pseudokirchnerilla subcapitata*	201	ISO 8692 SR 13328 EPA 1003.0	Algaltoxkit F (*Raphidocelis subcapitata, Selenastrum capricornutum, Chlorella vulgaris*)

No.	Toxicity tests	OECD	ISO/EPA	Microbiotests
Benthic organisms				
12.	Water - Sediment toxicity, *Chironomus riparius*	218, 219	-	Ostracodtokit(*Heterocypris incongruens*) Microtox (*Vibrio fischeri*)
Aquatic floating plants				
13.	Growth inhibition tests, *Lemna minor*	221	ISO 20079	-
Microorganisms				
14.	Inhibition of oxygen consumption by active sludge	209	ISO 8192	Microtox (*Vibrio fischeri, Photobacterium phosfophoreum*)
15.	Inhibition of nitrification of active sludge microorganisms	-	ISO 9509	Test ECHA (*Bacilus stearithermophilis*) Toxi-chromotest PAD, MetPAD, MetPLATE, FluoroMetPLATE, SOS – Chromotest, Toxi-chromotest (*Escherichia coli*)
16.	Bioluminescent bacteria inhibition test, *Vibrio fischeri*	-	ISO 11348-1,2,3	
17.	Bacteria growth inhibition test, *Pseudomonas aeruginosa*	-	ISO 10712	Muta- ChromoPlate (*Salmonella typhimurium mutant*) MARA test (*with 11 bacteria sp.*)

Table 2. OECD / ISO / EPA and microbiotests methods generally used in UE for aquatic toxicity assessment of chemicals / environmental samples

4. Laboratory experiments

4.1. Chemicals

To assess the ecotoxicity and risk assessment of cationic and amphoteric surfactants, seven compounds were selected:

- *cationic surfactants*: dialkylhydroxyethyl ammonium methasulphate (TEAQ) C16-C18, commercial name TETRANYL AT 7590, CAS: 93334-15-7, 1.017 meq/g, Kao Corporation S.A; Cetylpyridinium bromide, CAS: 140-72-7; benzenthonium chloride monohydrate, commercial name HYAMINE 1622, CAS: 121-54-0, >96% (Sigma-Aldrich).tow softeners base on TEA esterquats CAS 91995-81-2 and CAS 157905-74-3;

- *amphoteric surfactants:* laurilamidopropylbetaine / cocamidopropylbetaine – CAPB, commercial name AMFODAC LB, CAS: 4292-10-8, 34.6 %, Sasol Italy S.P.A; and a commercial toilet detergent base on CAPB

The ecotoxicity experiments were performed for individual surfactants, mixtures of the cationic with amphoteric surfactants and different products base on cationic and amphoteric compounds, in order to obtain a complex response of the surfactants toxicity.

Aquatic toxicity of surfactants
L(E)C50/ NOEC [mg/L]

CATIONIC SURFACTANTS

Esterquats [1, 11, 12, 21, 23, 25, 38 - 47]

Fish: 0.63-42 mg/L / 3.5 mg/L
Crustacean: 0.38 – 45 mg/L / 1 -3 mg/L
Bacteria: 10->130 mg/L / 0.9 -2.7 mg/L
Algae: 0.06 – 11 mg/L / 0.16 – 4.8 mg/L

Ammonium quaternary compounds [1, 10, 25, 48-51]

Fish: 0.62 -4.5 mg/L / 0.58 mg/L
Crustacean: 0.13 – 18 mg/L /0.18 -1.34 mg/L
Bacteria: 0.15- 6.9 mg/L
Algae: 0.05 – 18 mg/L / 0.12 mg/L

Alkyl dimethyl benzyl ammonium chloride [25, 52]

Fish: 0.28-2 mg/L / 0.004 – 0.03 mg/L
Crustacean: 0.004 – 0.006 mg/L
Algae: 0.67 – 1.8 mg/L

Alkyl trimethyl ammonium salts [25, 53 - 56]

Fish: 0.36 - 8.6 mg/L
Crustacean: 0.1 – 24 mg/L /0.43 -0.05 mg/L
Algae: 0.03 – 0.38 mg/L

Other cationic surfactants [51]

Fish: 0.07 – 24 mg/L
Crustacean: 0.07 - > 5 mg/L

AMPHOTERIC SURFACTANTS

Imidazoline derivatives [1]

Fish: 8.1 mg/L
Crustacean: 41 - 520 mg/L
Bacteria: 22 -900 mg/L

Coco alkyl derivatives [10, 25, 26, 30, 31, 57, 58]

Fish: 2 - 31 mg/L / 0.16 – 1.7 mg/L
Crustacean: 2.15 - 48 mg/L / 0.9 -1.6 mg/L
Bacteria: 5.2 - 78 mg/L
Algae: 0.09 – 48 mg/L / 0.09 - 10 mg/L

Table 3. Aquatic toxicity data for cationic and amphoteric surfactants

4.2. Analytical control

The methods used for qualitative and quantitative analytical control of surfactants are spectrometric, titrimetric and chromatographic [16, 59-64].

In our studies, the analytical control of cationic and amphoteric surfactants in the synthetic solutions used in ecotoxicity tests and also in the environmental samples (waste water and surface water) was performed according to the standard methods specified in Annex II of

Detergents Regulation (spectrometric methods: DIN/EN 38409/1989-20 for cationic surfactants and Orange II method - Boiteux 1984 for amphoteric surfactants). The performance parameters are shown in Table 4. According to other scientific studies [2, 65], our results were comparable.

Performance parameters of the methods	DIN 38409:1989, part 20 Cationic surfactants	Boiteux 1984 method, Amphoteric surfactants
	Hyamine 1622 (99.99%)	Cocamidopropylbetaine (34.6%)
Wavelength	628 nm	485 nm
Accuracy	96.8%	99%
Fidelity [CV (RSD)]	6.508 %	3.392 %
Repeatability (r)	0.088 mg/L	0.1128 mg/L
Intern reproducibility (R_L)	0.0115 mg/L	0.5161 mg/L
Calibration curve equation	x= 6.9108y –0.0069	x= 5.7837y –0.0925
Detection limits (LoD)	0.003 mg/L	0.002 mg/L
Quantification limits (LoQ)	0.035 mg/L	0.032 mg/L
Concentrations domain	0.003 - 4 mg/L	0.002 - 2 mg/L
Recuperation	80 % - 110 %	80 % - 110 %
Interferences	Small concentrations of anionic surfactants. This interference may be eliminated through the use of the ions exchange resins column.	Small concentration of cationic and anionic surfactants; this interference could be remove by pH adjustment at alkaline values or the use of the ions exchange resins column.

Table 4. Methods performance parameters for quantitative determination of cationic and amphoteric surfactants

Detection limits of spectrometric methods are 0.003 mg/L for cationic surfactants and 0.002 mg/L for amphoteric surfactants. The methods interferences are determined by the presence of other types of surfactants (anionic, cationic) and/or other organic substances, which react with the surfactant or with the color reagent to form stable compounds. These problems can be eliminated by using of ion exchange resins to separate the target surfactants.

Methods selectivity was ensured by using of standard curves performed for the main studied substances. For the selective detection of cationic compounds in environmental samples is recommended the use of standard HPLC techniques.

5. Biodegradability assessment

A significant request of Detergents Regulation is that each producer / distributor must to attest ultimate aerobic biodegradability of surfactants used in detergents. Our experiments target was to assess the primary and ultimate biodegradability for 2 surfactant raw materials (cationic - ammonium quaternary compounds and amphoteric – alkyl betaines), their mixture and 2 commercial cleaning products based on this type of surfactant.

For the compliance of the first criterion of aquatic risk assessment (biodegradability), was used OECD methodology specified in Annex III of Detergents Regulation (OECD 303A – similar with ISO 11733) –Simulation Test – Aerobic Sewage Treatment for primary biodegradability [3, 66]; OECD 301A (similar with ISO 7827) - DOC Die – Away Test [67] and OECD 301D (similar with ISO 10707) – Closed Bottle Test [68])

5.1. Primary biodegradability

OECD confirmatory test for primary biodegradability assessment of surfactants describes a small – activated sludge plant in continued flow (Figure 4), consisting in a vessel for synthetic sewage, an aeration vessel, a settling vessel, air-lift pumps to recycle the activated sludge and vessel for collecting the treated effluent. The degradation test was performed at 19-24°C and the duration of experiments was about 60 days. The monitored parameters of experimental equipment were the surfactants concentrations and chemical oxygen demand (COD) in influents and effluents, the content of dry mater in the activated sludge and oxygen concentration from aeration tank vessel.

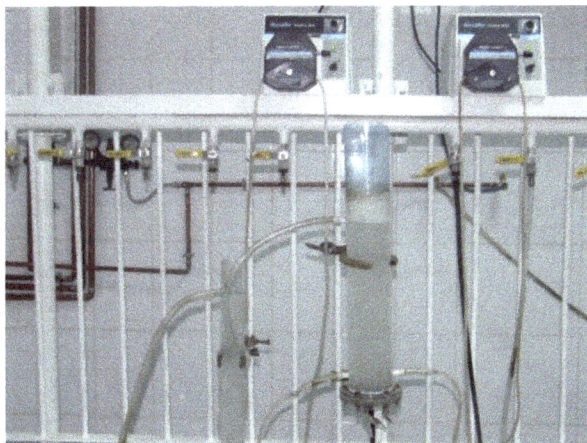

Figure 4. Laboratory simulation of aerobic sewage treatment

The efficiency of the biodegradation process (COD removal) and the percentage of biodegradability (surfactants degradation) were calculated (Table 5). Surfactants biodegradability was calculated as an arithmetic mean of daily removal efficiency values of surfactants, obtained in effective biodegradation period, during which degradation has been regular and the operation of the equipment trouble-free (Figure 5).

Result type	Tetranyl AT 7590	Hyamine 1622	CAPB	Hyamine 1622 + CAPB	Tetranyl AT 7590 + CAPB	Cetylpyridinium bromide
Test time (days)	36			30		36
Lag time (days)	10			12		10
Effective biodegradation time (days)	26			18		26
COD removal (%)	70 - 89		90	61	68	50
Cationic surfactant removal (%)	90	84	-	-	-	77 - 80
Amphoteric surfactant removal (%)	-	-	99	-	-	-
Total surfactants removal (cationic + amphoteric) %	-	-	-	80	90	-
Biodegradation (%)	91	84	97	90.7	97.6	80

Table 5. Primary biodegradability of cationic and amphoteric surfactants

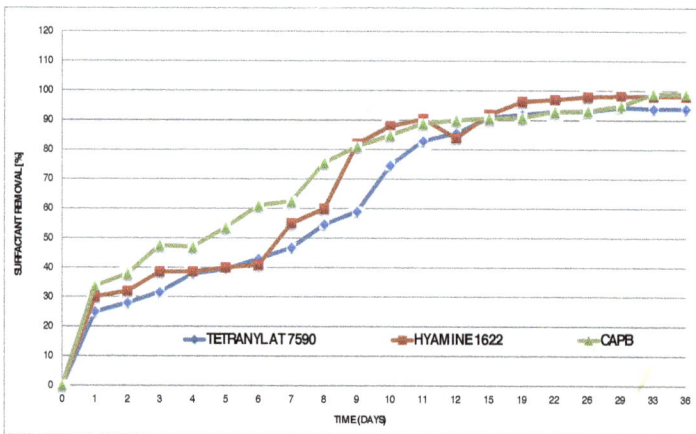

Figure 5. Primary biodegradability - individual surfactants removal

Considering the OECD confirmatory test, the level of primary biodegradability must be at least 80% in 21 days after the biological system initiation, for that the surfactant can be accepted as biodegradable and used as basic compound in commercial products. Our results showed a primary biodegradation > 90% for the cationic surfactant (TETRANYL AT 7590), amphoteric surfactant (cocamidopropylbetaine) and mixtures (cationic + amphoteric), while for the HYAMINE 1622 (cationic standard) and commercial product (cationic biocide - cetylpyridinium bromide) the primary biodegradation is in limit of 80-84%.

According to biodegradability criteria imposed by Regulation 648/2004, all the testing surfactants meet the conditions; the levels of biodegradation obtained were ≥ 80% [69-71]. The results were in line with the literature data on primary biodegradability of cationic and amphoteric surfactants (Table 5) [11, 22, 27].

5.2. Ultimate biodegradability

OECD 301A (ISO 7827) - DOC Die – Away Test, allowed the ultimate biodegradability assessment of substances / chemical products, in a given concentration, in a synthetic media, subject to aerobe microorganisms. According to this method, cleaning product – toilet detergent solution based on amphoteric surfactant was tested. The concentrations of DOC and amphoteric surfactant were determined and the percentage DOC / surfactant removal were calculated (Table 6). The obtained remove percentages were graphically represented in Figure 6.

Result type	Toilet detergent based on CAPB
Experimental period (days)	30
Maximum level of biodegradation (%)	91.43 – removal of DOC
Lag time (days)	3
Biodegradation time (days)	20
Amphoteric surfactant removal after 30 days (%)	72.85
Abiotic removal for DOC (%)	14

Table 6. Ultimate biodegradability test results for an amphoteric product

Biodegradability test performed considers that a substance is biodegradable if no significant abiotic removal was observed, the curves shows a typical form with lag and degradation phase and the DOC removal can be attributed to the biodegradation process of the substance. In conclusion, our results considered that:

• The total removal of dissolved organic carbon (DOC %) for the testing product (toilet cleaner based amphoteric surfactant - cocamidopropyl betaine) is ~ 92%, with an abiotic elimination of 14%;

• Effective biodegradation (the interval between the end of the lag time and the necessary time for the 90% DOC removal) is 20 days;

• Toilet cleaner commercial product base on amphoteric surfactant is biodegradable.

In line with the literature we estimated that 91.43 % of ultimate biodegradability obtained for cocamidopropilbetaine is within the range of 57% -100% specified for the same method or different methods recommended by OECD for ultimate biodegradability testing of amphoteric surfactants (see Table 1).

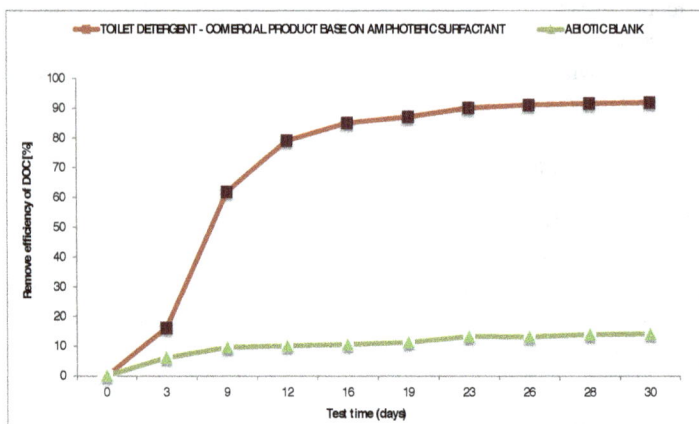

Figure 6. Ultimate biodegradability efficiency chart for the amphoteric product

OECD 301D (ISO 10707) – Closed Bottle Test, allows the ultimate biodegradability assessment of organic compounds present in a given concentration, subject to aerobe microorganisms, through biochemical oxygen demand (BOD) analyze. The test substance solution (which is the single source of carbon and energy), is inoculated with a little number of mixed aerobe microorganisms, incubated at dark, in closed and well filled recipients. The biodegradability indicator is dissolved oxygen concentration, parameter that is measured at regular intervals in a standard 28-days period. At the end of each time interval was calculated the oxygen removal, as a difference between the oxygen concentration of test substance and the control (Table 7). A biodegradability curve with the testing time on abscises and the biodegradability mean percentages on ordinate were plotted, for each time moment (Figure 7).

According with OECD 301D methodologies, an organic compound is biodegradable, when the biodegradation percentage is ≥60%, after 28 days of testing.

The experimental results obtained, have shown that the studied cationic surfactants and the cleaning product – laundry softeners are finally biodegraded with >70%. Considering other biodegradability studies for similar cationic compounds, our experimental results for TET-RANYL AT 7590 (78%) and fabric softeners based on it (77%, 85%) can be compared with the ultimate biodegradability values of esterquat and diesterquat cationic compounds, ranging from >60% - 79% using the same method and 75% - 92% using other OECD methods.

Regarding the ultimate biodegradability for the benzenthonium chloride, the literature, specify a range of 0-81% biodegradability using various OECD methods and >5% using the OECD 301D method (Table 1). Therefore, the percentage of biodegradation – 67% after 28 days, obtained in our laboratory experiments can be correlated with existing data.

Result type	Tetranyl AT 7590	Hyamine 1622	Laundry balm I	Laundry balm II	Aniline
Test time (days)			28		
Lag time (days)	7-10	7-10	7-10	7	7
Effective biodegradation time (days)	14	14	14	21	12-14
Remove of surfactants (%)	83.86	73.15	84.70	99.16	-
Biodegradability after 28 days (%)	78.37	67.23	77.34	85.80	95.23

Table 7. Ultimate biodegradability results for cationic surfactants

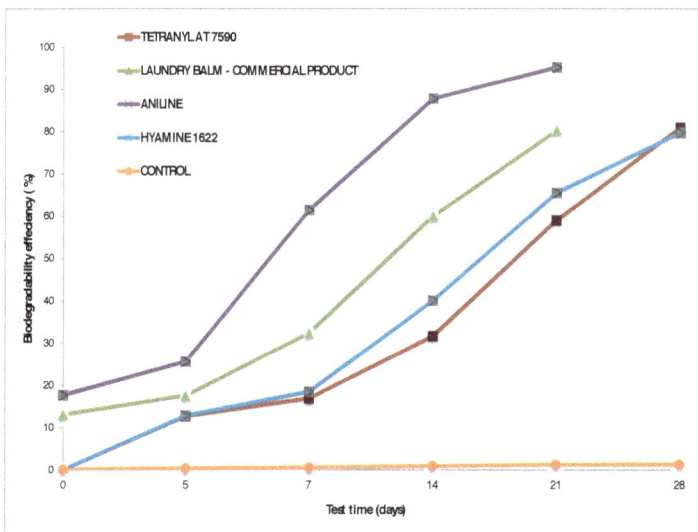

Figure 7. Ultimate biodegradation of cationic surfactants

6. Acute toxicity assessment

Considering the second criterion of ecotoxicological characterization / aquatic risk assessment, this chapter part aimed to evaluate the aquatic toxicity of surfactants on the most representative species of the Romanian surface waters. In accordance with the Europeans norms concerning surfactants and chemicals and OECD/ ISO/ ASTM testing methodology [72 - 75], the present study want to highlight the direct and indirect effects of cationic and amphoteric surfactants (benzenthonium chloride, dialkylhydroxyethyl ammonium methasulphate and cocamidopropylbetaine).

6.1. Direct toxicity

To evaluate the acute toxic effects of surfactants have conducted laboratory experiments in static and semi-static conditions, with distinct organisms from aquatic food chain, as followed:

- Acute lethal toxicity test with freshwater fish(1 year juvenile carp – Cyprinus carpio sp.) performed for determination of the mean Lethal Concentration which induce the death of half from the test organisms (fish) - LC50, according with OECD 203. The fish are exposed to testing surfactants, in different concentrations (0.5 mg/L – 100 mg/L), for 96h. The effect (mortality) is registered at each 24h and the concentration which kills 50% of fish at the final of test period is calculated.

- Acute toxicity test with water fleas Daphnia magna Status (Cladocera crustacea) performed for determination of Effective Concentration (EC50) which have a 50% impact on test organisms using Daphtoxkit FTM magna microbiotest, in accordance with OECD 202. The 24h to 48h EC50 bioassays was performed in disposable multiwall test plates starting from neonates of Daphnia magna, uniform in size and age, hatched from ephippia and exposed to different concentration of surfactant (0.05 mg/L – 50 mg/L) at 20oC, in darkness.

- Green algae growth inhibition test performed for determination of inhibitory / stimulatory concentration (EC50) with 50% effect on algae – Selenastrum capricornutum (Raphidocelis subcapitata or Pseudokirchneriella subcapitata) in accordance with OECD 201 and ISO/DIS 8692. This toxicity test was performed with Algaltoxkit FTM microbiotest which suppose the measurement of the algal growth (at 670 nm) in the long cells after 24h, 48h and 72h incubation (23oC) and calculation of inhibitory concentration in the test concentrations (0.05 – 10 mg/L surfactant) versus the growth in the control.

- Acute toxicity test with luminescent bacteria to estimate the toxic effect of surfactants on Vibrio fischeri sp, using the "BioFixLumi" equipment which respects criteria of DIN EN ISO 11348-3. The principle of method is: marine bacteria release luminescence as a metabolic product which can be affected by chemicals. With help of "BioFixLumi" system was measured the light intensity produced by bacteria, before and after 15 or 30 minute of incubation, in the presence of pollutant and against the control. The intensity difference between sample and control was associated with the effect of pollutants on microorganisms: inhibition or stimulation. The test concentrations of cationic surfactants were in the interval 0.05 – 10 mg/L and for amphoteric surfactant 3 – 80 mg/L.

- Microbial Assay for Risk Assessment (MARA) test– is a multi-species toxicity test based on responses of 11 microorganisms (prokaryote and eukaryote bacteria) to toxic compounds. The microbial growth is determined by a redox dye reduction which induces insoluble reaction products (red) which precipitate and form a pellet in the plate. The plate is scanned and the image is analyzed by MARA software for toxicity determination. The test was performed for 0.021 – 5 mg/L cationic standard solutions and 0.041 – 10 mg/L cationic raw material solutions.

The levels of toxicity class are drawn in accordance with international regulations EPA [72]and national legislative program (H.G. 1408/2008)[76], as followed: Highly toxic - LC_{50} / EC_{50} < 1mg/ L; Toxic - 1mg/L < LC_{50} / EC_{50} ≤ 10 mg/L; Harmful / hazardous for aquatic environment - 10 mg/L < LC_{50} / EC_{50} ≤ 100 mg/L; Very low toxic, non-toxic - LC_{50} / EC_{50} > 100 mg/L.

The final results concerning the acute effects of individual surfactants are summarized in the Table 8 and Figure 8.

Test organisms	HYAMINE 1622			TETRANYL AT 7590			CAPB		
	LC_{50}/EC_{50} mg/L	NOEC mg/L	LOEC mg/L	LC_{50}/EC_{50} mg/L	NOEC mg/L	LOEC mg/L	LC_{50}/EC_{50} mg/L	NOEC mg/L	LOEC mg/L
Cyprinus carpio	4.57 (1.94–9.77)	0.5	1	22.90 (11.22-33.65)	2	7	6.16 (2.81-11.74)	1	2
Daphnia magna	0.39 (0.15-0.48)	0.05	0.1	4.78 (3.05 -6.13)	0.05	0.1	9.54 (7.25–11.08)	1	5
Selenastrum capricornutum	0.56 (0.12 -1.25)	0.05	0.1	3.48 (1.67 -5.12)	0.05	0.1	5.55 (3.59 –7.21)	0.1	0.5
Vibrio fischeri	1.2	0.3	-	2.89	0.4	-	>100	0.4	-
Microbial toxicity	1.1	-	0.02	1.6	-	0.04	-	-	-
TOXICITY CLASS	HIGHLY TOXIC (for crustacean and algae) / TOXIC (for fish and bacteria)			TOXIC (for crustacean, algae, and bacteria) / HARMFUL/HAZARDOUS (for fish)			NON-TOXIC (for luminescent bacteria) / TOXIC (for fish, crustacean and algae)		
Literature toxicity data according to Table 3	Fish: LC_{50} – 0.28 – 42 mg/L; NOEC: 0.004 – 3.5 mg/L Crustacean: EC_{50} - 0.0059 – 78 mg/L; NOEC – 0.0041 – 3 mg/L Algae: EC_{50} – 0.09 -11 mg/L; NOEC – 0.16 -4.8 mg/L Bacteria: EC_{50} -0.5 - >130 mg/L.						Fish: LC_{50} - 2 – 31 mg/L; NOEC: 0.16 – 1.7 mg/L Crustacean: EC_{50} - 2.15 – >200 mg/L; NOEC – 0.9 – 1.6 mg/L Algae: EC_{50} – 0.55 -48mg/L; NOEC – 0.09 -10 mg/L Bacteria: EC_{50} – 5.2 - 900 mg/L.		

Table 8. Experimental toxicity data of studied surfactants

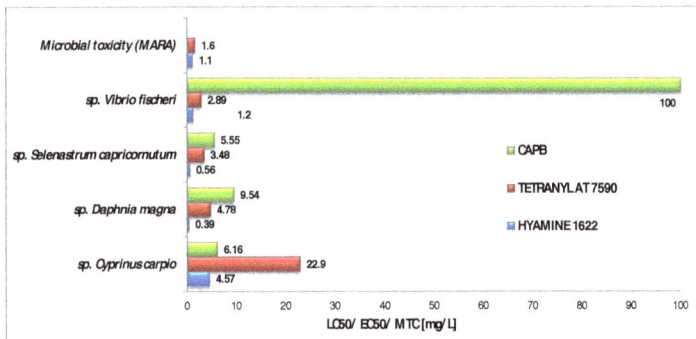

Figure 8. Toxicity quantum of cationic and amphoteric surfactants on aquatic organisms

Compared with other ecotoxicity studies (Table 3), the selected cationic surfactant Hyamine 1622 (benzenthonium chloride) with aromatic chains have a great toxic effect, while the TETRANYL AT 7590 (dialkylhydroxyethyl ammonium metasulphate) with linear alkyl chains have a toxic / harmful effect on aquatic organisms, indicating that toxicity was influenced by the chemical structure, which is also indicated in the literature [18, 32].

The highly toxic effect of benzenthonium chloride is caused by his biocide proprieties which damage the fish, algae, crustacean and bacteria, whit a great environmental risk potential in the most detrimental scenario. In case of cationic raw material (TETRANYL AT 7590), the acute toxicity effects on the testing organisms were smaller. This effect is due to the presence of the slight ester ties which are easily biodegraded by microorganisms and thus the substance biodisponibility to the target organisms is more reduced.

Because of intense foaming, the amphoteric surfactant, cause the exchange gases blocking in the gills and cell membranes, inducing the mortality / immobilization and growth inhibition in fish, water flea and algae. No toxic effect on bacteria was observed.

In accordance with Table 8 and Globally Harmonized System for Classification and Labelling of Chemicals (GHS) [77], we estimated that the cationic surfactant - benzenthonium chloride is classified as "Acute toxic, first class" because caused a highly toxic effect on the crustacean and algae at < 1mg/L. The cationic (diakylhydroxyethyl ammonium metasulphate) and amphorteric (cocamidpropylbetaine) surfactants were classified as "Acute toxic, second class" due to their toxic / harmful effects for the majority of test organisms at 1-10 mg/L.

Our toxicity results (LC50/ EC50 and NOEC) are in line with other toxicity values for this type of pollutants, and therefore we consider being scientifically relevant and can be used in aquatic risk assessment.

6.2. Indirect toxicity

To meet the requirements concerning the complementary aquatic risk of surfactants, toxicity bioassays (with fish, crustaceans, algae and bacteria) of effluents from biodegradation experiments were performed [5].

The toxicity evaluation of cationic surfactants effluents (benzenthonium chloride and dialkyl-hydroxyethyl ammonium metasulphte) resulted after ultimate biodegradability tests was performed according to "Toxicity Classification System for the discharged effluents into the aquatic environment" [37]. The principle is to determine and quantify the acute toxicity of effluents using a microbiotests battery. Effluent toxicity assessment is based on two types of values: an acute toxic value of effluent - transformed in toxicity units TU= $[1/L(E)C_{50}]$x100, that can fit in one of the 5 classes of toxicity and value of the weight score for each toxicity class.

The tests showed that biodegradation effluents have a toxic impact on target organisms and the level of toxicity varies depending of species. The algae and bacteria were the most sensible, which can be correlated with the effect caused by the original compounds of these species.

In Table 9 are presented the measured effects of the cationic biodegradation effluents and their toxicity classification. Experimental results have highlighted that benzenthoniu chloride

(Hyamine 1622) effluent was acutely toxic for all target organisms, while the Teranyl AT 7590 effluent determined a low toxic effect, and also, a greater influence on the algae and bacteria for both tested substances.

Organisms	Toxicity Unit (TU) calculated for each LC_{50} / EC_{50}		Classification System of discharged effluents in to natural aquatic environmental [32]
	Biodegradation effluent of HYAMINE 1622 (0.9 mg/L)	Biodegradation effluent of TETRANYL AT 7590 (0.25 mg/L)	
Cyprinus carpio	0 weight score 0	0 weight score 0	TU < 0.4 Class I – no acute toxicity
Daphnia magna	2 weight score 2	0 weight score 0	0.4<TU<1 Class II – small acute toxicity
Selenastrum capricornutum	3.32 weight score 2	1.2 weight score 2	1<TU<10 Class III – acute toxicity
Vibrio fischeri	4.98 weight score 2	2.24 weight score 2	10<TU<100 Class IV – high acute toxicity
TU for biotests battery / Toxicity Class	2.57 Class III – acute toxicity weight score 2	0.86 Class II – small acute toxicity weight score 1	TU>100 Class V – very high acute toxicity

Table 9. Toxicity classification of biodegradability effluents of the studied cationic surfactants

The toxic effects of surfactants biodegradation solutions, lead us to hypothesize of recalcitrant biodegradation metabolites occurrence with toxic effects potential on aquatic organisms, but their detection is not yet clarified. Other hypothesize is the persistence of testing surfactants, in case of benzenthonium chloride, for which was recorded the lowest ultimate biodegradability (67%), also confirmed by literature data.

For in situ extrapolation (surface water), the experimental toxicity values obtained will be reduced considerably, concerning the rivers dilution (100 fold to 1000 fold). Toxicity behavior of the surfactants depends on physical - chemical factors (pH, temperature, oxygen, microbial charge, climate change, the presence of other chemicals, etc.) that can affect the bioavailability.

Another toxicity experiment was performed in order to reveal the toxic effects of the cationic surfactants used as base ingredient in commercial products (eg. biocide - algaecide). In this case was estimated the toxicological behavior of this compound mixed with other ingredients. An acute growth inhibition test with algae was performed for a biocide product containing 50% of alkylbenzyldimethyl ammonium chloride C12-C16 (CAS 68424-85-1). In mixture with other ingredients (eg. ethylene glycol 2% and water 48%), cationic surfactants maintain his initial toxicity, but the level of effects depend of purpose of use and proportion of ingredients. According to international norms the product was highly toxic / very toxic to freshwater algae Selenastrum capricornutum, the estimated CE_{r50} value was <1 mg/L.

7. Aquatic risk assessment — Case study

The aim of surfactants aquatic risk assessment methodology was to establish the maximum allowable cationic and amphoteric surfactants (HYAMINE 1622 CAS 121-54-00 and cocami-dopropyl betaine CAS 4292-10-8) concentrations in surface water in order to avoid their negative impact on aquatic ecosystem and to assure the health of aquatic organisms in trophic chain.

The aquatic risk assessment involve the collection of literature data and laboratory testing results to estimate the predicted exposure concentrations of cationic and amphoteric surfac-tants in the water (PEC aquatic) and the no-effect concentration on organisms (PNEC aquatic). Comparison of these data allowed us to determine whether the studied substances have adverse effects in the aquatic environment, using the PEC / PNEC ratio, where the PEC value must be lower than the PNEC, so that the compounds not present risk to aquatic life. For individual substances PEC / PNEC must be <1, which indicates that there will not be necessary further researches to identify potentially risk.

Given the international methodologies, environmental risk studies [11, 52, 78-83] and the laboratory informations obtained in this work, the important steps of aquatic risk assessment strategy are presented in Figure 9.

In Table 10 and 11 are summarized the most important data of risk coefficients (PEC/PNEC ratio) for each studied surfactant class. We have selected several scenarios, considering the minimum and maximum of aquatic PEC values and the lowest acute (LC/EC$_{50}$) and chronic (NOEC) toxicity values, identified in the relevant literature studies and from our studies. In order to obtain the PNEC values, we used the lowest toxicity values and different application factors recommended at international level for risk assessment (OECD, EC and ECETOC).

The risk coefficients calculated for the studied surfactants were different. In case of benzen-thonium chloride (cationic surfactant class) from 15 scenarios of risk coefficients, the PEC/ PNEC rapport was <1 (Table 10), in the range of 2.56 - 512. The results suggest that this compound and its class homologues could have negative impact on the aquatic environment. This conclusion is sustained by hypothesis of complementary effects concerning the persis-tence or recalcitrant metabolites occurrence. As a result of risk data analysis and taking into consideration that monitoring and control of cationic surfactants concentrations are not imposed within national and international regulations on surface water quality, we recom-mend the value of benzenthonium chloride ≤0.002 mg/L as maximum allowed concentration in surface water (MATC), so that aquatic ecosystem life is not affected.

The risk coefficients of the amphoteric surfactant (cocamidopropylbetaine) were >1 and 10 different scenarios were analyzed in range of 0.036 – 0.38. In this case the studied amphoteric surfactant and its homologue class were safety for aquatic environment. Considering that amphoteric surfactants control and monitoring are not imposed, we estimate the value of 0.01 mg/L cocamidopropylbetaine as maximum allowed concentration (MATC) in surface water.

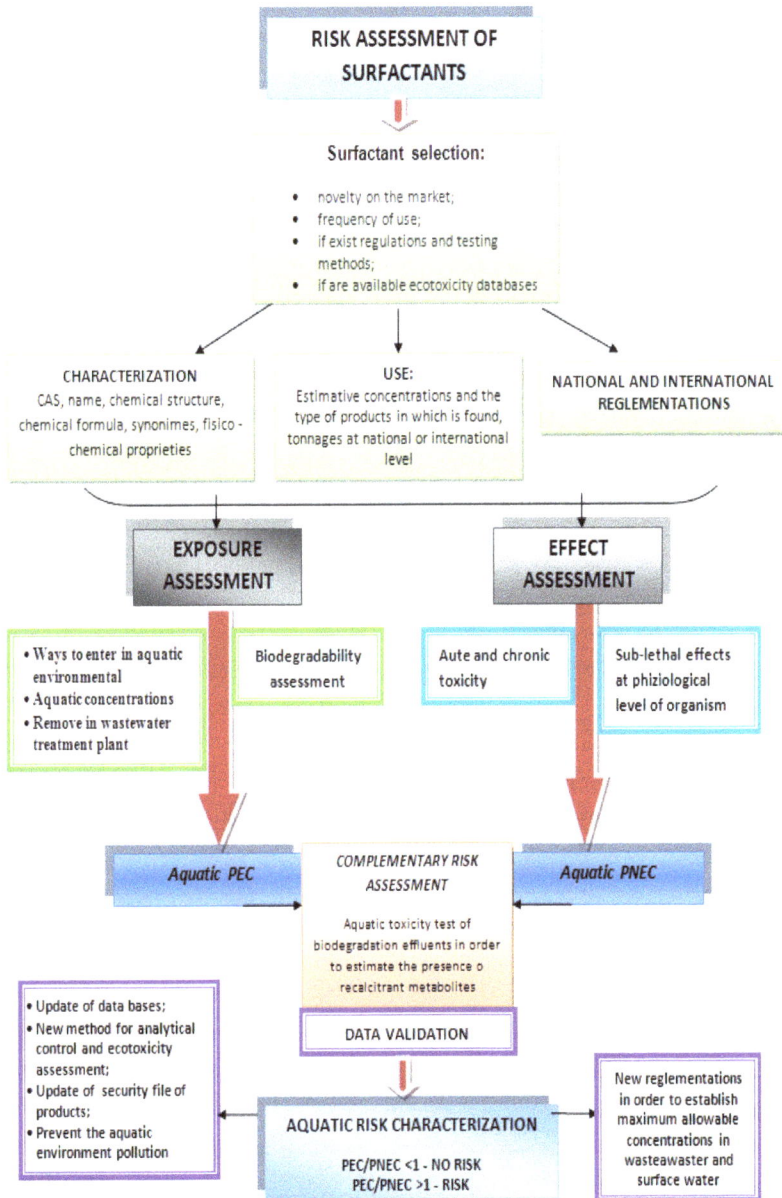

Figure 9. Aquatic risk assessment strategy plan for surfactants

PEC surface water	EC50/NOEC	Application factors for the lowest EC50 /NOEC	PNEC surface water	Risk coefficients (PEC/PNEC)
PEC$_{max}$= 0.2 mg/L [12]	0.39 mg/L (algae) [84]	100 (OECD)	0.0039	51.28
		1000 (EU)	0.00039	512.20
		200 (ECETOC)	0.00195	102.56
	0.004 mg/L (Daphnia) [12]	10 (OECD and EU)	0.0004	500
	0.05 mg/L (algae) [84]	100 (OECD)	0.0005	400
PEC$_{min}$=0.002 mg/L [12]	0.39 mg/L (algae) [84]	100 (OECD)	0.0039	0.51
		1000 (EU)	0.00039	5.12
		200 (ECETOC)	0.00195	1.025
	0.004 mg/L (Daphnia) [12]	10 (OECD and EU)	0.0004	5
	0.05 mg/L (algae) [84]	100 (OECD)	0.0005	4
PEC$_{max}$=0.01 mg/L (Danube River Romania)	0.39 mg/L (algae) [84]	100 (OECD)	0.0039	2.56
		1000 (EU)	0.00039	25.64
		200 (ECETOC)	0.00195	5.12
	0.004 mg/L (Daphnia) [12]	10 (OECD and EU)	0.0004	25
	0.05 mg/L (algae) [84]	100 (OECD)	0.0005	20

Table 10. Risk coefficients assessment for quaternary ammonium salts (eg. benzenthonium chloride / Hyamine 1622, CAS: 121-54-0) according to OECD, EC and ECETOC

PEC surface water	EC50/NOEC	Application factors for the lowest EC50 /NOEC	PNEC surface water	Risk coefficients (PEC/PNEC)
PEC$_{max}$=0.002 mg/L (Danube River Romania)	5.55 mg/L (algae) [84]	100 (OECD)	0.055	0.036
		1000 (EU)	0.0055	0.36
		200 (ECETOC)	0.027	0.074
	0.5 mg/L (algae) [84]	100 (OECD)	0.005	0.4
	0.09 mg/L (algae) [**] [30]	10 (OECD / EU)	0.009	0.22
PEC$_{min}$=0,0019 mg/L [13]	5.55 mg/L (algae) [84]	100 (OECD)	0.055	0.034
		1000 (EU)	0.0055	0.34
		200 (ECETOC)	0.027	0.07
	0.5 mg/L (algae) [84]	100 (OECD)	0.005	0.38
	0,09 mg/l (algae) [30]	10 (OECD / EU)	0.009	0.21

Table 11. Risk coefficients assessment for coco alkyl derives (eg. Cocamidopropylbetaine / AMFODAC LB, CAS: 4292-10-8) according to OECD, EC and ECETOC

8. Future challenges

The surfactants ecotoxicology domain remain open for new researches, because this compounds are in a dynamic change of molecular structure which can modified the level of biodegradability and toxicity. It is necessary to develop new control analytical methods for all type of surfactants (HPLC, LC / ELSD, LC–(ESI) MS). Some problems were highlighted concerning the strong absorption capacity of surfactants on the active sludge and also will be interesting to study the impact of cationic and amphoteric surfactants on sludge microorganisms.

There are still gaps in ecotoxicological and risk assessment databases of cationic and amphoteric surfactants, and also for the several nonionic surfactants. Also, an important subject in this research field is the study of biotic and abiotic factors influence on the bioavailability of surfactant compounds.

A great attention should be given to monitoring studies of the surfactants in the national and international surface waters, in order to underline the level of domestic and industrial pollution with this compounds and also to upgrade the current legislation or if is necessary to replace them.

In this field are significant gaps concerning the bioconcentration, bioaccumulation, acute and chronic sub lethal effects and the impact of surfactants on the metabolic pathways, whatever of surfactant type.

Another limitation of this research was the detection of metabolic compounds resulted after biodegradation process, which requires completion of equipment endowment and involve new expenses.

9. Conclusions

The aim of this chapter was the cationic and amphoteric surfactants ecotoxicological characterization according to European Regulation EC no. 648/2004 and risk assessment generated by them on the aquatic environment. Experimental researches were performed to establish the biodegradation level and aquatic toxicity, risk assessment and estimation of the maximum allowable limits in surface water.

Has been pointed that the cationic and amphoteric surfactants have a primary biodegradation >80% and a final removal >60%, noting that the cationic surfactants have registered the lowest values. In terms of acute aquatic toxicity was found that cationic surfactants are toxic for crustaceans, algae and bacteria ("Acute Toxicity, class 1") and amphoteric surfactants are toxic to fish, crustaceans and algae ("Acute Toxicity, class 2").

A complementary risk assessment study was performed for biodegradation liquids of cationic surfactants. The biodegradation effluents maintain the compounds toxicity on algae and bacteria in case of standard surfactant (Hyamine 1622), which means that in the surfactant biodegradation effluents the active substance was persistent or can arise recalcitrant metabolites.

Based on PEC / PNEC ratios, the aquatic risk assessment of cationic and amphoteric surfactants has been assessed: cationic surfactants PEC / PNEC > 1 - risk to aquatic organisms; amphoteric surfactant PEC / PNEC <1 - no risk to aquatic organisms.

Were estimated maximum allowable concentrations (MATC) of cationic surfactants (≤ 0.002 mg/L) and amphoteric (0.01 mg/L) in surface waters, so that the aquatic life in trophic chain, will not be affected.

The present study was relevant for the conformity control of market cleanup products to assure the human health and environment protection.

Abbreviations

AISE – International Association for Soaps, Detergents and Maintenance Products;

CEFIC – European Chemical Industry Council;

CESIO – European Committee of Organic Surfactants and their Intermediates;

EC - European Commission;

EC_{50} – lethal or inhibitory Effective Concentration with 50% effect on crustacean, algae or bacteria;

ECETOC – European Centre for Ecotoxicological & Toxicological Safety Assessment of Chemicals;

ECVAM – The European Centre for the Validation of Alternative Methods;

EPA - US Environmental Protection Agency;

GHS – Globally Harmonized System for Classification and Labeling of Chemicals;

HERA - Human and Environmental Risk Assessment on ingredients of European household cleaning products;

ICCVAM – Interagency Coordinating Committee on the Validation of Alternative Methods;

ISO – The International Standardization Organization;

IUCLID – International Uniform Chemical Information Database;

LC_{50} - Lethal Concentration with 50% effect on fish;

LOEC – Lowest Observed Effect Concentration

NICNAS – National Industrial Chemicals Notification and Assessment Scheme;

NOEC – No Observed Effect Concentration;

OECD – Organization for Economic Co – operation and Development;

Q(SAR) – Quantitative Structure Activity Relationship;

REACH – Regulation concerning Registration, Evaluation, Authorization and Restriction of Chemicals Substances;

SETAC - Society of Environmental Toxicology and Chemistry;

Acknowledgements

Authors thanks to National Research and Development Institute for Industrial Ecology - ECOIND and University of Bucharest for financial and technical support of this research.

Author details

Stefania Gheorghe*, Irina Lucaciu, Iuliana Paun, Catalina Stoica and Elena Stanescu

*Address all correspondence to: stefania_ivan2005@yahoo.com

National Research and Development Institute for Industrial Ecology — ECOIND Bucharest, Romania

References

[1] Berger, H. Environmentally compatible surfactants for the cosmetic industry, presented at the SCS Symposium Bristol 7-8 April 1997 Int. Journal of Cosmetic Science (1997). , 19, 227-237.

[2] Longman, G. F. The analysis of Detergents and Detergents Products; (1975).

[3] Regulation (EC) noof European Parliament and European Council, Official Journal of European Union, L104/1, April (2004).

[4] Regulation (EC) no the first Amendment of Regulation (2004). (648), 2006.

[5] Communities European CommissionCommission Recommendations from 23/XII/ 2005 concerning- the technical guidance for stratified approach of Regulation (CE) of implementation, Bruxelles, C (2005) 5677 final; (2005). (648)

[6] Regulation (EC) no. 66/2010 of the European Parliament and of the Council of 25 November 2009 concerning the EU ecological labeling.

[7] http://uk.cleanright.eu/ (accessed 12 November 2012).

[8]   http://www.cefic-lri.org (accessed 12 November 2012).

[9]   Ute MB and Erich J. Anaerobic Biodegradation of Detergent Surfactants, Materials 2009; 2,181-206.

[10] Tomislav I. and Hrenovic J. Surfactants in the Environment, Arh. Hig. Rada Toksicol 2010; 61, 95-110.

[11] HERA, Esterquats - Environmental Risk Assessment Report. Edition 1.0, March 2008.

[12] NICNAS, Full public report for Stepantex Esterquat, 28 January 2003.

[13] Kazuaki M, Naohiro N and Akiko Y. Aquatic environmental monitoring of detergent surfactants, Jurnal of Oleo Science 2008; 57(3)161-170.

[14] Detergent Products Environmental Project, Environmental and Health Assessment of Substances in Household Detergents and Cosmetic, 615; 2001.

[15] Séné C. Regulatory compliance of surfactants: Testing of surfactants and how the surfactant industry plans the implementation of Regulation. Workshop on the Implementation of Detergents Regulation (INFRA 32637), Bucharest, Romania; 2009.

[16] Gheorghe S, Lucaciu I, Grumaz R. Detergents legislative framework and ecotoxicological testing methodology. Journal of Environmental Protection and Ecology; 2011, book 3A(12)1525-1532.

[17] Domsch A and Klaus J. Biodegradability of amphoteric surfactants, in Karsa DR and Porter MR(eds.), Biodegradability of surfactants. Blackie Academic & Professional, Glasgow, United Kingdom, 231-254; 1995.

[18] Biffi E. Biodegradability Tests: Test Material: ANFODAC LB. Unpublished Report (Project number 96/200.A6), Biolab, 1996.

[19] Giolando ST, Rapaport RA, Larson RJ and Federle TW. Environmental fate and effects of DEEDMAC: a new rapidly biodegradable cationic surfactant for use in fabric softeners. Chemosphere 199; 530(6)1067-1083.

[20] Unilever.The ready and ultimate biodegradability of DEEDMAC in a manometric respirometry test. Report BD-MAN-10, 1991.

[21] Puchta R, Krings P and Sandkühler P, A new generation of softeners. Tenside Surf. Det. 1993; 30,186-191.

[22] Comber SDW, Painter H and Reynolds P. Cationic & Amphoteric Surfactant Primary Biodegradability Ring Test, European Union, ETD/98/502063, WRc Ref: CO 4909; 2000.

[23] Safty Datasheet STEPANTEX VL 90A, Stepan Europe; 2002.

[24] Games LM King JE and Larson RJ. Fate and distribution of a quaternary ammonium surfactant, octadecyltrimethylammonium chloride (OTAC), in wastewater treatment. Environ. Sci. Technol. 1982; 16,483-488.

[25] Torben M, Boyd HB, Nylen D, Pedersen AR, Petersen GI and Simonsen F. Environmental and Health Assessment of Substances in Household Detergents and Cosmetic Detergent Products, Environment Project no. 615, 2001.

[26] Garcia MT, Campos E, Marsal A, Ribosa I. Fate and effects of amphoteric surfactants in the aquatic environment, Environment International 2008,34,1001–1005.

[27] Brunner C, Baumann U, Pletscher E, Eugsterm, Total degradation or environmental experiment?. Tenside Surf. Det. 2000; 37, 276-280.

[28] IUCLID Dataset: European Commission-Chemicals Bureau, CAS 6178-40-0 (Cocamidopropyl betaine), CAS 911995-81-2 and CAS 93334-15-7 (Fatty acids, C10-20 and C16-18 –unsatd. reaction products with triethanolamine, di-Me sulfate-quaternized), February 2000.

[29] EPA, Index of Robust Summaries ACCFND Amides Chemical Category, Annex A; 2001.

[30] www.lenntech.com/aquatic/detergents (accessed 23 April 2012).

[31] Abel PD, Toxicity of synthetic detergents to fish and aquatic invertebrates. Journal of Fish Biology, 1974; 6,279–298.

[32] Masayuki Y and Takamasa T. Aquatic toxicity and biodegradability of advanced cationic surfactant APA – 22 compatible with the aquatic environment, Jurnal of Oleo Science 2008; 57(10)529-538.

[33] Zdenka S. Water quality and fish health, Eifac Technical Paper 54, 1993.

[34] Lewis MA. Chronic toxicities of surfactants and detergent builders to algae: A review and risk assessment, Ecotoxicology and Environmentally Safety 2004; 20(2)123-140.

[35] Pavlic Z, Zeljka VC, Dinko P. Toxicity of surfactants to green microalgae Pseudokirchneriella subcapitata and Scenedesmus subspicatus and to marine diatoms Phaeodactylum tricornutum and Skeletonema costatum, Chemosphere 2005; 61,1061–1068.

[36] Regulation (EC) no. 1907/2006 of the European Parliament and of the Council of 18 December 2006 concerning the Registration, Evaluation, Authorization and Restriction of Chemicals (REACH).

[37] Persoone G, Marsalek B, Blinova I, Törökne A, Zarina D, Manusadzianas L, Grzegorz NJ, Tofan L, Stepanova N, Tothova L, Kolar B. A practical and user-friendly toxicity classification system with microbiotests for natural waters and wastewaters, Environmental Toxicology 2003;18(6)395 – 402.

[38] Kao Corporation SA, Safety Datasheet for TETRANYL AT-7590; 2009.

[39] Unilever.The acute toxicity of Stepantex degradation intermediate to Daphnia magna, Report no.: AT/R202/01,1989b.

[40] Unilever. The acute toxicity of Hamburg ester quat to Daphnia magna. Report no.: AT/K5/08, 1990b.

[41] Unilever.The chronic toxicity of Hamburg ester quat to Daphnia magna in ASTM medium. Report no.: CT/K5/06,1991c.

[42] Unilever. The chronic toxicity of Hamburg ester quat to Oncorhynchus mykiss 28 day growth test. Report no.: CT/K05/07,1991d.

[43] Unilever. The acute toxicity of DEEDMAC to Zebra fish. Report no.: AT/S057/0, 1996a.

[44] Unilever. The toxicity of DEEDMAC monoester to Scenedesmus subspicatus. Report no.: AL/S052/02, 1997c.

[45] Procter & Gamble, Acute toxicity of SS0138.01 on Selenastrum capricornutum printz. Report no.: 40390, ABC Laboratories, Inc. Columbia, Missouri, 1993a.

[46] Procter & Gamble, Chronic toxicity of SS0565.01 to Daphnia magna using natural surface water under flow-through conditions, ABC Laboratories, Inc. Columbia, Missouri, report no. 44855,1999a.

[47] Henkel KGaA. Stepantex VS 90: Algal toxicity, Department of Ecology, Düsseldorf, Germany; 1989a.

[48] Grzegorz N, Jawecki C, Ska-Sota EG and Narkiewicz P. The toxicity of cationic surfactants in four bioassays, Guidance Document Methodology, Human & Environmental Risk Assessment on Ingredients of Household; 2005.

[49] Ying GG. Fate, behaviour and effects of surfactants and their degradation products in the environment, Environ. Int. 2006; 32(3)417-31.

[50] Stterlin H, Alexy R, Kmmerer K. The toxicity of the quaternary ammonium compound benzalkonium chloride alone and in mixture with other anionic compounds to bacteria and test systems with Vibrio fischeri and Pseudomonas putida, Ecotoxicol. Environ. Saf. 2008; 71, 498-505.

[51] Roghair CJ, Buijze A and Schoon HNP. Ecotoxicological risk evaluation of the cationic fabric softener DTDMAC. I. Ecotoxicological effects. Chemosphere 1992; 24 (5) 599-609.

[52] EPA, Alkyl Dimethyl Benzyl Ammonium Chloride (ADDAC) Preliminary Risk Assessment Office of Pesticide Programs. Antimicrobials Division; 2006.

[53] Lewis MA and Hamm BG. Environmental modification of the photosynthetic response of lake plankton to surfactants and significance to a laboratory – field comparison. Water Res. 1986; 20, 1575-1582.

[54] Lewis RJ. Sax´s Dangerous Properties of Industrial Materials.9th ed. Van Nostrand Reinhold, New York, United States, 1996.

[55] Utsunomiya A, Watanuki T, Matsushita K and Tomita I. Toxic effects of linear alkylbenzene sulfonate, quaternary alkylammonium chloride and their complexes on Dunaliella sp. and Chlorella pyrenoidosa. Environ. Toxicol. Chem. 1997; 16, 1247-1254.

[56] Toxicology/Regulatory Services, Inc., Fatty Nitrogen Derived Cationics Category High Production Volume (HPV) Chemicals Challenge, Test Plan, 2001.

[57] Berol N. Amphoteen 24, Produkt information (in Swedish); 1993.

[58] Brøste P, Brugsanvisning for Dehyton AB 30 (in Danish). Materials safety data sheet for Dehyton AB 39. P. Brøste A/S, Lyngby, Denmark; 1998.

[59] Iancu V. Study of HPLC analitical determination of cationic and amphoteric surfactants in aqueous solutions. In house paper, June 2007.

[60] Ciurcanu I, Lucaciu I, Mihaila E. Experimental research to elaborate and verify a spectrophotometer method for the determination of cationic surfactants concentration in aqueous solution, proceedings of the International Symposium "Environment and Industry" 29 – 31 October 2003, Bucharest, Romania; 2003.

[61] Bratucu L, Ciurcanu I, Lucaciu I, Mihaila E. Laboratory assimilation and implementation of analytical methods used for cationic and amphoteric surfactants concentrations assessment from aqueous solutions,, proceedings of The XXIX National Conference of Chemistry, 04-06 October 2006, Ramnicu Valcea, Romania; 2006.

[62] Levsen K, Emmrich M and Behnert S. Determination of dialkyldimethylammonium compounds and other cationic surfactants in sewage water and activated sludge, Fresenius'Journal of Analytical Chemistry 2004; 732-737.

[63] Yong Q, Gaoyong Z, Baoan K, Yumei Z. Primary aerobic biodegradation of cationic and amphoteric surfactants, Journal of Surfactants and Detergents 2005; 8(1)55-58.

[64] German Standard Methods for the investigation of water, wastewater and sludge. Collective parameters for effects and substances (Group H). Determination of disulphine blue active substances (H20). DIN 38409 – Ausgabe:1989-07; Part 20.

[65] Idouhar M and Tazerouti A. Spectrophotometric determination of cationic surfactants using patent blue V: application to the wastewater industry in Algiers. Journal of Surfactants and Detergents, 2008; 11(4)263-267.

[66] OECD, Chemicals Group. Revised guidelines for tests for ready biodegradability, Paris, 1993a.

[67] OECD 301A. Dissolved organic carbon analyze method – DOC method – Biodegradability test.

[68] OECD 301D. Assessment of ultimate aerobe biodegradability in aqueous media of organic compounds. The biochemical oxygen remove analyze method (the closed tubes test).

[69] Gheorghe S, Lucaciu I, Pascu L. Biodegradability assessment of cationic and amphoteric raw materials, Journal of Environmental Protection and Ecology 2012;13 (1)155-164.

[70] Gheorghe S, Lucaciu I, Pascu L. Removal of surfactants from household cleaning products and/or cosmetic detergents during the ready biodegradability tests performed in conformity with the new European regulations, proceedings of the International Symposium „The environment and industry" (SIMI) 28-30 October 2009, Bucharest, Romania, EstFalia; 2009.

[71] Gheorghe S, Lucaciu I, Pascu L. Ultimate biodegradability assessment of cationic and amphoteric surfactants, proceedings of the International Conference on fishery and aquaculture - a view point upon the sustainable management of the water resources in the Balkan area, 26-28.05.2010, Galati, Romania; 2010.

[72] American Society for Testing and Materials: Standard methods for conducting acute toxicity tests with fishes, macro invertebrates, and amphibians, in Annual Book of ASTM Standards. Designation E 729-88a, Philadelphia: ASTM, 1992a; 11,403-422.

[73] OECD, Guidelines for the testing of chemicals, 2006.

[74] OECD, Manual for Investigation of Hpv Chemicals, 2006.

[75] Official Journal of the European Communities: Methods for the determination of ecotoxicity, L133, pp.88-127, Office for Official Publications. L2985, Luxembourg, 1988.

[76] Governmental Decision 1408/2008 concerning the classification, labeling and packing of dangerous chemicals.

[77] Globally Harmonized System for Classification and Labeling of Chemicals (GHS), revision 4, United Nations, New York and Geneva, 2011; www.unece.org (accessed 12 May 2012).

[78] CESIO, Environmental Risk Assessment of Detergent Chemicals, Proceedings of the AISE/CESIO Limelette III Workshop, 28–29 November 1995, Brussels; 1996.

[79] HERA, Cocoamidopropyl Betaine (CAPB).Edition 1.0, June 2005.

[80] Ivan (Gheorghe) S., Determination of cationic and amphoteric surfactants ecotoxicity and their aquatic risk assessment, PhD thesis. Bucharest University and INCD ECOIND Bucharest;2012.

[81] Leeuwen K, Roghair C, Nijs T and Greef J. Ecotoxicological risk evaluation of the cat-
      ionic fabric softener DTDMAC. III. Risk assessment, Chemosphere 1992; 24
      (5)629-639.

[82] Notox BV. Safety and Environmental Research, New Detergent Regulation published
      in the Official Journal, Effective on 8th October 2005.

[83] Technical Guidance Document on Risk Assessment, European Chemicals Institute
      for Health and Consumer Protection Bureau, Part II, 2003.

[84] Gheorghe S, Lucaciu I, Grumaz R, Stoica C. Acute toxicity assessment of several cati-
      onic and amphoteric surfactants on aquatic organisms, Journal of Environmental
      Protection and Ecology 2012;13(2) 541-553.

Chapter 5

Biodegradable Polymers

Babak Ghanbarzadeh and Hadi Almasi

Additional information is available at the end of the chapter

1. Introduction

In developing countries, environmental pollution by synthetic polymers has assumed dangerous proportions. Petroleum-derived plastics are not readily biodegradable and because of their resistance to microbial degradation, they accumulate in the environment. In addition in recent times oil prices have increased markedly. These facts have helped to stimulate interest in biodegradable polymers. Biodegradable plastics and polymers were first introduced in 1980s. Polymers from renewable resources have attracted an increasing amount of attention over the last two decades, predominantly due to two major reasons: firstly environmental concerns, and secondly the realization that our petroleum resources are finite. There are many sources of biodegradable plastics, from synthetic to natural polymers. Natural polymers are available in large quantities from renewable sources, while synthetic polymers are produced from non-renewable petroleum resources. Biodegradation of polymeric biomaterials involves cleavage of hydrolytically or enzymatically sensitive bonds in the polymer leading to polymer erosion. A vast number of biodegradable polymers have been synthesized recently and some microorganisms and enzymes capable of degrading them have been identified.

The objective of this chapter is to classification of biodegradable polymers. The chemical structure, sources, production and synthesis methods, physical properties (mechanical, barrier and thermal properties) and applications of most important biodegradable polymers would be discussed.

2. Classification and properties of biodegradable polymers

The biodegradable polymers can be classified according to their chemical composition, origin and synthesis method, processing method, economic importance, application, etc. In the

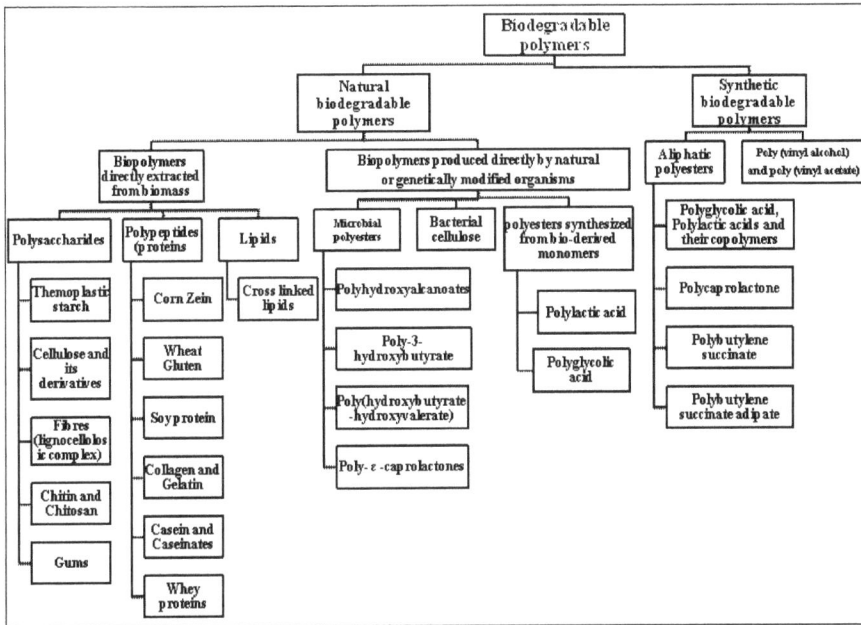

Figure 1. Schematic presentation of biobased polymers based on their origin and method of production

present chapter, biodegradable polymers classified according to their origin into two groups: natural polymers which obtained from natural resources and synthetic polymers which produced from oil. An overview of these categories is given in Fig. 1.

2.1. Natural biodegradable polymers

Biopolymers are polymers formed in nature during the growth cycles of all organisms; hence, they are also referred to as natural polymers. Their synthesis generally involves enzyme-catalyzed, chain growth polymerization reactions of activated monomers, which are typically formed within cells by complex metabolic processes.

2.1.1. Biopolymers directly extracted from biomass

2.1.1.1. Polysaccharides

For materials applications, the principal polysaccharides of interest are cellulose and starch, but increasing attention is being given to the more complex carbohydrate polymers produced by bacteria and fungi, especially to polysaccharides such as xanthan, curdlan, pullulan and hyaluronic acid. These latter polymers generally contain more than one type of carbohydrate unit, and in many cases these polymers have regularly arranged branched structures. Because of this difference, enzymes that catalyze hydrolysis reactions during the biodegradation of each kind of polysaccharides are different and are not interchangeable.

2.1.1.1.1. Thermoplastic starch

Starch is the major polysaccharide reserve material of photosynthetic tissues and of many types of plant storage organs such as seeds and swollen stems. The principal crops used for its production include potatoes, corn and rice. In all of these plants, starch is produced in the form of granules, which vary in size and somewhat in composition from plant to plant (Chandra and Rustgi, 1998). The starch granule is essentially composed of two main polysaccharides, amylose and amylopectin with some minor components such as lipids and proteins. Amylose is essentially a linear molecule of $(1\rightarrow4)$-linked α-D-glucopyranosyl units with some slight branches by $(1\rightarrow6)$-α-linkages (Fig. 2). Typically, amylose molecules have molecular weights ranging from 10^5 to 10^6 gmol^{-1} (Buleon et al., 1998). Amylopectin is a highly branched molecule composed of chains of α-D- glucopyranosyl residues linked together mainly by $(1\rightarrow4)$-linkages but with $(1\rightarrow6)$ linkages at the branch points. Amylopectin consist of hundreds of short chains of $(1\rightarrow4)$-linked α-D-glucopyranosyl interlinked by $(1\rightarrow6)$-α-linkages (Fig. 2). It is an extremely large and highly branched molecule (molecular weights ranging from 10^6 to 10^8 gmol^{-1}).

There are three types of crystallinity in starch. They are the 'A' type mainly cereal starches such as maize, wheat, and rice; 'B' type such as tuber starches (potato, sago); and finally the 'C' type crystallinity which is the intermediate between A and B type crystallinity, normally found in bean and other root starches (Blanshard, 1987). Another type of crystallinity is the Vh-type, which is the characteristic of amylose complexed with fatty acids and monoglycerides.

Starch granules contain alternating 120-400 nm amorphous and semi-crystalline layers or growth rings (Buleon et al., 1998). The semi-crystalline growth rings are composed of alternating amorphous and crystalline lamellae. The sum of one amorphous and one crystalline lamella is around 9-10 nm in size. Amylopectin is often presumed to support the framework of the semi- crystalline layers in the starch granule. The short chains with polymerization degrees ranging between 15 and 18 form a double helical conformation (Buleon et al., 1998) and associating into clusters. These clusters pack together to produce a structure of alternating crystalline and amorphous lamellar composition. The side chains clusters which are predominantly linear and form double helices are responsible for the crystalline lamellae while the branching regions of the amylopectin molecule are responsible for the amorphous lamellae.

Thermoplastic starch is plasticized starch that has been processed (typically using heat and pressure) to completely destroy the crystalline structure of starch to form an amorphous thermoplastic starch. Thermoplastic starch processing typically involves an irreversible order-disorder transition termed gelatinization. Starch gelatinization is the disruption of molecular organization within the starch granules and this process is affected by starch-water interactions. Fig. 3 highlights the gelatinization process diagrammatically (Lai and Kokini, 1991). This figure shows raw starch granules made up of amylose (linear) and amylopectin (branched) molecules (step (a)). Then the addition of water breaks up crystallinity and disrupts helices (step (b)). Addition of heat and more water causes granules to swell and amylose diffuses out

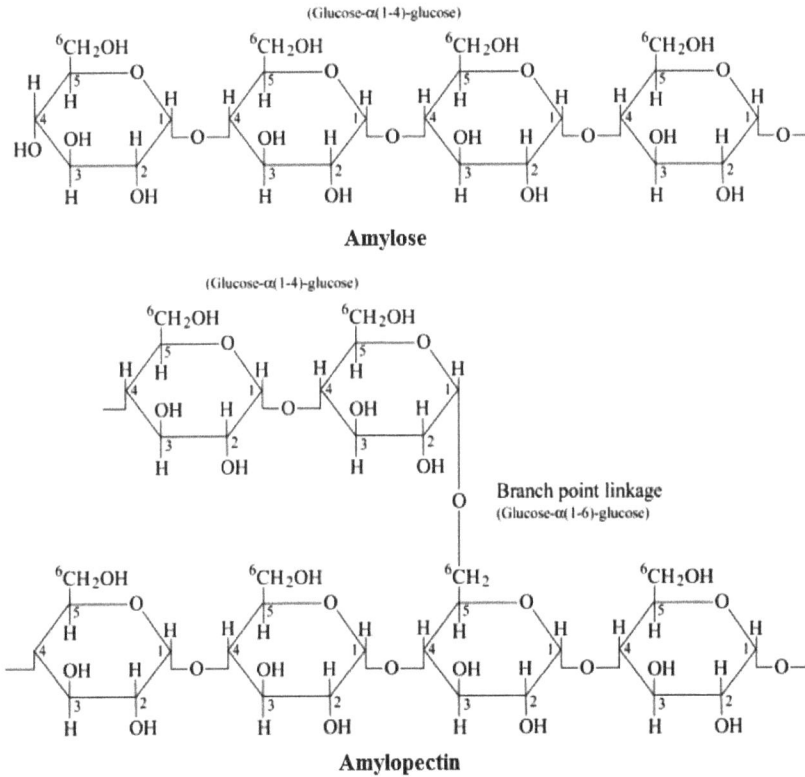

Figure 2. Chemical structure of amylose and amylopectin (Buleon et al., 1998)

of the granule (step (c)). Granules, mostly containing amylopectin are collapsed and held in a matrix of amylose (step (d)).

Thermoplastic starch is produced using dry native starch with a swelling or plastifying agent in compound extruders without adding water. In extrusion, starch is converted by application of both thermal and mechanical energy, and basically three phenomena occur at different structural levels: fragmentation of starch granules; hydrogen bond cleavage between starch molecules, leading to loss of crystallinity; and partial depolymerization of the starch polymers (Fang and Fowler, 2003). Furthermore, the extrusion process ensures the very intimate mixing of the polymers and any additives. By introduction of mechanical energy and heat in a temperature range of 120-220 °C, crystal starch, is homogenized and melted in an extrusion process with a plasticizer which lowers the melting point of the starch. With this process, a permanent conversion of the molecular structure to thermoplastic starch is performed (Lorcks,

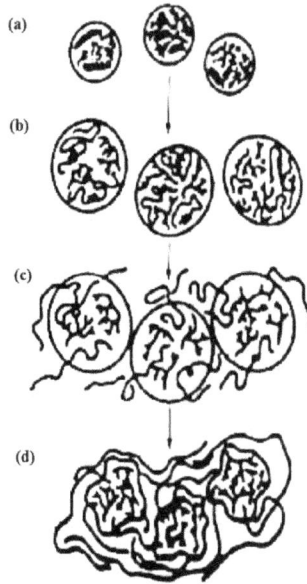

Figure 3. Starch gelatinization process (Lai and Kokini, 1991)

1998). The thermoplastic starch is free of crystalline fractions. Molecules such as polyglycols (e.g. glycerol, sorbitol, etc), amides and amines serve as non-volatile plasticizers for starch (Wiedersheim and Strobel, 1991).

Depending on the starch source and processing conditions, a thermoplastic material may be obtained with different properties suitable for various applications. Starch has been widely used as a raw material in film production because of increasing prices and decreasing availability of conventional film-forming resins (Chandra and Rustgi, 1998). Potential applications of starch films include production of disposable food serviceware, food packaging, purchase bags, composting bags and loose fill products (Xiong et al., 2008). Starch is also used in hygiene and cosmetics. Moreover, starch has been used for many years as an additive to plastic for various purposes. Starch was added as filler to various resin systems to make films that were impermeable to water but permeable to water vapour (Pedroso and Rosa, 2005). Starch as biodegradable filler in LDPE was reported (Nakamura et al., 2005). A starch-filled polyethylene film was prepared which becomes porous after the extraction of the starch. This porous film can be readily invaded by microorganisms and rapidly saturated with oxygen, thereby increasing polymer degradation by biological and oxidative pathways. Starch is also useful for making agricultural mulch films because it degrades into harmless products when placed in contact with soil microorganisms. Starch is also used in medical applications. For example, starch-based thermoplastic hydrogels for use as bone cements or drug-delivery carriers have been developed through blending starch with cellulose acetate (Pereira et al., 1998; Espigares et al., 2002).

Important properties of thermoplastic starch based materials include (Lorcks, 1998):

- compostable in accordance with DIN 54900
- high water vapour permeability
- good oxygen barrier
- not electrostatically chargeable
- low thermal stability

In general, the low resistance to water and the variations in mechanical properties under humid conditions affect the use of starch for various applications. As water has a plasticizing power, the material behavior changes according to the relative humidity of the air (Averous, 2002). Strong hydrophilic character (water sensitivity) and poor mechanical properties compared to conventional synthetic polymers are the most important disadvantages of starch which make it unsatisfactory for some applications such as packaging purposes. Generally, many approaches are suggested to mitigate these shortcomings. One approach is the modification of starch. Cross-linking can be produce low water sensitive and high strength materials (Ghanbarzadeh et al., 2010). According to Takore et al., (2001), the esterification of starch allows the increase of its thermoplastic characteristics, as well as its thermal stability. Other approach to improve the functional properties of the starch films is to blend starch with other polymers. Mao et al. (2000) examined the extrusion of thermoplastic cornstarch- glycerol-polyvinyl alcohol (PVOH) blends and noted the effect of PVOH to improve mechanical properties and slow biodegradation. Development of the polymer nanocomposites is one of the latest revolutionary steps of the polymer technology. In terms of nanocomposite reinforcement of thermoplastic starch polymers there has been much exciting new developments. Dufresne and Cavaille (1998) and Angles and Dufresne (2000) highlight work on the use of microcrystalline whiskers of starch and cellulose as reinforcement in thermoplastic starch polymer and synthetic polymer nanocomposites. They find excellent enhancement of properties, probably due to transcrystallisation processes at the matrix/fibre interface. Almasi et al. (2010) examined the use of nanoscale montmorillonite into starch/carboxymethyl cellulose blends and finds excellent improvements in film impermeability and tensile properties.

2.1.1.1.2. Cellulose and its derivatives

At present, cellulose is the most abundant polymer available worldwide with an estimated annual natural production of 1.5×10^{12} tons and considered as an almost inexhaustible source of raw materials (Cao et al., 2009). Cellulose is composed of polymer chains consisting of unbranched β $(1 \rightarrow 4)$ linked D- glucopyranosyl units (anhydroglucose unit) (Fig. 4). The length of these β $(1 \rightarrow 4)$ glucan chains depends on the source of cellulose. As the main component of cell wall, cellulose is predominantly located in the secondary wall. In the primary cell wall, cellulose consists of roughly 6000 glucose units. Three hydroxyl groups, placed at the positions C2 and C3 (secondary hydroxyl groups) and C6 (primary hydroxyl groups) can form intra-

and intermolecular hydrogen bonds (Abdul Khalil et al., 2012). Because of the strong tendency for intra- or intermolecular hydrogen bonding, bundles of cellulose molecules aggregate to microfibrils, which form either highly ordered (crystalline) or less ordered (amorphous) regions (Hamad, 2006). Microfibrils are further aggregated to fibrils and finally to cellulose fibers.

Cellulose is the main constituent of cell wall in lignocellulosic plant, and its content depends on the plant species, growing environment, position, growth, and maturity. Generally, cellulose content in lignocellulosic plant is 23–53% on a dry-weight basis, less than that in cotton, which is almost made of pure fibrous cellulose. In most straw species, approximately 35–45% of the dry substance is cellulose (Knauf and Moniruzzaman, 2004).

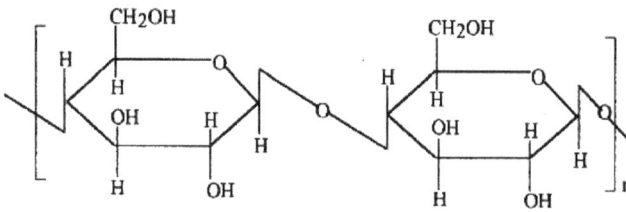

Figure 4. Chemical structure of cellulose

In the lignocellulosic materials, cellulose is embedded in a gel matrix composed of hemicelluloses, lignins, and other carbohydrate polymers. Cellulose was isolated for the first time some 150 years ago (Smith, 2005). The combination of the chemical and the mechanical treatments is necessary for the dissolution of lignins, hemicelluloses, and other noncellulosic substances. A protocol based on acidified sodium chlorite is frequently applied to delignify woody materials as an initial step in the isolation of cellulose. Alkali extraction to dissolve hemicelluloses before or after delignification is the common method. The presence of high amounts of lignin in isolated cellulose fibers after delignification affects the structure and properties of the cellulose fibers. Fibers with high amounts of lignin are coarse and stiff, and have a brownish color. Therefore, it is challenging to obtain fibers that are relatively free of bound lignin. To achieve this aim, chemical bleaching, which is used to obtain fibers with higher cellulose content from delignified and unbleached fibers, is usually considered as a continuation of delignification process to isolate cellulose from woody raw materials (Brendel et al., 2000). Nowadays, there are various procedures for extraction of cellulose microfibrils (e.g. pulping methods, acid hydrolysis, steam explosion, atc.)(Abdul Khalil et al., 2012).

Many useful properties stem from unique functional characteristics related to the chemical structure of cellulose. These structural properties include an extended, planar chain conformation and oriented, parallel-chain packing in the crystalline state. The absence of branches in this 100% linear polymer contributes to efficient chain packing in the native crystalline state, resulting in stiff, dimensionally stable fibers (Smith, 2005). Cellulose fibers thus exhibit a high degree of crystallinity (upwards of 70%) when isolated and purified. However, cellulose fibers present in native woody biomass exhibit approximately 35% crystallinity, due to the presence of other lignocellulosic components (Abdul Khalil et al., 2012). The crystal nature (monoclinic sphenodic) of naturally occurring cellulose is known as cellulose I. Cellulose is resistant to strong alkali (17.5 wt%) but is easily hydrolyzed by acid to water-soluble sugars. Cellulose is relatively resistant to oxidizing agents (John and Thomas, 2008). The tight fiber structure created by hydrogen bonds results in the typical material properties of cellulose, such as high tensile strength and insolubility in most solvents. There are significant differences between the properties of straw cellulose, wood cellulose, and cotton cellulose. The cellulose crystallites are longer in straw pulps than in wood pulps, but they are not as long as in cotton cellulose. In addition, the degree of crystallinity of straw pulps appears to be less than that of wood cellulose. Low crystallinity can be useful when a cellulose derivative is to be manufactured (Sun and Tomkinson, 2000).

Cellulose has received more attention than any other polymer since it is attacked by a wide variety of microorganisms. The biodegradation of cellulose is complicated, because cellulose exists together with lignin however, it is fortunate that pure cellulose does decompose readily (Chandra and Rustgi, 1998). Fermentation of cellulose has been suggested as a source of chemicals such as ethanol and acetic acid, but this has not achieved any commercial importance to date.

The most significant cellulosic applications are in the paper, wood product, textile, film, and fiber industries but recently it has also attracted significant interest as a source of biofuel production (Mantia and Morreale, 2011). The natural cellulosic carbon skeleton can be utilized in two major applications on an industrial scale. The first is as regenerated or mercerized cellulose (cellulose II, Rayon), which is not moldable and is used only for film and fiber production. The second represents a broader class of applications, which employs chemically modified celluloses, principally the cellulose esters (Chandra and Rustgi, 1998).

As mentioned before, in all forms, cellulose is a very highly crystalline, high molecular weight polymer, which is infusible and insoluble in all but the most aggressive, hydrogen bond-breaking solvents such as N-methylmorpholine-N-oxide. Because of its infusibility and insolubility, cellulose is usually converted into derivatives to make it more processable. All of the important derivatives of cellulose are reaction products of one or more of the three hydroxyl groups, which are present in each glucopyranoside repeating unit, including (Chandra and Rustgi, 1998):

• ethers, e.g. methylcellulose, hydroxypropyl methyl cellulose and hydroxylethyl cellulose;

• esters, e.g. cellulose acetate, carboxymethyl cellulose and cellulose xanthate, which is used as a soluble intermediate for processing cellulose into either fibre or film forms, during which the cellulose is regenerated by controlled hydrolysis;

• acetals, especially the cyclic acetal formed between the C2 and C3 hydroxyl groups and butyraldehyde.

These modified forms of cellulose can be tailored to exhibit particular physical and chemical properties by varying the pattern and degrees of substitution within the cellulose backbone. Industrial applications are numerous and widespread for cellulose derivatives owing to rigidity, moisture vapor permeability, grease resistance, clarity, and appearance (Edgar et al., 2001). Esterification of the cellulose backbone provides structural changes that allow for a greatly expanded range of applications, not available to the parent polysaccharide. Commercially available forms of cellulose acetate have degrees of substitution between 1.7 and 3.0 and are currently used in high volume applications ranging from fibers, to films, to injection moulding thermoplastics, to low solids solvent-borne coatings for metal and automobile industries (Mohanty et al., 2000). Methylcellulose exhibits thermal gelation and has excellent film-forming properties. It has been widely used to prepare edible films (Debeaufort and Voilley, 1997; Peressini et al., 2003). Carboxymethyl cellulose is also widely used in the pharmaceutical and food industries. It has good film forming properties (Ghanbarzadeh et al., 2011). Carboxymethyl cellulose based film is a very efficient oxygen, carbon dioxide, and lipid barrier. However, it has poor resistance to water vapor transmission (Ghanbarzadeh and Almasi, 2011).

The chemically modified celluloses are degradable only under certain circumstances, as more recalcitrant, hydrophobic ester groups replace the native glucopyranosyl hydroxyls (to varying degrees) in the esterification procedure. Structurally, the degrees of substitution and C-2 hydroxyl substitution patterns are important criteria in predicting biodegradation patterns for these polymers (Amass et al., 1998). Biodegradation rates of cellulose esters generally increase with decreasing degrees of acetate substitution.

2.1.1.1.3. Fibers (Lignocellolosic complex)

Plant fibers include bast (or stem or soft sclerenchyma) fibers, leaf or hard fibers, seed, fruit, wood, cereal straw, and other grass fibers. All these plant based natural fibers are lignocellulosic in nature and are composed of cellulose, hemicelluloses, lignin, pectin and waxy substances (Kabir et al., 2012). Lignocellulosic biomass comprises approximately 50% of the global biomass and is by far the most abundant renewable organic resource on earth. This woody material is comprised of 30-50% cellulose, 20-50% hemicellulose, and 15-35% lignin, dependent upon the plant species and environmental (growing) conditions (Galbe and Zacchi, 2002). Fig. 5 presents the model of the structural organization of the three major structural constituents of the fiber cell wall (Madsen, 2004). Hemicellulose molecules are hydrogen bonded with cellulose fibrils and they form cementing materials for the fiber structure. Lignin and pectin are coupled with the cellulose–hemicellulose network and provides an adhesive quality to hold the molecules together. This adhesive quality is the cause for the strength and stiffness properties of the fiber.

Hemicellulose is not a form of cellulose and the name is a misnomer. They comprise a group of polysaccharides composed of a combination of 5- and 6-carbon ring sugars (Fig. 6a). Hemicellulose differs from cellulose in three aspects. Firstly, they contain several different sugar units whereas cellulose contains only 1,4-β-D-glucopyranose units. Secondly, they exhibit a considerable degree of chain branching containing pendant side groups giving rise to its non crystalline nature, whereas cellulose is a linear polymer. Thirdly, the degree of polymerization of native cellulose is 10–100 times higher than that of hemicelluloses (John and Thomas, 2008). The degree of polymerization (DP) of hemicellulose is around 50–300. Hemicellulose is very hydrophilic, soluble in alkali and easily hydrolyzed in acids.

Lignin is a complex hydrocarbon polymer with both aliphatic and aromatic constituents (Fig. 6b). They are totally insoluble in most solvents and cannot be broken down to monomeric units. Lignin is totally amorphous and hydrophobic in nature. It is the compound that gives rigidity to the plants. It is thought to be a complex, three-dimensional copolymer of aliphatic and aromatic constituents with very high molecular weight. Hydroxyl, methoxyl and carbonyl groups have been identified. Lignin has been found to contain five hydroxyl and five methoxyl groups per building unit. It is believed that the structural units of lignin molecule are derivatives of 4-hydroxy-3-methoxy phenylpropane (John and Thomas, 2008). The main difficulty in lignin chemistry is that no method has been established by which it is possible to isolate lignin in its native state from the fiber. Lignin is considered to be a thermoplastic polymer exhibiting a glass transition temperature of around 90 °C and melting temperature of around 170 °C (Olesen and Plackett, 1999). It is not hydrolyzed by acids, but soluble in hot alkali, readily oxidized, and easily condensable with phenol.

Pectins are a collective name for heteropolysaccharides. They give plants flexibility. Pectin, is a complex anionic polysaccharide composed of β-1,4-linked D-galacturonic acid residues, where in the uronic acid carboxyls are either fully (HMP, high methoxy pectin) or partially (LMP, low methoxy pectin) methyl esterified (Tharanathan, 2003) (Fig. 6c).

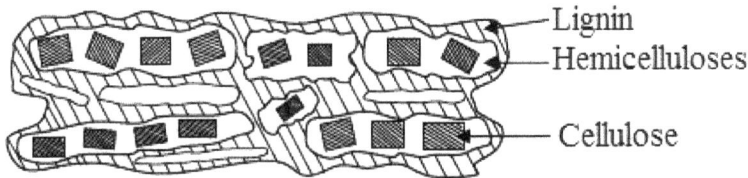

Figure 5. Structural organization of the three major constituents in the fiber cell wall (Kabir et al., 2012).

It has been proposed various techniques for separation of these components from lignocellulosic complex. The 'Clean Fractionation Process', developed and patented by the National Renewable Energy Laboratory (Golden, CO, USA), is one example of an organic solvent-based system used to separate and purify the three major feedstocks present in lignocellulosic biomass (Smith, 2005). Lignin and hemicellulose are disrupted and solubilized in the solvent mixture composed of water, methyl-isobutyl-ketone (MIBK), ethanol, and sulfuric acid (H_2SO_4), following a steam explosion treatment catalyzed by the acidic conditions created within the reactor due to the added sulfuric acid and endogenous acetic acid released during the hydrolysis. This environmentally benign process selectively separates cellulose, hemicellulose, and lignin with a high degree of purity, substantial energy savings, and lessened production cost (Kulesa, 1999).

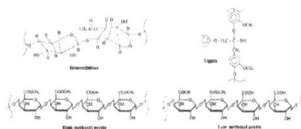

Figure 6. Chemical structure of hemicelluloses, lignin and pectins (Kabir et al., 2012; Tharanathan, 2003).

These lignocellulosic materials have the potential to be utilized as a feedstock for the production of a wide variety of industrial and commodity products, ranging from paper, lumber, and platform chemicals to a variety of fuels and advanced materials, including biodegradable polymers (Smith, 2005). For example, HMP forms excellent films. Plasticized blends of citrus pectin and high amylase starch give strong, flexible films, which are thermally stable up to 180°C. Pectin is also miscible with poly(vinyl alcohol) in all proportions. Potential commercial uses for such films are water soluble pouches for detergents and insecticides, flushable liners and bags, and medical delivery systems and devices (Tharanathan, 2003).

Hemicellulose can be utilized in microbial fermentations for the production of a variety of value-added products. Detoxified hemicellulosic hydrolysates have been used as xylose-rich feedstocks in a variety of biotechnological applications including the microbial production of ethanol, xylitol, and biodegradable polyhydroxyalkanoate (PHA) polymers (Smith, 2005). Production of PHAs based on renewable, bio-based substrates could make PHA-derived thermoplastic products more economically competitive with petroleum-based plastics, as the major costs in PHA production are the carbon source and the separation process. In the next section of this chapter we will describe the properties of this family of degradable microbial polyesters.

2.1.1.1.4. Chitin and chitosan

Chitin is a polysaccharide found in the shells of crabs, lobsters, shrimps and insects or can be generated via fungal fermentation processes. Chitosan is the deacylated derivative of chitin and forms the exoskeleton of arthropod. Structurally chitosan is a linear polysaccharide consisting of β (1-4) linked D-glucosamine with randomly located N-acetylglucosamine groups depending upon the degree of deacetylation of the polymer (Nair and Laurencin, 2007). Fig. 7 shows the structure of chitin and chitosan.

Chitin is insoluble in its native form but chitosan, is water soluble. Chitosan is soluble in weakly acidic solutions resulting in the formation of a cationic polymer with a high charge density and can therefore form polyelectrolyte complexes with wide range of anionic polymers (Pachence et al., 2007). Chitosan has been found to be non-toxic after oral administration in humans and is an FDA approved food additive (Nair and Laurencin, 2007).

Enzymes, such as chitosanase, lysozyme and papain are known to degrade chitosan *in vitro*. The *in vivo* degradation of chitosan is primarily due to lysozyme and takes place through the hydrolysis of the acetylated residues. The rate of degradation of chitosan inversely depends on the degree of acetylation and crystallinity of the polymer (Shi et al., 2006). The highly deacetylated form exhibits the lowest degradation rates and may last several months *in vivo*. Apart from this, chemical modification of chitosan can significantly affect its solubility and degradation rate. Chemical modification of chitosan produces materials with a variety of physical and mechanical properties.

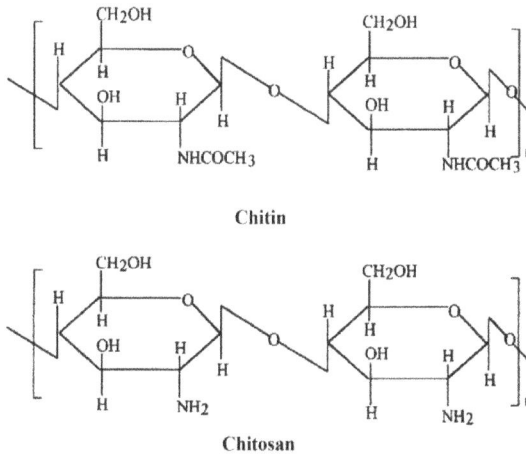

Figure 7. Chemical structure of chitin and chitosan (Nair and Laurencin, 2007).

These biopolymers are biocompatible and have antimicrobial activities as well as the ability to absorb heavy metal ions. They also find applications in the cosmetic industry because of their water-retaining and moisturizing properties (Chandra and Rustgi, 1998). Chitosan has been formed into membranes and matrices suitable for several tissue-engineering applications (Shalaby et al., 2004). Chitin derivatives can also be used as drug carriers. Chitosan was used to develop injectable thermo-sensitive carrier material for biomedical applications. Due to the mild gelling conditions, the hydrogel has been found to be a potential delivery vehicle for growth factors, small molecular weight drugs and cells for localized therapy (Nair et al., 2006). The high chemical reactivity of chitosan, has also led to several chitosan-drug conjugates for cancer therapy (Onishi et al., 2001). Chitosan gels, powders, films, and fibers have been formed and tested for many applications such as encapsulation, membrane barriers, contact lens materials, cell culture, and inhibitors of blood coagulations (Pachence et al., 2007). Chitosan has good film forming properties and therefore can be used as a food packaging material (Ghanbarzadeh et al., 2008; Suyatama et al., 2005).

2.1.1.1.5. Gums

Gums are a group of polysaccharides that can form gels in solution upon the introduction of counterions. The degree of cross-linking is dependent on various factors such as pH, type of counterion, and the functional charge density of these polymers (Chandra and Rustgi, 1998). The common types of these polysaccharides will be discussed here.

Alginic acid present within the cell walls and intercellular spaces of brown algae and has a structural role in giving flexibility and strength to marine plants. Alginate is a non-branched, binary copolymer of (1-4) glycosidically linked β-D-mannuronic acid and α-L-guluronic acid monomers. The composition of alginate (the ratio of the two uronic acids and their sequential arrangements) varies with the source (Nair and Laurencin, 2007). Alginates are extracted from the algae using a base solution and then reacted with acid to form alginic acid. They are high molecular weight polymers having molecular weights up to 500 kDa (Augst et al., 2006). Alginic acid forms water-soluble salts with monovalent cations, low molecular weight amines, and quaternary ammonium compounds. Due to its non-toxicity, alginate has been extensively used as a food additive and a thickener in salad dressings and ice creams (Nair and Laurencin, 2007). Alginate gels have been used widely in controlled release drug delivery systems. Alginates have been used to encapsulate various herbicides, microorganisms and cells. Even though alginates have been extensively investigated as biomaterials, one of the main disadvantages of using alginate-based materials is their inability to undergo enzymatic degradation by mammals (Bouhadir et al., 2001).

2.1.1.2. Polypeptides (Proteins)

Proteins can be defined as natural polymers able to form amorphous three-dimensional structures stabilized mainly by noncovalent interactions. The functional properties of these materials are highly dependent on structural heterogeneity, thermal sensitivity, and hydrophilic behavior of proteins. Numerous vegetable and animal proteins are commonly used as biodegradable polymers.

2.1.1.2.1. Corn zein

Zein comprises a group of alcohol-soluble proteins (prolamins) found in corn endosperm. It accounts for 50% or more of total endosperm protein, and its only known role is the storage of nitrogen for the germinating embryo (Gennadios, 2002). It can be extracted with aqueous alcohol and dried to a granular powder. Based on solubility differences, zein consists of three protein fractions, i.e., α-zein, β-zein, and γ-zein. α-Zein accounts for 75 to 85% of the total protein and is dominated by two groups of proteins, Z19 and Z22, according to sodium dodecyl sulfate-polyacrylamide gel electrophoresis (SDS PAGE) (Argos et al., 1982). Z19 and Z22 consist of 210 and 245 amino acids, respectively.

Due to the hydrophobic nature of zein, water sorption is extremely low in the low water activity (a_w) range. Water sorption, though, increased exponentially with a_w (Watt, 1983). Zein, although not soluble in water, is readily plasticized by it. Zein films are brittle at ambient temperature and need plasticization to become flexible. Plasticization increases polymer mobility, decreases glass transition temperature (T_g), and markedly changes rheological properties (Ghanbarzadeh, et al., 2006a; 2006b; 2007a). Common zein plasticizers include glycerol, glyceryl monoesters, poly ethylene glycol, and fatty acids.

The film-forming properties of zein have been recognized for decades, and they are the basis for its commercial utilization (Lai et al., 1997). Coating films are formed on hard surfaces by covering them with zein solutions and allowing the solvent to evaporate off. The dried zein residue forms hard and glossy, scuff-proof, protective coatings that also are resistant to microbial attack (Reiners et al., 1973). Zein coatings are used as oxygen, lipid, and moisture barriers for nuts, candies, confectionery products, and other foods. Rice fortified with vitamins and minerals has been coated with zein/stearic acid/wood resin mixtures to prevent vitamin and mineral losses during washing in cold water. Pharmaceutical tablets are zein-coated for controlled ingredient release and protection (Gennadios, 2002). Use of zein-based coatings has been suggested for reducing oil uptake by deep-fat fried foods, for protecting active ingredients in chewing gum, for achieving controlled release of active ingredients in pharmaceutical tablets and for masking the taste of orally administered drugs (Gennadios, 2002). Zein, upon casting from aqueous aliphatic alcoholic solutions forms tough, glossy and grease resistant films (Ghanbarzadeh et al., 2006c; 2007b). By cross-linking the tensile strength of the films is further improved. Zein films have water vapor permeability (WVP) values lower than or similar to those of other protein films, cellulose ethers, and cellophane (Krochta, 1992). However, their WVP is notably higher than that of LDPE or ethylene-vinyl alcohol copolymer (EVOH) (Gennadios, 2002).

2.1.1.2.2. Wheat gluten

Whereas dry wheat flour comprises 9–13% protein and 75–80% starch, wheat gluten consists mainly of wheat storage protein (70–80%, dry matter basis) with traces of starch and non - starch polysaccharides (10–14%), lipids (6–8%), and minerals (0.8–1.4%). Osborne distinguished four wheat protein classes based on their solubility in different solvents, namely, albumins, globulins, gliadins, and glutenins. The albumins and globulins (15–22% of total protein), which are, respectively, water- and salt-soluble, are removed with starch granules during gluten processing. In contrast, the gliadins, which are alcohol-soluble, and the glutenins, which are soluble (or at least dispersible) in dilute acid or alkali solutions, are being collected into gluten (Wrigley and Bietz, 1988). Gliadin molecules may interact together or with glutenin molecules via hydrophobic interactions and hydrogen bonds. In the fully hydrated state, gliadin exhibits viscous flow properties without significant elasticity. For cereal technologists, gliadin accounts for the extensibility of wheat flour dough and acts as a filler diluting glutenin interactions. Contrary to gliadin, which is comprised of distinct polypeptide chains, glutenin consists of polymers made from polypeptide chains (also named subunits) linked end-to-tail by SS bonds (Kasarda, 1999). Vital wheat gluten is the cohesive and elastic mass that is leftover after starch is washed away from wheat flour dough. Commercially, it is an industrial by-product of wheat starch production via wet milling (Gennadios, 2002).

Wheat gluten is suitable for numerous food and nonfood uses. Its main application is in the bakery industry, where it is used to strengthen weak flours rendering them suitable for bread baking (Gennadios, 2002). The other potential applications of gluten are very diverse: windows in envelopes, surface coatings on paper, biodegradable plastic films for agricultural uses, water-soluble bags with fertilizers, detergents, cosmetics, cigarette filters and additives and molded objects (Cuq et al., 1998). Wheat gluten-based materials are homogeneous, transparent, mechanically strong, and relatively water resistant. They are biodegradable and a priori biocompatible, apart from some wheat gluten-specific characteristics such as allergenicity (Guilbert et al., 1996).

The moisture barrier properties of wheat gluten-based films are relatively poor as compared to synthetic films, such as LDPE. The gas (O_2, CO_2, and ethylene) barrier properties of wheat gluten-based films are highly interesting, as they are exceptionally good at low relative humidity (RH) conditions. Water and other plasticizers can lower the glass transition temperature (T_g) of the wheat gluten and enable processing at temperatures below those that lead to protein decomposition, which means that protein-based films can be formed by using techniques that are conventionally used with synthetic polymers (e.g., extrusion, injection, and molding) (Gennadios, 2002).

2.1.1.2.3. Soy protein

The major use of soybean in the food industry is as a source of oil, while soy protein concentrate and isolate are readily available as co-products of the oil processing industry (Pszczola, 1998). Soy protein is a complex mixture of proteins with widely different molecular properties. The major soybean proteins have molecular weights ranging from 200 to 600 kDa. Most soy proteins (~90%) are globulins, which can be fractionated into 2S, 7S, 11S and 15S according to their sedimentation coefficients. The 7S and 11S fractions, the main fractions making up about 37% and 31% of the total extractable protein, have the capability of polymerization (Wolf, 1972). Soy protein used in the food industry is classified as soy flour, concentrate, or isolate based on the protein content. Soy flour contains 50–59% protein and is obtained by grinding defatted soy flakes. Soy protein concentrate contains 65–72% protein and is obtained by aqueous liquid extraction or acid leaching process. Soy protein isolate contains more than 90% protein and is obtained by aqueous or mild alkali extraction followed by isoelectric precipitation (Soy Protein Council, 1987).

Soy protein is an abundant and relatively cheap ingredient source for various food applications. The functional properties that make soy protein useful in foods include cohesiveness, adhesiveness, emulsification, dough and fiber formation, whippability, solubility, and foaming (Gennadios, 2002). Soy protein also is used in infant formulas and in baked, meat, and dairy products (Witherly, 1990). The use of soy protein as a film-forming agent can add value to soybeans by creating new channels for marketing soy proteins (Cao et al., 2007). Soy protein is a viable and renewable resource for producing edible and environmentally friendly biodegradable films. Soy protein films are flexible, smooth, transparent, and clear compared to other films from plant proteins (Cho and Rhee, 2004). These films have good mechanical properties but they are generally slightly water-resistant (Cuq et al., 1998). Soy protein films are typically prepared by drying thin layers of cast film-forming solutions (Gennadios, 2002).

Biodegradable plastics were also produced from soy isolate and concentrate by a thermo-molding process (Jane et al., 1994).

2.1.1.2.4. Collagen and gelatin

Collagen is an abundant protein constituent of connective tissue in vertebrate (about 50% of total human protein) and invertebrate animals. Similar to cellulose in plants, collagen molecules support mechanical stresses transferred to them by a low-modulus matrix (Gennadios, 2002). Collagen is a rod-type polymer nearly 300 nm long with a molecular weight of 300,000. There have been more than twenty two different types of collagen identified so far in the human body, with the most common being Type I–IV. Type I collagen is the single most abundant protein present in mammals and is the most thoroughly studied protein. The Type I collagen is composed of three polypeptide subunits with similar amino acid compositions. Each polypeptide is composed of about 1050 amino acids, containing approximately 33% glycine, 25% proline and 25% hydroxyproline with a relative abundance of lysine (Nair and Laurencin, 2007). Collagen is a hydrophilic protein because of the greater content of acidic, basic, and hydroxylated amino acid residues than lipophilic residues. Therefore, it swells in polar liquids with high solubility parameters.

Collagen undergoes enzymatic degradation within the body via enzymes, such as collagenases and metalloproteinases, to yield the corresponding amino acids. Due to their enzymatic degradability, unique physico-chemical, mechanical and biological properties collagen has been extensively investigated for various applications. Collagen is mostly soluble in acidic aqueous solutions and can be processed into different forms such as sheets, tubes, sponges, foams, nanofibrous matrices, powders, fleeces, injectable viscous solutions and dispersions. Studies have also shown that the degradation rate of collagen used for biomedical applications can be significantly altered by enzymatic pre-treatment or cross-linking using various cross-linking agents (Nair and Laurencin, 2007).

Type I collagen is found in high concentrations in tendon, skin, bone, and fascia, which are consequently convenient and abundant sources for isolation of this natural polymer. The major sources of collagen currently used for industrial applications are bovine or porcine skin or bovine or equine Achilles tendons (Pachence et al., 2007). Thermal or chemical dissociation of collagen polypeptide chains forms products known as gelatin. Insoluble collagen is converted to soluble gelatin by acid or alkaline/lime (mild and slow) processing. Two processes are mainly used for commercial production of gelatin. In the first process, the collagen in hide or demineralized bone is partly depolymerized by prolonged liming that breaks down covalent cross-links. The occurring hydrolysis results in extensive release of collagenous material, which is solubilized at near neutral pH at temperatures of 60–90 °C (Type B gelatin). The acid process (Type A gelatin) involves soaking skin or bone in a dilute acid followed by extraction at acid pH (Johnston-Banks, 1990).

The properties of collagen and gelatin are of great interest to various fields, such as surgery (implantations; wound dressings), leather chemistry (tanning), pharmacy (capsule production; tablet binding), and food science (gels; edible films) (Arvanitoyannis et al., 1998). Reportedly, about 65% of gelatin manufactured worldwide is used in foods, 20% in photographic applications, 10% in pharmaceutical products, and 5% in other specialized and industrial applications (Slade and Levine, 1987). Collagen has been extensively investigated for the localized delivery of low molecular weight drugs including antibiotics (Gruessner et al., 2001). Collagen films have traditionally been used for preparing edible sausage casing (Hood, 1987). Gelatin has been successfully used to form films that are transparent, flexible, water-resistant, and impermeable to oxygen (Hebert and Holloway 1992). These films were made by cooling and drying an aqueous film-forming solution based on gelatin. Gelatin is also used as a raw material for photographic films, and to microencapsulate aromas, vitamins, and sweeteners (Balassa and Fanger 1971).

2.1.1.2.5. Casein and caseinates

Casein is the main protein of milk, representing 80% of the total milk proteins, it is a phosphoprotein that may be separated into various electrophoretic fractions, α_{s1}-casein, α_{s2}-casein, β-casein and κ-casein which differ in primary, secondary and tertiary structure and molecular weight. These four different types of casein are found in bovine milk in the approximate ratio of 4:1:4:1 respectively (Dalgleish, 1997). Casein exists in the form of micelles containing all four casein species complexed with colloidal calcium phosphate. The casein micelles are stable to most common milk processes such as heating, compacting, and homogenization. Micellar integrity is preserved by extensive electrostatic and hydrogen bonding, and hydrophobic interactions (Gennadios, 2002).

Two principal methods have been established for the production of casein on commercial scale, i.e., isoelectric precipitation (acid casein) and enzymatic coagulation (rennet casein). The preparation of acid casein from skim milk is quite simple. The essential steps involve acidification to about pH 4.6 (isoelectric point of casein) to induce the coagulation of casein, adjustment of temperature to between 30 and 40 °C for better handling properties of the product, washing, pressing or centrifuging the curd to remove excessive water, and finally drying and grinding. Hydrochloric acid is usually used for both laboratory scale and industrial preparation of casein. The key difference in producing rennet casein from producing acid casein is the means of coagulation. A proteolytic enzyme, such as chymosin (rennin), cleaves the -casein fraction to release a glycomacropeptide, thus destabilizing the casein micelles and promoting coagulation of casein in the presence of calcium cations (Gennadios, 2002). Though water-insoluble casein has some applications, most food application would require casein with high water solubility. This is achieved by dispersing the casein in water and adjusting the pH to between 6.5 and 7.0 with an alkali. The most commonly used soluble caseinate is sodium caseinate. It is normally manufactured by dissolving fresh acid casein curd in sodium hydroxide followed by spray drying. Other soluble caseinates prepared in a similar manner include potassium, calcium, magnesium, and ammonium caseinates (Fox and McSweeney, 1998).

Its relative simple isolation and the useful properties of casein as an industrial material and food ingredient have led to commercial production of casein and caseinates since the 19th century (Muller, 1982). Casein and caseinates are suitable for numerous food and nonfood uses such as in industrial applications (especially in glues, paper coatings, paints, leather finishing, textile fibers, and plastics), and in various food products. The end-uses of casein and caseinates have gradually shifted from industrial to food applications. About 70 to 80% of the casein produced worldwide is used as a food ingredient (Gennadios, 2002). Film-forming properties of caseins have been used to improve the appearance of numerous foods, to produce water-soluble bags, and to produce origin or quality identification labels inserted under precut cheeses, to ensure the surface retention of additives on intermediate-moisture foods, and to encapsulate polyunsaturated lipids for animal feeds (Cuq et al., 1998). Casein-based edible films are attractive for food applications due to their high nutritional quality, excellent sensory properties and potential to adequately protect food products from their surrounding environment (Fabra et al., 2009). The mechanical properties of casein and caseinate films, being neither too tough nor too fragile, also make them suitable for edible purposes. Though more permeable to water vapor than plastic films, they are capable of retarding moisture transfer to some degree (Buonocore et al., 2003). Casein and caseinate films dissolve nearly instantaneously in water and this is desirable for many food applications.

2.1.1.2.6. Whey proteins

Whey proteins are those proteins that remain in milk serum after pH/rennet coagulation of casein during cheese or casein manufacture (Gennadios, 2002). Whey protein, which represents approximately 20% of total milk proteins is a mixture of proteins with diverse functional properties. The five main proteins are α-lactalbumins, β-lactoglobulins, bovine serum albumin, immunoglobulins, and proteose peptones. α-lactalbumins, β-lactoglobulins and bovine serum albumin comprise 57, 19 and 7% of the total whey protein. The immunoglobulins and proteose-peptone fractions represent the remainder of the whey protein. Whey proteins are globular and soluble at pH 4.6 (Dybing and Smith, 1991).

The industrial processes used for whey protein recovery are ultrafiltration, reverse osmosis, gel filtration, electrodialysis, and ion exchange chromatography. Fractionated whey constituents of various degrees of concentration can be obtained by combining two or more of the above recovery processes (Sienkiewicz and Riedel, 1990). Whey protein products can be classified according to their composition. In particular, they are divided according to their protein content. Whey protein concentrate (WPC) contains 25–80% protein. Whey protein isolate (WPI) is nearly all protein (>90%) (Gennadios, 2002).

Whey protein, a byproduct of the cheese industry, has excellent nutritional and functional properties and the potential to be used for human food and animal feed. The film-forming properties of whey proteins have been used to produce transparent, flexible, colorless, and odorless films, such as those produced from caseins (Cuq et al., 1998). The use of whey proteins to make an edible packaging film material brings several environmental advantages because of the film's biodegradability and its capacity to control moisture, carbon dioxide, oxygen, lipid, flavor and aroma transfer (Ozdemir and Floros, 2008). These properties offer the potential to extend the shelf-life of many food products, avoiding quality deterioration (Gounga et al., 2007).

2.1.1.2.7. Other proteins

Prevalent types of plant and animal proteinous biopolymers have been discussed previously. Nevertheless, these are not the only available biopolymers from this category. There are other proteins, which have potential to use as biopolymeric materials for different applications. Most important of them includes elastin (a major protein component of vascular and lung tissue), egg albumins, fish myofibrillar protein and wool keratin (Gennadios, 2002).

An interesting property of elastin is its ability to undergo folding when the temperature is increased above 25 °C. This is due to its transition from a disordered form to an or-dered form at higher temperature called inverse temperature transition (ITT). Due to the unique thermal transition properties of elastin, it has been extensively investigated as a smart, injectable drug delivery system (Mithieux et al., 2004). Use of egg albumins to en-capsulate organic hydrophobic compounds in cosmetics and foods has been proposed in many patents (Gennadios et al., 1999). The potential use of myofibrillar proteins for the preparation of films and coatings was proposed to identify new practical applications for fish proteins (Shiku et al., 2004). Yamauchi et al (1996) developed water-insoluble films based on keratin by casting and drying alkaline dispersions. The large amount of cystine in keratin favors formation of many disulfide bonds that could stabilize the proteic net-work. However, because of their unpleasant mouthfeel, edible coatings based on keratin have not found many applications (Yamauchi and Khoda, 1997).

Other proteins have been also used for various purposes, including proteins from rye, pea, barley, sorghum, rice, sunflower, pistachio and peanut (Gennadios, 2002).

2.1.2. Biopolymers produced directly by natural or genetically modified organisms

2.1.2.1. Microbial polyesters

The microbial polyesters are produced by biosynthetic function of a microorganism and readily biodegraded by microorganisms and within the body of higher animals, including humans. In the field of medicine, they can be used as implanting material and a drug carriers.

2.1.2.1.1. Polyhydroxyalcanoates (PHAs)

Polyhydroxyalkanoates (PHAs) are a family of intracellular biopolymers synthesized by many bacteria as intracellular carbon and energy storage granules (Fig. 8). Depending on growth conditions, bacterial strain, and carbon source, the molecular weights of these polyesters can range from tens into the hundreds of thousands (Pachence et al., 2007). Bacterially synthesized PHAs have attracted attention because they can be produced from a variety of renewable resources and are truly biodegradable and highly biocompatible thermoplastic materials.

The plastic-like properties and biodegradability of PHAs offer an attraction as a potential replacement for non-degradable polyethylene and polypropylene. Many efforts have been made to produce PHA as environmentally degradable thermoplastics (Chen et al., 2001; Chen and Page, 1997; Byrom, 1992). PHAs have a promising potential for food packaging applica-tions. However, due to exorbitant production costs, few suppliers exist in the market.

Over 90 different types of PHA consisting of various monomers have been reported and the number is increasing. Some PHAs behave similarly to conventional plastics such as polyethylene and polypropylene, while others are elastomeric (Yang et al., 2002). The most representative member of this family is poly(3-hydroxybutyrate) (PHB).

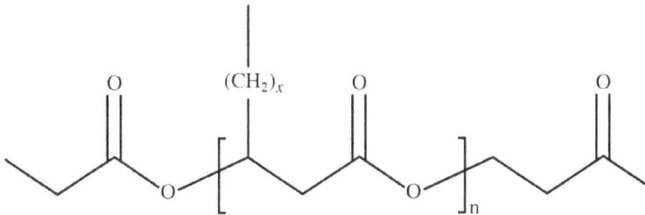

Figure 8. Chemical structure of polyhydroxyalkanoate (Pachence et al., 2007).

2.1.2.1.2. Poly-3-hydroxybutyrate (PHB)

Among the PHA family, poly-3-hydroxybutyrate (PHB) is the most common member; it belongs to the short chain length PHA with its monomers containing 4-5 carbon atoms (Smith, 2005). ICI developed a biosynthetic process for the manufacture of PHB, based on the fermentation of sugars by the bacterium *Alcaligenes eutrophus*. PHB homopolymer, like all other PHA homopolymers, is highly crystalline, extremely brittle, and relatively hydrophobic. Consequently, the PHA homopolymers have degradation times *in vivo* on the order of years (Holland et al., 1987; Miller and Williams, 1987).

PHB has the poorest mechanical properties compared with its copolymers. Efforts have been made to improve the mechanical properties of PHB; Iwata et al. (2003) prepared uniaxially oriented films of PHB, with sufficient strength and flexibility by cold-drawing from an amorphous preform at a temperature below, but near to, the glass transition temperature. Melt crystallized and solvent-cast films of PHB are usually quite brittle, and the orientation is critical and difficult to reproduce consistently.

Since current production technology is still unable to produce any PHB that is competitive with conventional plastics such as polyethylene, polypropylene or polystyrene which are manufactured on a large scale, the application of PHB as environmentally friendly packaging materials is still unrealistic. Therefore, increasing research is focused on the biosynthesis of PHB with unconventional structures that may bring new properties and new applications for PHB (Eggink et al., 1995; Lutke-Eversloh, et al., 2001; Kim, et al., 1996).

PHB has been found to have low toxicity, in part due to the fact that it degrades *in vivo* to d3-hydroxybutyric acid, a normal constituent of human blood. Applications of these polymers previously tested or now under development include controlled drug release, artificial skin, and heart valves as well as such industrial applications as paramedical disposables (Doi et al., 1990; Sodian et al., 2000).

2.1.2.1.3. Poly (Hydroxybutyrate-Hydroxyvalerate) (PHB/HV)

Blends of the PHB family are usually compatible and co-crystallization is enhanced. Yoshie et al., (2004) studied solid-state structures and crystallization kinetics of poly (3-hydroxybutyrate-co-3-hydroxyvalerate) (PHB/HV) blends. It was found that PHB and HV can co-crystallize.

The copolymers of PHB with hydroxyvaleric acid (PHB/HV) (Fig. 9) are less crystalline, more flexible, and more readily processible, but they suffer from the same disadvantage of being too hydrolytically stable to be useful in short-term applications when resorption of the degradable polymer within less than one year is desirable.

PHB and its copolymers with up to 30% of 3-hydroxyvaleric acid are now commercially available under the trade name Biopol. It was found previously that a PHA copolymer of PHB/HV with a 3- hydroxyvalerate content of about 11%, may have an optimum balance of strength and toughness for a wide range of possible applications (Pachence et al., 2007).

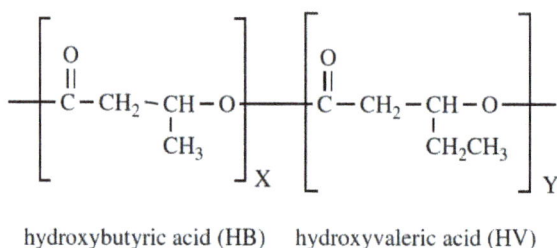

$$
\left[\begin{array}{c} O \\ \| \\ C-CH_2-CH-O \\ | \\ CH_3 \end{array}\right]_X \left[\begin{array}{c} O \\ \| \\ C-CH_2-CH-O \\ | \\ CH_2CH_3 \end{array}\right]_Y
$$

hydroxybutyric acid (HB)   hydroxyvaleric acid (HV)

Figure 9. Poly(b-hyroxybutyrate) and copolymers with hydroxyvaleric acid. For a homopolymer of HB, Y = 0; commonly used copolymer ratios are 7, 11, or 22 mole percent of hydroxyvaleric acid. (Pachence et al., 2007)

2.1.2.1.4. Poly- ε-Caprolactones (PCL)

Poly (ε-caprolactone) (PCL) (Fig. 10) is aliphatic polyester and is of great interest as it can be obtained by the ring opening polymerization of a relatively cheap monomeric unit 'ε-caprolactone'. This polyester is highly processible as it is soluble in a wide range of organic solvents (Nair and Laurencin, 2007).

PCL exhibits several unusual properties not found among the other aliphatic polyesters. Most noteworthy are its exceptionally low glass transition temperature of -62°C and its low melting temperature of 57°C. Another unusual property of PCL is its high thermal stability. Whereas other tested aliphatic polyesters had decomposition temperatures (T_d) between 235 and 255°C, PCL has a T_d of 350°C (Pachence et al., 2007). PCL has low tensile strength (approximately 23MPa) but an extremely high elongation at breakage (4700%) (Gunatillake et al., 2006). In addition, ε-caprolactone can be copolymerized with numerous other monomers (e.g., ethylene oxide, chloroprene, THF, δ-valerolactone, 4-vinylanisole, styrene, methyl methacrylate, vinylacetate). Particularly noteworthy are copolymers of ε-caprolactone and lactic acid that have been studied extensively (Feng et al., 1983).

PCL undergoes hydrolytic degradation due to the presence of hydrolytically labile aliphatic ester linkages; however, the rate of degradation is rather slow (2–3 years). PCL is a semicrystalline polymer with a low glass transition temperature of about -60°C. Thus, PCL is always in a rubbery state at room temperature. Among the more common aliphatic polyesters, this is an unusual property, which undoubtedly contributes to the very high permeability of PCL for many therapeutic drugs (Pitt et al., 1987). Due to the slow degradation, high permeability to many drugs and non-toxicity, PCL was initially investigated as a long-term drug/vaccine delivery vehicle. Extensive research is ongoing to develop various micro- and nano-sized drug delivery vehicles based on PCL (Sinha et al., 2004). Due to its excellent biocompatibility, PCL has also been extensively investigated as scaffolds for tissue engineering.

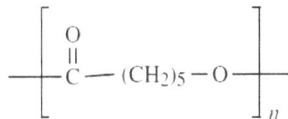

Figure 10. Chemical structure of polyhydroxyalkanoate (Pachence et al., 2007)

2.1.2.2. Bacterial Cellulose (BC)

Bacterial cellulose (BC) belongs to specific products of primary metabolism and is mainly a protective coating, whereas plant cellulose (PC) plays a structural role. Cellulose is synthesized by bacteria belonging to the genera *Acetobacter*, *Rhizobium*, *Agrobacterium*, and *Sarcina* (Bielecki, 2004). Its most efficient producers are Gram-negative, acetic acid bacteria *Acetobacter xylinum* (reclassified as *Gluconacetobacter xylinus*), which have been applied as model microorganisms for basic and applied studies on cellulose (Pamela et al., 1992).

Extensive research on BC revealed that it is chemically identical to PC, but its macromolecular structure and properties differ from the latter (Fig. 11). Nascent chains of BC aggregate to form subfibrils, which have a width of approximately 1.5 nm and belong to the thinnest naturally occurring fibers, comparable only to subelemental fibers of cellulose detected in the cambium of some plants and in quinee mucous. BC subfibrils are crystallized into microfibrils, these into bundles, and the latter into ribbons (Bielecki, 2004). Dimensions of the ribbons are 3-4 (thickness) × 70-80 nm (width), whereas the width of cellulose fibers produced by pulping of birch or pine wood is two orders of magnitude larger (1.4-4.0×10^{-2} and 3.0-7.5×10^{-2} mm, respectively) (Iguchi et al., 2000).

BC is also distinguished from its plant counterpart by a high crystallinity index (above 60%) and different degree of polymerization (DP), usually between 2000 and 6000, but in some cases reaching even 16,000 or 20,000, whereas the average DP of plant polymer varies from 13,000 to 14,000 (Iguchi et al., 2000).

One of the most important features of BC is its chemical purity, which distinguishes this cellulose from that from plants, usually associated with hemicelluloses and lignin, removal of which is inherently difficult (Bielecki, 2004). Because of the unique properties, resulting from the ultrafine reticulated structure, BC has found a multitude of applications in paper, textile, and food industries, and as a biomaterial in cosmetics and medicine (Hu et al., 2011). Wider application of this polysaccharide is obviously dependent on the scale of production and its cost. Therefore, basic studies run together with intensive research on strain improvement and production process development.

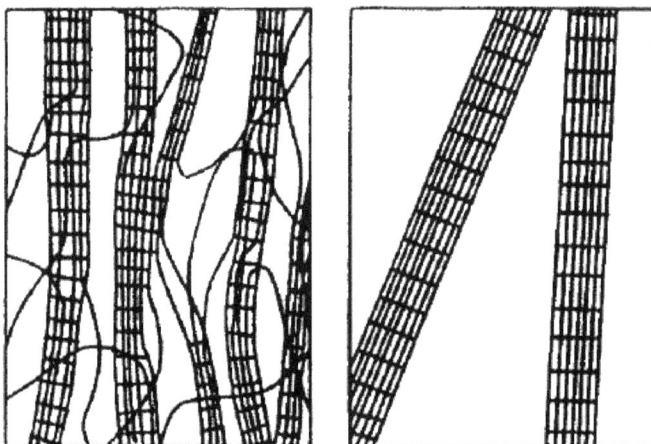

Figure 11. Schematic model of BC microfibrils (Right) drawn in comparison with the `fringed micelles'; of PC fibrils (Left)(Iguchi et al.,2000)

2.1.2.3. Biopolymers (Polyesters) synthesized from bio-derived monomers

This category of biopolymers belongs to biodegradable polyesters and produced by polycondensation or ring-opening polymerization of biologically derived monomers.

2.1.2.3.1. Polylactic Acid or polylactide (PLA)

Among the family of biodegradable polyesters, polylactides (i.e. PLA) have been the focus of much attention because they are produced from renewable resources such as starch, they are biodegradable and compostable, and they have very low or no toxicity and high mechanical performance, comparable to those of commercial polymers (Yu et al., 2006).

PLA or poly-lactide was discovered in 1932 by Carothers. He was only able to produce a low molecular weight PLA by heating lactic acid under vacuum while removing the condensed water. The problem at that time was to increase the molecular weight of the products; and, finally, by ring-opening polymerization of the lactide, high-molecular weight PLA was synthesized (Jamshidian et al., 2010). PLA was 1st used in combination with polyglycolic acid (PGA) as suture material and sold under the name Vicryl in the U.S.A. in 1974 (Mehta et al., 2005).

Lactic acid (2-hydroxy propionic acid), the single monomer of PLA (Fig. 12), is produced via fermentation or chemical synthesis. Its 2 optically active configurations, the L(+) and D(−) stereoisomers are produced by bacterial (homofermentative and heterofermentative) fermentation of carbohydrates. The homofermentative method is preferably used for industrial production because its pathways lead to greater yields of lactic acid and to lower levels of by-products. The general process consists of using species of the *Lactobacillus* genus such as *Lactobacillus delbrueckii, L. amylophilus, L. bulgaricus, and L. leichmanii*, a pH range of 5.4 to 6.4, a temperature range of 38 to 42 °C, and a low oxygen concentration. Generally, pure L-lactic acid is used for PLA production (Mehta et al., 2005).

The polymerization of racemic (D,L)-lactide and mesolactide, results in the formation of amorphous polymers. Among these monomers, L-lactide is the naturally occurring isomer. Similar to polyglycolide, poly(L-lactide) (PLLA) is also a crystalline polymer (~37% crystallinity) and the degree of crystallinity depends on the molecular weight and polymer processing parameters. It has a glass transition temperature of 60–65 °C and a melting temperature of approximately 175 °C (Nair et al., 2007).

Poly (L-lactide) is a slow-degrading polymer compared to polyglycolide, has good tensile strength, low extension and a high modulus (approximately 4.8 GPa) and hence, has been considered an ideal biomaterial for load bearing applications, such as orthopaedic fixation devices. It is classified as generally recognized as safe (GRAS) by the United State Food and Drug Administration (FDA) and is safe for all food packaging applications (Madhavan Nampoothiri et al., 2010; FDA 2002).

Polylactides undergo hydrolytic degradation via the bulk erosion mechanism by the random scission of the ester backbone. It degrades into lactic acid a normal human metabolic by-product, which is broken down into water and carbon dioxide via the citric acid cycle (Maurus and Kaeding, 2004).

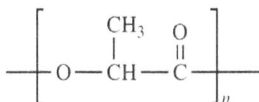

Figure 12. Chemical structure of polylactic acid (Jamshidian et al., 2010)

2.1.2.3.2. Polyglycolic Acid (PGA)

Polyglycolide or Polyglycolic acid (PGA) is a biodegradable, thermoplastic polymer and the simplest linear, aliphatic polyester (Fig. 13).

PGA has been known since 1954 as a tough fiber-forming polymer (Pachence et al., 2007).

Polyglycolide has a glass transition temperature between 35-40 °C and its melting point is reported to be in the range of 225-230 °C. PGA also exhibits an elevated degree of crystallinity, around 45-55%, thus resulting in insolubility in water. The solubility of this polyester is somewhat unique, in that its high molecular weight form is insoluble in almost all common organic solvents (acetone, dichloromethane, chloroform, ethyl acetate, tetrahydrofuran), while low molecular weight oligomers sufficiently differ in their physical properties to be more soluble. However, polyglycolide is soluble in highly fluorinated solvents like hexafluoroiso-propanol (HFIP) and hexafluoroacetone sesquihydrate, that can be used to prepare solutions of the high molecular weight polymer for melt spinning and film preparation. Fibers of PGA exhibit high strength and modulus (7 GPa) and are particularly stiff (Nair et al., 2007).

Polyglycolide is characterized by hydrolytic instability owing to the presence of the ester linkage in its backbone. The degradation process is erosive and appears to take place in two steps during which the polymer is converted back to its monomer glycolic acid: first water diffuses into the amorphous (non-crystalline) regions of the polymer matrix, cleaving the ester bonds; the second step starts after the amorphous regions have been eroded, leaving the crystalline portion of the polymer susceptible to hydrolytic attack. Upon collapse of the crystalline regions the polymer chain dissolves (Tian et al., 2012).

The traditional role of PGA as a biodegradable suture material has led to its evaluation in other biomedical fields. Implantable medical devices have been produced with PGA, including anastomosis rings, pins, rods, plates and screws. It has also been explored for tissue engineering or controlled drug delivery. Tissue engineering scaffolds made with polyglycolide have been produced following different approaches, but generally most of these are obtained through textile technologies in the form of non-woven meshes. The Kureha Corporation has announced its commercialization of high molecular weight polyglycolide for food packaging applications under the trade name of Kuredux®. Production is at Belle, West Virginia, with an intended capacity of 4000 annual metric tons, according to a Chemicals Technology report. Its attributes as a barrier material result from its high degree of crystallization. A low molecular weight version (approximately 600 amu) is available from the DuPont Co. and is purported to be useful in oil and gas applications (Chandra and Rustgi, 1998).

Owing to its hydrolytic instability, however, its use has initially been limited. Currently polyglycolide and its copolymers (poly(lactic-co-glycolic acid) with lactic acid, poly(glycolide-co-caprolactone) with ε-caprolactone, and poly (glycolide-co-trimethylene carbonate) with trimethylene carbonate) are widely used as a material for the synthesis of absorbable sutures and are being evaluated in the biomedical field.

2.2. Synthetic biodegradable polymers

2.2.1. Aliphatic polyesters

Polyester is a category of polymers which contain the ester functional group in their main chain. Aliphatic polyesters are biodegradable but often lack in good thermal and mechanical

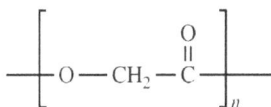

Figure 13. Chemical structure of polyglycolic acid (Pachence et al., 2007)

properties. Vice versa, aromatic polyesters, like Polyethylene terephthalate (PET), have excellent material properties, but are resistant to microbial attack. Among biodegradable polymers, aliphatic polyester-based polymeric structures are receiving special attention because they are all more or less sensitive to hydrolytic degradation, a feature of interest when compared with the fact that living systems function in aqueous media. Only some of these aliphatic polyesters are enzymatically degradable. A smaller number is biodegradable, and an even more limited number is biorecyclable.

2.2.1.1. Polyglycolic Acid (PGA), Polylactic Acids (PLA) and their copolymers

Poly (glycolic acid) (PGA), poly (lactic acid) (PLA), and their copolymers are the most widely used synthetic degradable polymers in medicine. In section 2.1.2.3., structure and properties of these polymers has been explained and in this section synthetic production of them are presented. PGA can be obtained through several different processes starting with different materials (Nair et al., 2007):

1. polycondensation of glycolic acid;

2. ring-opening polymerization of glycolide;

3. solid-state polycondensation of halogenoacetates

Polycondensation of glycolic acid is the simplest process available to prepare PGA, but it is not the most efficient because it yields a low molecular weight product. The most common synthesis used to produce a high molecular weight form of the polymer is ring-opening polymerization of "glycolide", the cyclic diester of glycolic acid.

In spite of its low solubility, this polymer has been fabricated into a variety of forms and structures. Extrusion, injection and compression molding as well as particulate leaching and solvent casting, are some of the techniques used to develop polyglycolide-based structures for biomedical applications (Gunatillake et al., 2006).

The high rate of degradation, acidic degradation products and low solubility however, limit the biomedical applications for PGA. Therefore, several copolymers containing glycolide units are being developed to overcome the inherent disadvantages of PGA.

Due to its hydrophilic nature, surgical sutures made of PGA tend to lose their mechanical strength rapidly, typically over a period of two to four weeks post-implantation. In order to adapt the materials properties of PGA to a wider range of possible applications, researchers undertook an intensive investigation of copolymers of PGA with the more hydrophobic PLA.

Alternative sutures composed of copolymers of glycolic acid and lactic acid is currently marketed under the trade names Vicryl and Polyglactin 910 (Pachence et al., 2007).

Due to the presence of an extra methyl group in lactic acid, PLA is more hydrophobic than PGA. The hydrophobicity of high-molecular-weight PLA limits the water uptake of thin films to about 2% and results in a rate of backbone hydrolysis lower than that of PGA (Reed and Gilding, 1981). In addition, PLA is more soluble in organic solvents than is PGA.

Industrial lactic acid production utilizes the lactic fermentation process rather than synthesis because the synthetic routes have many major limitations, including limited capacity due to the dependency on a by-product of another process, inability to only make the desirable L-lactic acid stereoisomer, and high manufacturing costs (Datta and Henry 2006). Three ways are possible for the polymerization of lactic acid (Jamshidian et al., 2010):

1.   Direct condensation polymerization;

2.   Direct poly-condensation in an azeotropic solution (an azeotrope is a mixture of 2 or more chemical liquids in such a ratio that its composition cannot be changed by simple distillation. This occurs because, when an azeotrope is boiled, the resulting vapor has the same ratio of constituents as the original mixture);

3.   Polymerization through lactide formation.

Polymerization through lactide formation is being industrially accomplished for high molecular weight PLA production.

Recently, PLA, PGA, and their copolymers have been combined with bioactive ceramics such as Bioglass particles and hydroxyapatite that stimulate bone regeneration while greatly improving the mechanical strength of the composite material (Rezwan et al., 2006). Bioglass particles combined with D,L-PLA-co-PGA have also been shown to be angiogenic, suggesting a novel approach for providing a vascular supply to implanted devices. PLA, PGA, and their copolymers are also being intensively investigated for a large number of drug delivery applications.

2.2.1.1.1. Polybutylene Succinate (PBS)

Polybutylene succinate (PBS), chemically synthesized by polycondensation of 1,4-butanedial with succinic acid (Fig. 14), is a chemosynthetic polyester with a relatively high melting temperature ($T_m \sim 113°C$) and favorable mechanical properties, which are comparable with those of such widely used polymers as polyethylene and polypropylene (Chen et al., 2011). PBS is thermoplastic, aliphatic polyester with many interesting properties including biodegradability, melt processability, and thermal and chemical resistance. In addition, its excellent processability in the field of both textiles into melt blown, multifilament, monofilament, nonwoven, flat, and split yarn fabrics and plastics into injection-molded products, makes it a promising polymer for various potential engineering applications (Lim et al., 2011). However, other properties of the PBS, such as its melt viscosity, melt strength, softness, and gas barrier characteristics are still regarded to be insufficient for various end-use applications.

PBS has a relatively low biodegradation rate because of its high crystallization rate and high crystallinity. To promote the physical properties, extend the application field, and increase the biodegradability of PBS, numerous approaches have been used, such as physical blending, copolymerization, or formation of composites (Okamoto et al., 2003).

Owing to the excellent processability of PBS, it can be processed using conventional polyolefin equipment in the range 160-200 ºC. Injection, extrusion or blow moulding is suitable for processing PBS. Its applications include mulch film, cutlery, containers, packaging film, bags and flushable hygiene products.

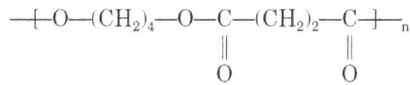

$$\text{---}\!\!\left[\!\text{O---(CH}_2)_4\text{---O---}\overset{\displaystyle\|}{\underset{\displaystyle O}{C}}\text{---(CH}_2)_2\text{---}\overset{\displaystyle\|}{\underset{\displaystyle O}{C}}\text{---}\right]_n$$

Figure 14. Chemical structure of polybutylene succinate (Okamoto et al., 2003)

2.2.1.1.2. Polybutylene Succinate Adipate (PBSA)

Because of the relatively low degradation rates, PBS can be copolymerized by adipate in order to increase the biodegradability. Poly (butylene succinate-*co*-adipate) (PBSA) is synthesized by the reaction of glycols with aliphatic dicarboxylic acids (Fig. 15) and is available for use in a variety of applications including films, laminations, sheet extrusion, monofilaments, multifilaments, blow-molded containers, injection molded cutlery, and foam cushions (Steeves et al., 2007). The succinic acid which is used to prepare this polymer is created by fermentation of sugar extracted from sugarcane or corn, therefore classifying it as a biobased material.

PBSA film has properties very similar to linear low-density polyethylene (LLDPE) and relatively high biodegradability, and is therefore suitable for a composting bag of kitchen waste (Ren et al., 2005).

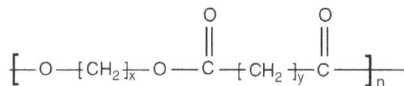

$$\left[\!\text{O---}(\text{CH}_2)_x\text{---O---}\overset{\displaystyle O}{\overset{\displaystyle\|}{C}}\text{---}(\text{CH}_2)_y\text{---}\overset{\displaystyle O}{\overset{\displaystyle\|}{C}}\text{---}\right]_n$$

Figure 15. Chemical structure of polybutylene succinate adipate, where x = 4, y = 2, 4 (Steeves et al., 2007)

2.2.2. Poly (Vinyl Alcohol) (PVOH) and Poly (Vinyl Acetate) (PVA)

Poly (vinyl alcohol) (PVOH) is the most readily biodegradable of vinyl polymers. It is readily degraded in waste-water-activated sludges. Unlike many vinyl polymers, PVOH is not prepared by polymerization of the corresponding monomer. The monomer, vinyl alcohol, almost exclusively exists as the tautomeric form, acetaldehyde. PVOH instead is prepared by partial or complete hydrolysis of polyvinyl acetate to remove acetate groups (Fig. 16). PVOH has a melting point of 180 to 190°C. It has a molecular weight of between 26,300 and 30,000, and a degree of hydrolysis of 86.5 to 89% (Ramaraj, 2007).

PVOH is an odorless and tasteless, translucent, white or cream colored granular powder. It is used as a moisture barrier film for food supplement tablets and for foods that contain inclusions or dry food with inclusions that need to be protected from moisture uptake. PVOH belongs to the water soluble polymers. In the context of the application, solubility and speed of solution are important characteristics.

PVOH has excellent film forming, emulsifying and adhesive properties. It is also resistant to oil, grease and solvents. It has high tensile strength and flexibility, as well as high oxygen and aroma barrier properties. However these properties are dependent on humidity, in other words, with higher humidity more water is absorbed. The water, which acts as a plasticizer, will then reduce its tensile strength, but increase its elongation and tear strength. PVOH is fully degradable and dissolves quickly (Vercauteren and Donners, 1986).

PVOH is the largest synthetic water-soluble polymer produced in the world. The prominent properties of PVOH may include its biodegradability in the environment. The generally accepted biodegradation mechanism occurs via a two-step reaction by oxidation of hydroxyl group followed by hydrolysis. The biodegradation of PVOH is influenced by the stereo-chemical configuration of the hydroxyl groups of PVOH. The isotactic material of PVOH preferentially degraded. The microbial degradation of PVOH has been studied, as well as its enzymatic degradation by secondary alcohol peroxidases isolated from soil bacteria of the *Pseudomonas* strain (Jansson et al., 2006).

PVOH has been studied extensively because of its good biodegradability and mechanical properties. These properties have made PVOH as attractive material for disposable and biodegradable plastic substitutes. Its water solubility, reactivity, and biodegradability make it a potentially useful material in biomedical, agricultural, and water treatment areas, e.g. as a flocculant, metal-ion remover, and excipient for controlled release systems.

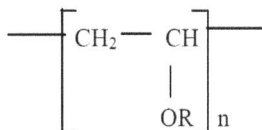

Figure 16. Chemical structure of poly (vinyl alcohol), Where R= H or COCH$_3$ (Ramaraj, 2007)

Polyvinyl acetate, (PVA), is a rubbery synthetic polymer. It is a type of thermoplastic and belongs to the polyvinyl esters family. Polyvinyl acetate is prepared by polymerization of vinyl acetate monomer (free radical vinyl polymerization of the monomer vinyl acetate) (Fig. 17). The degree of polymerization of polyvinyl acetate typically is 100 to 5000. The ester groups of the polyvinyl acetate are sensitive to base hydrolysis and will slowly convert PVA into polyvinyl alcohol and acetic acid. Under alkaline conditions, boron compounds such as boric acid or borax cause the polymer to cross-link, forming tackifying precipitates or slime (Dionisio et al., 1993).

PVA reportedly undergoes biodegradation more slowly. Copolymers of ethylene and vinyl acetate were susceptible to slow degradation in soil-burial tests. The weight loss in a 120-day period increased with increasing acetate content. Because PVOH is obtained from the hydrolysis of PVA, which can be controlled easily in terms of the extent of hydrolysis and the sequence of PVA and PVOH, a controlled hydrolysis of PVA followed by controlled oxidation should provide degradation materials having a wide range of properties and degradability.

Polyvinyl acetate is a component of a widely used glue type, commonly referred to as wood glue, white glue, carpenter's glue, or PVA glue. The stiff homopolymer PVA, mostly the more soft copolymer a combination of vinyl acetate and ethylene, vinyl acetate ethylene (VAE), is used also in paper coatings, paint and other industrial coatings, as binder in nonwovens in glass fibers, sanitary napkins, filter paper and in textile finishing (Chandra and Rustgi, 1998).

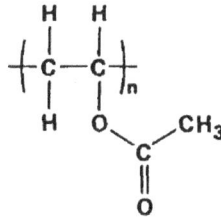

Figure 17. Chemical structure of poly (vinyl acetate) (Chandra and Rustgi, 1998)

3. Factors affecting biodegradation

Several factors affect extent of polymer biodegradation that most impotents of them are polymer structure, polymer morphology, molecular weight, Radiation and chemical treatments.

3.1. Polymer structure

Natural macromolecules, e.g. protein, cellulose, and starch are generally degraded in biological systems by hydrolysis followed by oxidation. It is not surprising, then, that most of the reported synthetic biodegradable polymers contain hydrolyzable linkages along the polymer chain; for example, amide enamine, ester, urea, and urethane linkages are susceptible to biodegradation by microorganisms and hydrolytic enzymes. Since many proteolytic enzymes specifically catalyze the hydrolysis of peptide linkages adjacent to substituents in proteins, substituted polymers containing substituents such as benzyl, hydroxy, carboxy, methyl, and phenyl groups have been prepared in the hope that an introduction of these substituents might increase biodegradability (Savenkova et al., 2000).

Since most enzyme-catalyzed reactions occur in aqueous media, the hydrophilic–hydrophobic character of synthetic polymers greatly affects their biodegradabilities. A polymer containing both hydrophobic and hydrophilic segments seems to have a higher biodegradability than those polymers containing either hydrophobic or hydrophilic structures only. A series of poly(alkylene tartrate)s was found to be readily assimilated by *Aspergillus niger*.

However, the polymers derived from C_6 and C_8 alkane diols were more degradable than the more hydrophilic polymers derived from C_2 and C_4 alkane diols or the more hydrophobic polymers derived from the C_{10} and C_{12} alkane diols.

In order for a synthetic polymer to be degradable by enzyme catalysis, the polymer chain must be flexible enough to fit into the active site of the enzyme. This most likely accounts for the fact that, whereas the flexible aliphatic polyesters are readily degraded by biological systems, the more rigid aromatic poly (ethylene terephthalate) is generally considered to be bioinert (Chandra and Rustgi, 1998).

3.2. Polymer morphology

One of the principal differences between proteins and synthetic polymers is that proteins do not have equivalent repeating units along the polypeptide chains. This irregularity results in protein chains being less likely to crystallize. It is quite probable that this property contributes to the ready biodegradability of proteins. Synthetic polymers, on the other hand, generally have short repeating units, and this regularity enhances crystallization, making the hydrolyz-able groups inaccessible to enzymes. It was reasoned that synthetic polymers with long repeating units would be less likely to crystallize and thus might be biodegradable; indeed, a series of poly (amide-urethane)s were found to be readily degraded by subtilisin (Zilberman et al., 2005).

Selective chemical degradation of semicrystalline polymer samples shows certain character-istic changes. During degradation, the crystallinity of the sample increases rapidly at first, then levels off to a much slower rate as the crystallinity approaches 100%. This is attributed to the eventual disappearance of the amorphous portions of the sample. The effect of morphology on the microbial and enzymatic degradation of PCL, a known biodegradable polymer with a number of potential applications, has been studied. Scanning electron microscopy (SEM) has shown that the degradation of a partially crystalline PCL film by filamentous fungi proceeds in a selective manner, with the amorphous regions being degraded prior to the degradation of the crystalline region. The microorganisms produce extracellular enzymes responsible for the selective degradation. This selectivity can be attributed to the less-ordered packing of amorphous regions, which permits easier access for the enzyme to the polymer chains. The size, shape and number of the crystallites all have a pronounced effect on the chain mobility of the amorphous regions and thus affect the rate of the degradation. This has been demon-strated by studying the effects of changing orientation via stretching on the degradation (Chandra and Rustgi, 1998).

Biodegradation proceeds differently from chemical degradation. Studies on the degradation by solutions of 40% aqueous methylamine have shown a difference in morphology and molecular weight changes and in the ability of the degrading agents to diffuse into the substrate. Also, it was found that the differences in degradation rates between amorphous and crystalline regions are not same. The enzyme is able to degrade the crystalline regions faster than can methylamine. Quantitative GPC (gel permeation chromatography) analysis shows that methylamines degrade the crystalline regions, forming single and double transverse length products. The enzyme system, on other hand, shows no intermediate molecular weight material and much smaller weight shift with degradation. This indicates that although degradation is selective, the crystalline portions are degraded shortly after the chain ends are made available to the exoenzyme. The lateral size of the crystallites has a strong effect on the rate of degradation because the edge of the crystal is where degradation of the crystalline material takes place, due to the crystal packing. A smaller lateral crystallite size yields a higher crystallite edge surface in the bulk polymer. Prior to the saturation of the enzyme active sites, the rate is dependent on available substrate; therefore, a smaller lateral crystallite size results in a higher rate of degradation. The degradation rate of a PCL film is zero order with respect to the total polymer, but is not zero order with respect to the concentrations of the crystallite edge material. The drawing of PCL films causes an increase in the rate of degradation, whereas annealing of the PCL causes a decrease in the rate of degradation. This is probably due to opposite changes in lateral crystallite sizes.

In vitro chemical and enzymatic degradations of polymers, especially polyesters, were analyzed with respect to chemical composition and physical properties. It was found quite often that the composition of a copolymer giving the lowest melting point is most susceptible to degradation (Vert, 2005). The lowest packing order, as expected, corresponds with the fastest degradation rate.

3.3. Radiation and chemical treatments

Photolysis with UV light and the γ-ray irradiation of polymers generate radicals and/or ions that often lead to cleavage and crosslinking. Oxidation also occurs, complicating the situation, since exposure to light is seldom in the absence of oxygen. Generally this changes the material's susceptibility to biodegradation. Initially, one expects the observed rate of degradation to increase until most of the fragmented polymer is consumed and a slower rate of degradation should follow for the crosslinked portion of the polymer. A study of the effects of UV irradiation on hydrolyzable polymers confirmed this. Similarly, photooxidation of polyalkenes promotes (slightly in most cases) the biodegradation. The formation of carbonyl and ester groups is responsible for this change (Miller and Williams, 1987).

Processes have been developed to prepare copolymers of alkenes containing carbonyl groups so they will be more susceptible to photolytic cleavage prior to degradation. The problem with this approach is that negligible degradation was observed over a two year period for the buried specimens. Unless a prephotolysis arrangement can be made, the problem of plastic waste disposal remains serious, as it is undesirable to have open disposal, even with constant sunlight exposure.

As expected, γ-ray irradiation greatly affects the rate of *in vitro* degradation of polyesters. For polyglycolide and poly(glycolide-*co*-lactide), the pH of the degradation solution decreased as the process proceeded. The change-time curves exhibit sigmoidal shapes and consist of three stages: early, accelerated, and later; the lengths of these three regions were a function of γ-ray irradiation. Increasing radiation dosage shortens the time of the early stage. The appearance of the drastic pH changes coincides with loss of tensile breaking strength. Similar effects via enzymatic and microbial degradation remain to be demonstrated (Chandra and Rustgi, 1998).

3.4. Molecular weight

There have been many studies on the effects of molecular weight on biodegradation processes. Most of the observed differences can be attributed to the limit of detecting the changes during degradation, or, even more often, the differences in morphology and hydrophilicity–hydrophobicity of polymer samples of varying molecular weight. Microorganisms produce both exoenzymes [degrading polymers from terminal groups (inwards)] and endoenzymes (degrading polymers randomly along the chain). One might expect a large molecular effect on the rate of degradation in the ease of exoenzymes and a relatively small molecular weight effect in the case of endoenzymes. Plastics remain relatively immune to microbial attack as long as their molecular weight remains high. Many plastics, such as poly ethylene, poly propylene and poly styrene do not support microbial growth. Low molecular weight hydrocarbons, however, can be degraded by microbes. They are taken in by microbial cells, 'activated' by attachment to coenzyme-A, and converted to cellular metabolites within the microbial cell. However, these processes do not function well (if at all) in an extracellular environment, and the plastic molecules are too large to enter the cell. This problem does not arise with natural molecules, such as starch and cellulose, because conversions to low molecular weight components by enzyme reactions occur outside the microbial cell. Photodegradation or chemical degradation may decrease molecular weight to the point that microbial attack can proceed, however (Chandra and Rustgi, 1998).

The upper limits of molecular weight, beyond which uptake and intracellular degradation do not occur, have not been established for all alkane-derived materials. Very slow degradation of paraffins, PE glycols, and linear alkyl benzene sulphonates occurs when the length of the polymer chain exceeds 24–30 carbon atoms. It could be concluded from these amply documented results that alkane-based plastics with molecular weights exceeding 400–500 daltons (i.e. greater than 30 carbon atoms) must be degraded into smaller molecules by photodegradation, chemical or other biological means before biodegradation. LDPE with a molecular weight average of $Mw = 150\ 000$ contains about 11 000 carbon atoms. Decreasing molecules of this size to biologically acceptable dimensions requires extensive destruction of the PE matrix. This destruction can be partly accomplished in blends of PE and biodegradable natural polymers by the action of organisms, such as arthropods, millipedes, crickets, and snails (Santerre et al., 2005).

4. Future strategy

Synthetic polymers are gradually being replaced by biodegradable materials especially those derived from replenishable, natural resources. Bioplastics development is just beginning; until now it covers approximately 5-10% of the current plastic market, about 50,000 t in Europe. More than the origin, the chemical structure of the biopolymer that determines its biodegradability. Use of such biopackagings will open up potential economic benefits to farmers and agricultural processors. The principal field regards the use of packaging films for food products, loose films used for transport packaging, service packaging like carry bags, cups, plates and cutlery, biowaste bags, in agricultural and horticultural fields like bags and compostable articles. Bilayer and multicomponent films resembling synthetic packaging materials with excellent barrier and mechanical properties need to be developed. Cross-linking, either chemically or enzymatically, of the various biomolecules is yet another approach of value in composite biodegradable films. Sustained multidisciplinary research efforts by chemists, polymer technologists, microbiologists, chemical engineers, environmental scientists and bureaucrats are needed for a successful implementation and commercialization of biopolymer-based eco-friendly packaging materials. Undoubtedly, biodegradation offers an attractive route to environmental waste management. Their development costs are high and yet they do not have the benefit of economic scale. It was shown that polyolefins present the same oxo-biodegradability of biopolymers, but they are more economical and effecting during use. Bio-based polymers have already found important applications in medicine field, where cost is much less important than function. It seems very unlikely that biodegradable oil based polymers will be displaced from their current role in packaging application, where cost is more important for the consumer market than environmental acceptability. Biopolymers fulfill the environmental concerns but they show some limitations in terms of performance like thermal resistance, barrier and mechanical properties, associated with the costs. Then, this kind of packaging materials needs more research, more added value like the introduction of smart and intelligent molecules (which is the nanotechnology field) able to give information about the properties of the material inside the package (quality, shelf-life, safety) and nutritional values. It is necessary to make researches on this kind of material to enhance barrier properties, to incorporate intelligent labelling, to give to the consumer the possibility to have more detailed product information than the current system.

Author details

Babak Ghanbarzadeh and Hadi Almasi

University of Tabriz, Iran

References

[1] Almasi, H, Ghanbarzadeh, B, & Entezami, A. A. (2010). Physicochemical properties of starch-CMC-nanoclay biodegradable films. International Journal of Biological Macromolecules. , 46, 1-5.

[2] Amass, W, Amass, A, & Tighe, B. (1998). A review of biodegradable polymers: Uses, current developments in the synthesis and characterization of biodegradable polyesters, blends of biodegradable polymers, and recent advances in biodegradation studies. Polymer International. , 47, 89-144.

[3] Angles, M. N, & Dufresne, A. (2000). Plasticized starch/tunicin whiskers nanocomposites. Macromolecules. , 33(22), 8344-8353.

[4] Argos, P, Pederson, K, Marks, M. D, & Larkins, B. A. (1982). Structure Model for Maize Zein Proteins. Journal of Biological Chemistry. , 257, 9984-9990.

[5] Arvanitoyannis, I, Nakayama, A, & Aiba, S. (1998). Chitosan and Gelatin Based Edible Films: State Diagrams, Mechanical and Permeation Properties. Carbohydrate Polymers. , 37, 371-382.

[6] Augst, A. D, Kong, H. J, & Mooney, D. J. (2006). Alginate hydrogels as biomaterials. Macromolecular Bioscience. , 6(8), 623-33.

[7] Averous, L. (2002). Starch based biodegradable materials suitable for thermoforming packaging. Starch/Starke. , 53, 368-371.

[8] Balassa, L. L, & Fanger, G. O. (1971). Microencapsulation in the food industry. CRC Critical Reviews in Food Technology. , 2, 245-265.

[9] Bielecki, S, Krystynowicz, A, Turkiewicz, L, & Kalinowska, H. (2004). Bacterial cellulose. In: Biopolymers. ed. Steinbüchel, A., (Wiley-VCH., 3-7.

[10] Blanshard, J. M. V. (1987). Starch granule structure and function: a physicochemical approach. In Starch: Properties and potential. Galliard, T., ed (Chichester: John Wiley & Sons., 16-54.

[11] Bouhadir, K. H, Lee, K. Y, Alsberg, E, Damm, K. L, Anderson, K. W, & Mooney, D. J. (2001). Degradation of partially oxidized alginate and its potential application for tissue engineering. Biotechnology Progress. , 17, 945-50.

[12] Brendel, O, Iannetta, P. P. M, & Stewart, D. and simple method to isolate pure alpha-cellulose. Phytochemical Analysis. , 11, 7-10.

[13] Buleon, A, Colonna, P, Planchot, V, & Ball, S. (1998). Starch granules: structure and biosynthesis. International Journal of Biological Macromolecules. , 23, 85-112.

[14] Buonocore, G. G. Del Nobile, M.A., Martino, C.D., Gambacorta, G., La Notte, E., and Nicolais, L. ((2003). Modeling the water transport properties of casein-based edible coating. Journal of Food Engineering. , 60, 99-106.

[15] Byrom, D. (1992). Production of poly-β-hydroxybutyrate and poly-β-hydroxyvalerate copolymers. FEMS Microbiology Reviews. , 103, 247-250.

[16] Cao, Y, Wu, J, Zhang, J, Li, H. Q, Zhang, Y, & He, J. S. (2009). Room temperature ionic liquids (RTILs): a new and versatile platform for cellulose processing and derivatization. Chemical Engineering Journal. , 147, 13-21.

[17] Caoa, N, Fua, Y, & Hea, J. (2007). Preparation and physical properties of soy protein isolate and gelatin composite films. Food Hydrocolloids. , 21, 1153-1162.

[18] Chandra, R, & Rustgi, H. (1998). Biodegradable polymers. Progress in Polymer Science. , 23, 1273-1335.

[19] Chen, C. H, Yang, C. S, Chen, M, Shih, Y. C, Hsu, H. S, & Lu, S. F. (2011). Synthesis and characterization of novel poly(butylene succinate-co-2-methyl-1,3-propylene succinate)s. eXPRESS Polymer Letters. , 5(4), 284-294.

[20] Chen, G. Q, & Page, W. J. (1997). Production of poly-beta-hydroxybutyrate by Azotobacter vinelandii UWD in a two-stage fermentation process. Biotechnological Biotechniques. , 11, 347-350.

[21] Chen, G. Q, Zhang, G, Park, S. J, & Lee, S. (2001). Industrial Production of Poly(hydroxy- butyrate-co- hydroxyhexanoate). Applied Microbiological Biotechnology. , 57, 50-55.

[22] Choa, S. Y, & Rhee, C. (2004). Mechanical properties andwater vapor permeability of edible films made from fractionated soy proteins with ultrafiltration. LWT. , 37, 833-839.

[23] Cuq, B, Gontard, N, & Guilbert, S. (1998). Proteins as Agricultural Polymers for Packaging Production. Cereal Chemistry. , 75(1), 1-9.

[24] Dalgleish, D. G. (1997). Structure-Function Relationships of Caseins. In Food Proteins and Their Applications. eds., S. Damodaran and A. Paraf. (New York: Marcel Dekker., 199-223.

[25] Datta, R, & Henry, M. (2006). Lactic acid: recent advances in products, processes and technologies: a review. Journal of Chemical Technology and Biotechnology. , 81, 1119-1129.

[26] Debeaufort, F, & Voilley, A. (1997). Methylcellulose-based edible films and coatings: 2. Mechanical and thermal properties as a function of plasticizer content. Journal of Agricultural and Food Chemistry. , 45(3), 685-689.

[27] Dionisio, M. S, Moura-ramos, J. J, & Williams, G. (1993). Molecular motion in poly (vinyl acetate) and in poly (vinyl acetate) / ϱ-nitroaniline mixtures. Polymer. , 34(19), 4105-4113.

[28] Doi, Y, Kanesawa, Y, Kunioka, M, & Saito, T. (1990). Biodegradation of microbial copolyesters: poly(3-hydroxy-butyrate-co-3-hydroxyvalerate) and poly(3-hydroxybutyrate-co-4-hydroxyvalerate). Macromolecules. , 23, 26-31.

[29] Dufresne, A, & Cavaille, J. Y. (1998). Clustering and percolation effects in microcrystalline starch reinforced thermoplastic. Journal of Polymer Science: Part B. , 36(12), 2211-2224.

[30] Dybing, S. T, & Smith, D. E. (1991). Relation of Chemistry and Processing Procedures to Whey Protein Functionality: A Review. Cultured Dairy Products Journal. , 26, 4-12.

[31] Edgar, K. J, Buchanan, C. M, Debenham, J. S, Rundquist, P. A, Seiler, B. D, Shelton, M. C, & Tindall, D. (2001). Advances in cellulose ester performance and application. Progress in Polymer Science. , 26, 1605-1688.

[32] Eggink, G, De Waard, P, & Huijberts, G. N. M. (1995). Formation of novel poly(hydroxy- alkanoates) from long-chain fatty acids. Canadian Journal of Microbiology. , 41(1), 14-21.

[33] Espigares, I, Elvira, C, Mano, J. F, & Vazquez, B. San Roman, J., and Reis, R.L. ((2002). New partially degradable and bioactive acrylic bone cements based on starch blends and ceramic fillers. Biomaterials. , 23, 1883-95.

[34] Fabra, M, Talens, P, & Chiralt, A. (2009). Microstructure and optical properties of sodium caseinate films containing oleic acid-beeswax mixtures. Food Hydrocolloids. , 23, 676-683.

[35] Fang, J, & Fowler, P. (2003). The use of starch and its derivatives as biopolymer sources for packaging materials. Food, Agriculture & Environment. , 1, 82-84.

[36] FDA (2002). Inventory of Effective Food Contact Substance (FCS) Notifications http://www.accessdata.fda.gov/scripts/fcn/fcnDetailNavigation.cfm?rpt=fcsListing&id=178. (178)

[37] Feng, X. D, Song, C. X, & Chen, W. Y. (1983). Synthesis and evaluation of biodegradable block copolymers of E-caprolactone and d,l-lactide. Journal of Polymer Science. , 21, 593-600.

[38] Fox, P. F, & Mcsweeney, P. L. H. (1998). Dairy Chemistry and Biochemistry. London: Blackie Academic & Professional.

[39] Galbe, M, & Zacchi, G. (2002). A review of the production of ethanol from softwood. Applied Microbiological Biotechnology. , 59, 618-628.

[40] Gennadios, A. (2002). Protein- based films and coatings. (CRC Press LLC., 134-149.

[41] Ghanbarzadeh, B, Almasi, H, & Entezami, A. A. (2011). Improving the barrier and mechanical properties of corn starch-based edible films: Effect of citric acid and carboxymethyl cellulose. Industrial Crops and Products. , 33, 229-235.

[42] Ghanbarzadeh, B, Almasi, H, & Entezami, A. A. (2010). Physical properties of edible modified starch/carboxymethyl cellulose films. Innovative Food Science and Emerging Technologies. , 11, 697-702.

[43] Ghanbarzadeh, B, & Almasi, H. (2011). Physical properties of edible emulsified films based on carboxymethyl cellulose and oleic acid. International Journal of Biological Macromolecules. , 48, 44-49.

[44] Ghanbarzadeh, B, Musavi, M, Oromiehie, A. R, & Rezayi, K. Razmi Rad, E., and Milani, J. ((2007). Effect of plasticizing sugars on water vapor permeability, surface energy and microstructure properties of zein films. LWT. , 40, 1191-1197.

[45] Ghanbarzadeh, B, Oromiehie, A, Musavi, M, & Rezayi, K. Razmi Rad, E., and Milani, J. ((2006c). Investigation of water vapour permeability, hydrophobicity and morphology of zein films plasticized by polyols. Iranian Polymer Journal. , 15(7), 34-42.

[46] Ghanbarzadeh, B, Oromiehie, A. R, Musavi, M, Emam, D-J. o. m. e. h, Razmi, Z, Rad, E, & Milani, J. (2006a). Effect of plasticizing sugars on rheological and thermal properties of zein resins and mechanical properties of zein films. Journal of Food Research International. , 39, 882-890.

[47] Ghanbarzadeh, B, Oromiehie, A. R, & Musavi, M. Massimiliano Falcone, P., Emam D-Jomeh, Z., and Razmi Rad, E. ((2007b). Study of Mechanical Properties, Oxygen Permeability and AFM Topography of Zein Films Plasticized by Polyols. Packaging Technology and Science. , 20, 155-163.

[48] Ghanbarzadeh, B, & Oromiehie, A. R. Musavi, Razmi Rad, E., and Milani, J. ((2006b). Effect of Polyolic Plasticizers on Rheological and Thermal Properties of Zein Resins. Iranian Polymer Journal. , 15(10), 779-787.

[49] Ghanbarzadeh, B, Orumiei, A. R, & Saboonchi, S. H. (2008). Corona treatment: effect on properties of chitosan coated corn prolamin films. International chemical engineering congress. Kish island; January., 2-5.

[50] Gounga, M. E, Xu, S, & Wang, Z. (2007). Whey protein isolate-based edible films as affected by protein concentration, glycerol ratio and pullulan addition in film formation. Journal of Food Engineering. , 83, 521-530.

[51] Gruessner, U, Clemens, M, Pahlplatz, P. V, Sperling, P, Witte, J, & Sperling, P. (2001). Improvement of perineal wound healing by local administration of gentamicin-impregnated collagen fleeces after abdominoperineal excision of rectal cancer. American Journal of Surgery. , 182, 502-9.

[52] Guilbert, S, Gontard, N, & Gorris, L. G. M. (1996). Prolongation of the Shelf-life of Perishable Food Products Using Biodegradable Films and Coatings. Food Science and Technology. , 29, 10-17.

[53] Gunatillake, P, Mayadunne, R, & Adhikari, R. (2006). Recent developments in biodegradable synthetic polymers. Biotechnology Annual Reviews. , 12, 301-347.

[54] Hall, P. E, Sanderson, S. M, Johnston, D. M, & Cannon, R. E. (1992). Transformation of Acetobacter xylinum with plasmid DNA by electroporation Plasmid. , 28(3), 194-200.

[55] Hamad, W. (2006). On the Development and Applications of Cellulosic Nanofibrillar and Nanocrystalline Materials. The Canadian Journal of Chemical Engineering. , 84, 513-519.

[56] Handa, A, Gennadios, A, Froning, G. W, Kuroda, N, & Hanna, M. A. (1999). Tensile, Solubility, and Electrophoretic Properties of Egg White Films as Affected by Surface Sulfhydryl Groups. Journal of Food Science. , 64(1), 68-76.

[57] Hebert, G. D, & Holloway, O. E. (1992). Product and process of coating nuts with edible protein. U.S. patent 5,149,562.

[58] Holland, S. J, Jolly, A. M, Yasin, M, & Tighe, B. J. (1987). Polymers for biodegradable medical devices II. Hydroxybutyrate-hydroxyvalerate copolymers: hydrolytic degradation studies. Biomaterials. , 8, 289-295.

[59] Hood, L. L. (1987). Collagen in Sausage Casings. In: Advances in Meat Research, eds., A.M. Pearson, T.R. Dutson and A.J. Bailey, (New York: Van Nostrand Reinhold., 4, 109-129.

[60] Hu, W, Chen, S, Xu, Q, & Wang, H. (2011). Solvent-free acetylation of bacterial cellulose under moderate conditions. Carbohydrate Polymers. , 83, 1575-1581.

[61] Iguchi, M, Yamanaka, S, & Budhiono, A. (2000). Bacterial cellulose-a masterpiece of nature arts. Journal of Materials Science. , 35, 261-270.

[62] Iwata, T, Tsunoda, K, Aoyagi, Y, Kusaka, S, Yonezawa, N, & Doi, Y. (2003). Mechanical properties of uniaxially cold-drawn films of poly(β-3-hydroxybutyrate). Polymer Degradation and Stability. , 79, 217-224.

[63] Jacob JohnM., and Thomas, S. ((2008). Biofibres and biocomposites. Carbohydrate Polymers. , 71, 343-364.

[64] Jamshidian, M. Arab Tehrany, E., Imran, M., Jacquot, M., and Desobry, S. ((2010). Poly-Lactic Acid: Production, Applications, Nanocomposites, and Release Studies. Comprehensive Reviews in Food Science and Food Safety. , 9, 552-572.

[65] Jane, J, Lim, S, Paetau, I, Spence, K, & Wang, S. (1994). Biodegradable plastics made from agricultural biopolymers. In: Polymers from Agricultural Coproducts. M. L. Fishman, R. B. Friedman, and S. J. Huang, eds. (p.p. 92-100). ACS Symposium Series 575: Chicago.

[66] Jansson, A, Järnström, A, Rättö, P, & Thuvander, F. (2006). Physical and Swelling Properties of Spray-Dried Powders made from Starch and Poly(vinyl alcohol). Starch/Stärke. , 58, 632-641.

[67] Johnston-banks, F. A. (1990). Gelatine. In Food Gels. ed., P. Harris. (New York. Elsevier Applied Science., 233-289.

[68] Kabir, M. M, Wang, H. K. T, & Cardona, F. (2012). Chemical treatments on plant-based natural fibre reinforced polymer composites: An overview. Composites: Part B. In Press Paper.

[69] Kasarda, D. D. (1999). Glutenin Polymers: The In Vitro to In Vivo Transition. Cereal Foods World. , 44, 566-572.

[70] Kim, Y. B, Rhee, Y. H, Han, S. H, Heo, G. S, & Ki, J. S. (1996). Poly-3-hydroxyalka-noates produced from Pseudomonas oleovorans grown with !-phenoxyalkanoates. Macromolecules, , 29, 3432-3435.

[71] Knauf, M, & Moniruzzaman, M. (2004). Lignocellulosic biomass processing: a perspective. International Sugar Journal. , 106, 147-150.

[72] Krochta, J. M. (1992). Control of Mass Transfer in Foods with Edible Coatings and Films. In Advances in Food Engineering. eds, R.P. Singh and M.A. Wirakartakasu-mah, Boca Raton, FL. (CRC Press, Inc., 517-538.

[73] Kulesa, G. (1999). Clean fractionation-inexpensive cellulose for plastics production. http://

[74] La MantiaF.P., and Morreale, M. ((2011). Green composites: A brief review. Composites: Part A. , 42, 579-588.

[75] Lai, H. M, & Padua, G. W. (1997). Properties and Microstructure of plasticized zein films. Cereal Chemistry. , 74(6), 771-775.

[76] Lai, L. S, & Kokini, J. L. (1991). Physiochemical changes and rheological properties of starch during extrusion (a review). Biotechnology Progress. , 7, 251-266.

[77] Lim, S, Lee, J, Jang, S, Lee, S, Lee, K, Choi, H, & Chin, J. (2011). Synthetic Aliphatic Biodegradable Poly(butylene succinate)/Clay Nanocomposite Foams with High Blowing Ratio and Their Physical Characteristics. Polymer Engineering and Science. , 123, 1316-1325.

[78] Lircks, J. (1998). Properties and applications of compostable starch-based plastic material.

[79] Lutke-eversloh, T, Bergander, K, Luftman, H, & Steinbuchel, A. (2001). Identification of a new class of biopolymer: bacterial synthesis of a sulfur-containing polymer with thioester linkages. Microbiology. , 147, 11-19.

[80] Madsen, B. (2004). Properties of plant fiber yarn polymer composites. PhD thesis, BY-GDTU, Technical University of Denmark; 8-77877-145-5

[81] Mao, L. J, Imam, S, Gordon, S, Cinelli, P, & Chiellini, E. (2000). Extruded cornstarch-glycerol- polyvinyl alcohol blends: Mechanical properties, morphology, and bio degradability. Journal of Polymers and the Environment. , 8(4), 205-211.

[82] Maurus, P. B, & Kaeding, C. C. (2004). Bioabsorbable implant material review. Oper-
      ation Techniques and Sport Medicine. , 12, 158-60.

[83] Mehta, R, Kumar, V, Bhunia, H, & Upadhyay, S. N. (2005). Synthesis of poly (lactic
      acid): a review. Journal of Macromolecular Science and Polymer Review. , 45, 325-49.

[84] Miller, N. D, & Williams, D. F. (1987). On the biodegradation of poly-β-hydroxybuty-
      rate (PHB) homopolymer and poly-b-hydroxybutyrate- hydroxyvalerate copolymers.
      Biomaterials. , 8, 129-137.

[85] Mithieux, S. M, Rasko, J. E. J, & Weiss, A. S. (2004). Synthetic elastin hydrogels de-
      rived from massive elastic assemblies of selforganized human protein monomers. Bi-
      omaterials. , 25, 4921-4927.

[86] Mohanty, A. K, Misra, M, & Hinrichsen, G. (2000). Biofibres, biodegradable poly-
      mers, and biocompostites: an overview. Macromolecular Materials Engineering. ,
      276, 1-24.

[87] Muller, L. L. (1982). Manufacture of Casein, Caseinates and Coprecipitates. In Devel-
      opments in Dairy Chemistry-1. Proteins. ed., P.F. Fox (London: Applied Science
      Publishers., 315-337.

[88] Nair, L. S, Bijoux, C, Trevor, S, & Laurencin, C. T. (2006). Development of injectable
      thermogelling chitosan-inorganic phosphate solution for biomedical application. So-
      cial Biomaterial Meeting.

[89] Naira, L. S, & Laurencin, C. T. (2007). Biodegradable polymers as biomaterials. Prog-
      ress in Polymer Science. , 32, 762-798.

[90] Nakamura, E. M, Cordi, L, Almeida, G. S. G, Duran, N, & Mei, L. H. I. (2005). Study
      and development of LDPE/starch partially biodegradable compounds. Journal of
      Materials Processing Technology. , 162, 236-241.

[91] Nampoothiri, K. M, Nair, N. R, & John, R. P. (2010). An overview of the recent devel-
      opments in polylactide (PLA) research. Bioresource Technology. , 101, 8493-8501.

[92] Okamoto, K, Ray, S. S, & Okamoto, M. (2003). New Poly (butylene succinate)/
      Layered Silicate Nanocomposites. II. Effect of Organically Modified Layered Silicates
      on Structure, Properties, Melt Rheology, and Biodegradability. Journal of Polymer
      Science: Part B: Polymer Physics. , 41, 3160-3172.

[93] Olesen, P. O, & Plackett, D. V. (1999). Perspectives on the performance of natural
      plant fibres presented at natural fibres performance forum.Copenhagen, May http://
      www.ienica.net/fibreseminar/olesen.pdf/>., 27-28.

[94] Unishi, H, Takahashi, H, Yoshiyasu, M, & Machida, Y. (2001). Preparation and in vi-
      tro properties of N-Succinylchitosan or carboxymethylchitin-mitomycin C conjugate
      microparticles with specified size. Drug Delivery Industry in Pharmaceutical. , 27,
      659-67.

[95] Ozdemir, M, & Floros, J. D. (2008). Optimization of edible whey protein films con-
      taining preservatives for mechanical and optical properties. Journal of Food Engi-
      neering. , 84, 116-123.

[96] Pachence, J. M, Bohrer, M. P, & Kohn, J. (2007). Biodegradable Polymers. In Princi-
      ples of Tissue Engineering, Lanza, B., Langer, C., and Vacanti, P., edr. 3rd Edition.
      (Elsevier, Inc, 323.

[97] Pedroso, A. G, & Rosa, D. S. (2005). Mechanical, thermal and morphological charac-
      terization of recycled LDPE/corn starch blends. Carbohydrate Polymers. , 59, 1-9.

[98] Pereira, C. S, Cunha, A. M, Reis, R. L, & Vazquez, B. and San Roman, J. ((1998). New
      starch-based thermoplastic hydrogels for use as bone cements or drug-delivery carri-
      ers. Journal of Material Science. , 12, 825-33.

[99] Peressini, D, Bravin, B, Lapasin, R, Rizzotti, C, & Sensidoni, A. (2003). Starch-methyl-
      cellulose based edible films: rheological properties of film-forming dispersions. Jour-
      nal of Food Engineering. , 59, 25-32.

[100] Pitt, C. G, Andrady, A. L, Bao, Y. T, & Sarnuei, N. K. P. (1987). Estimation of the rate
      of drug diffusion in polymers. In: Controlled-Release Technology, Pharmaceutical
      Applications. eds.P. I. Lee and W. R. Good, (American Chemical Society, Washing-
      ton, DC., 49-77.

[101] Pszczola, D. E. (1998). Addressing Functional Problems in Fortified Foods. Food
      Technology. 52(7):38, 40-41, 44, 46.

[102] Ramaraj, B. (2007). Crosslinked Poly(vinyl alcohol) and Starch Composite Films. II.
      Physicomechanical, Thermal Properties and Swelling Studies. Journal of Applied
      Polymer Science. , 103, 909-916.

[103] Reed, A. M, & Gilding, D. K. (1981). Biodegradable polymers for use in surgery-
      poly(glycolic)/poly(lactic acid) homo and copolymers: 2. In vitro degradation. Poly-
      mer. , 22, 494-498.

[104] Reiners, R. A, Wall, J. S, & Inglett, G. E. (1973). Corn Proteins: Potential for Their In-
      dustrial Use. In Industrial Uses of Cereals, ed., Y. Pomeranz, St. Paul, MN. (Ameri-
      can Association of Cereal Chemists., 285-298.

[105] Ren, M, Song, J, Song, C, Zhang, H, Sun, X, Chen, Q, Zhang, H, & Mo, Z. (2005).
      Crystallization kinetics and morphology of poly(butylene succinate-co-adipate).
      Journal of Polymer Science Part B: Polymer Physics. , 43(22), 3231-3241.

[106] Rezwan, K, Chen, Q. Z, Blaker, J. J, & Boccaccini, A. R. (2006). Biodegradable and bio-
      active porous polymer/inorganic composite scaffolds for bone tissue engineering. Bi-
      omaterials. , 27, 3413-3431.

[107] Santerre, J. P, Woodhouse, K, Laroche, G, & Labow, R. S. (2005). Understanding the biodegradation of polyurethanes: from classical implants to tissue engineering materials. Biomaterials. , 26, 7457-7470.

[108] Savenkova, L, Gercberga, Z, Nikolaeva, V, Dzene, A, Bibers, I, & Kalmin, M. (2000). Mechanical properties and biodegradation characteristics of poly-(hydroxy butyrate)-based films. Process Biochemistry. , 35, 573-579.

[109] Shalaby, S. W. DuBose, J.A., and Shalaby, M. ((2004). Chitosan-based systems. In Absorbable and Biodegradable Polymers. Shalaby, S. W. B., and Karen, J. L., eds. (CRC Press, Boca Raton, FL., 77.

[110] Shi, C, Zhu, Y, Ran, X, Wang, M, Su, Y, & Cheng, T. (2006). Therapeutic potential of chitosan and its derivatives in regenerative medicine. Journal of Surgery Research. , 133, 185-92.

[111] Shiku, Y, Hamaguchi, P. Y, Benjakul, S, Visessanguan, W, & Tanaka, M. (2004). Effect of surimi quality on properties of edible films based on Alaska Pollack. Food Chemistry. , 86, 493-499.

[112] Sienkiewicz, T, & Riedel, C. L. (1990). Utilization of Whey. In Whey and Whey Utilization, Second Edition. eds., T. Sienkiewicz and C-L. Riedel, Gelsenkirchen-Buer (Germany: Verlag Th. Mann., 92-130.

[113] Sinha, V. R, Bansal, K, Kaushik, K, Kumria, R, & Trehan, A. (2004). Poly-ε-caprolactone microspheres and nanospheres: an overview. International Journal of Pharmaceutical. 278: 1 23.

[114] Slade, L, & Levine, H. (1987). Polymer-Chemical Properties of Gelatin in Foods. In Advances in Meat Research, eds., A.M. Pearson, T.R. Dutson and A.J. Bailey (New York: Van Nostrand Reinhold Company., 4, 251-266.

[115] Smith, R. (2005). Biodegradable polymers for industrial applications. (Woodhead Publishing., 223-226.

[116] Sodian, R, Hoerstrup, S. P, Sperling, J. S, Martin, D. P, Daebritz, S, Mayer, J. E, & Vacanti, J. P. (2000). Evaluation of biodegradable, three-dimensional matrices for tissue engineering of heart valves. ASAIO Journal. , 46, 107-110.

[117] Soy Protein Council (1987). Soy Protein Products: Characteristics, Nutritional Aspects, and Utilization. Washington, DC.

[118] Steeves, D. M, Farrell, F, & Ratto, J. A. (2007). Investigation of Polybutylene Succinate-co-Adipate (PBSA)/Montmorillonite Layered Silicate (MLS) Melt-Processed Nanocomposites. Journal of Biobased Materials and Bioenergy. , 1, 94-108.

[119] Sun, R. C, & Tomkinson, J. (2000). Essential guides for isolation/purification of polysaccharides. Encylopaedia Seperation Science. , 6, 4568-4574.

[120] Suyatama, N. E, Tighzert, L, & Copinet, A. (2005). Effects of hydrophilic plasticizers on mechanical, thermal, and surface properties of chitosan Films. Journal of Agricultural and Food Chemistry. , 53, 3950-3957.

[121] Thakore, I. M, Desai, S, Sarawadade, B. D, & Devi, S. (2001). Studies on biodegradability, morphology and thermomechanical of LDPE/modified starch blends. European Polymer Journal. , 37, 151-160.

[122] Tharanathan, R. N. (2003). Biodegradable films and composite coatings: past, present and future. Trends in Food Science & Technology. , 14, 71-78.

[123] Tian, H, Tang, Z, Zhuang, X, Chen, X, & Jing, X. (2012). Biodegradable synthetic polymers: Preparation, functionalization and biomedical application. Progress in Polymer Science. , 37, 237-280.

[124] Vercauteren, F, & Donners, W. A. B. (1986). A C nuclear magnetic resonance study of the microstructure of poly (vinyl alcohol). Polymer. , 27, 567-578.

[125] Vert, M. (2005). Aliphatic polyesters: great degradable polymers that cannot do everything. Biomacromolecules. , 6, 538-546.

[126] Watt, I. C. (1983). The Theory of Water Sorption by Biological Materials. In Physical Properties of Foods. eds., R. Jowitt, F. Escher, B. Hallstrom, H.F.T. Meffert,W.E.L. Spiess and G. Vos. (London, England: Applied Science Publishers., 27-41.

[127] Wiedersheim, W, & Strobel, E. (1991). Compounding of thermoplastic starch with twin screw extruders. Starch/Stärke. , 43, 138-145.

[128] Witherly, S. A. (1990). Soy Formulas Are Not Hypoallergenic. American Journal of Clinical Nutrition. , 51, 705-706.

[129] Wolf, W. J. (1972). Purification andproperties of the proteins. In Soybeans: Chemistry and technology. eds. A.K. Smith, and S.J. Circle. (Westport, CN: The Avi Publishing Company, Inc., 93-143.

[130] Wrigley, C. W, & Bietz, J. A. (1988). Protein and Amino Acids. In Wheat Chemistry and Technology. ed.,Y. Pomeranz, St. Paul,MN. (American Association of Cereal Chemists., 159-275.

[131] Xiong, H. Tang, Sh., Tang, H. and Zou, The structure and properties of a starch-based biodegradable film. Carbohydrate Polymers. 71: 263-268., 2008.

[132] Yamauchi, K, & Khoda, A. (1997). Novel Proteinous Microcapsules from Wool Keratins. Colloids and Surfaces B: Biointerfaces. , 9, 117-119.

[133] Yamauchi, K, Yamauchi, A, Kusunoki, T, Khoda, A, & Konishi, Y. (1996). Preparation of Stable Aqueous Solution of Keratins, and Physicochemical and Biodegradational Properties of Films. Journal of Biomedical Materials and Resourses. , 31, 439-444.

[134] Yang, X, Zhao, K, & Chen, G. Q. (2002). Effect of surface treatment on the biocompat-
       ibility of microbial polyhydroxyalkanoates. Biomaterials. , 23, 1391-1397.

[135] Yoshie, N, Saito, M, & Inoue, Y. (2004). Effect of chemical compositional distribution
       on solid-state structure and proeprties of poly(3-hydroxybutyrate-co-3-hydroxyval-
       erate). Polymer. , 45, 1903-11.

[136] Yu, L, Dean, K, & Li, L. (2006). Polymer blends and composites from renewable re-
       sources. Progress in Polymer Science. , 31, 576-602.

[137] Zilberman, M, Nelson, K. D, & Eberhart, R. C. (2005). Mechanical properties and in
       vitro degradation of bioresorbable fibers and expandable fiber-based stents. Journal
       of Biomedical Material Research. Part B: Appled Biomaterials. , 74, 792-799.

Chapter 6

Creation of Novel Green Surfactants Containing Carbonate Linkages

Taisuke Banno, Taro Toyota and Shuichi Matsumura

Additional information is available at the end of the chapter

1. Introduction

1.1. What is green chemistry?

Chemistry has a key role to play in maintaining and improving our quality of life, such as in medicine, materials and electronics. However, it has also caused damage to human health and the natural environment. To make chemistry compatible with human health and the environment, Anastas and Warner have proposed 12 principles of green chemistry, which help to explain what it means in practice [1]. The principles cover a wide range of concepts, such as the molecular design and synthetic routes of product and the best means of waste disposal. In recent years, the establishment of a new field of green chemistry has been recognized as a necessary goal for sustainable development. The greening of chemistry will be realized by the discovery and development of new synthetic routes using renewable resources, reaction conditions and catalysts for improved selectivity and energy minimization, and the design of bio-/environmentally compatible chemicals. On the basis of these concepts, green polymer chemistry [2-4], synthetic organic chemistry using environmentally friendly processes [5,6] and technology for the production of bio-based product [7] have so far been developed and improved.

1.2. Requirements for green surfactants

Surfactants are widely used in large quantities in industrial fields including fibers, pharmaceutical agents and foods, and are also used as household detergents. Owing to the water-solubility of surfactants, they are generally difficult to recover or reuse; therefore, they are discharged as drainage into the environment if they are not biodegradable. On the other hand, surfactants should be chemically recycled, particularly after industrial use, in order to conserve

energy resources in addition to reducing their environmental impact. Therefore, the development of surfactants with improved biodegradability and chemical recyclability using renewable resources by an environmentally benign process is now needed with respect to the
establishment of green chemistry. Also, increasing the performance of surfactants is one way
to reduce their consumption. Figure 1 shows a simplified conceptual scheme of green surfactants.

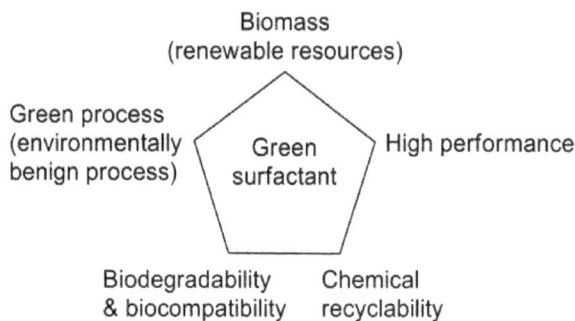

Figure 1. Simplified conceptual scheme of green surfactants.

Some green surfactants have been synthesized using renewable resources, such as amino acids,
sugars and organic acids, and they have been used in human life because of their low toxicity
and high biodegradability [8]. Surfactants containing the sugar moiety as a hydrophilic group
have been synthesized by many researchers and used as detergents, emulsifiers and cosmetics
[9]. The syntheses and properties of anionic, cationic and amphoteric surfactants containing
amino acid moieties, such as arginine [10,11], tryptophan [12], lysine [13], cysteine [14],
glutamate and aspartate [15] have also been reported. Microbially producible polycarboxylates
may become versatile starting materials for the production of next-generation green surfactants. The polycarboxylates applicable to a biomass refinery system include fumaric acid,
maleic acid, itaconic acid and aconitic acid [16]. The production of these polycarboxylates has
been extensively studied with the aim of establishing a biomass refinery system as well as
establishing green chemistry. Also, fatty acids and alcohols from plant oils occupy a concrete
position as the hydrophobic group of a green surfactant. Furthermore, sugar-based biosurfactants produced by a variety of microorganisms exhibit unique properties, such as mild
production conditions, multifunctionality and high environmental compatibility [17]. The
numerous advantages of biosurfactants have led to their application not only in the food,
cosmetic and pharmaceutical industries but also in environmental protection and energy-
saving technology. These features of biosurfactants should broaden their range of applications
in new advanced technologies.

It is known that ester- or amide-containing linear-type surfactants consisting of one hydrophobic group and one hydrophilic group are rapidly biodegraded by environmental microbes
[18,19]. This is due to the high biodegradability of the degradation compounds produced by

the enzymatic hydrolysis of ester or amide linkage. However, ester-containing surfactants are generally labile to hydrolysis, particularly under alkaline conditions. Thus, more hydrolytically stable and biodegradable surfactants are needed. Also, chemodegradable surfactants containing a labile linkage that is rapidly cleaved by a chemical stimulus have attracted considerable attention as biodegradable surfactants [20]. Acetal- or cyclic acetal-containing surfactants exhibit good surfactant properties and rapid hydrolyzability under acidic conditions [21,22]. On the other hand, the ester linkage of cationic surfactants is hydrolyzed to produce a fatty acid and a quaternary-ammonium alcohol at a high pH [23]. Furthermore, photolabile sulfonate-type anionic surfactants [24], ozone-cleavable anionic surfactants [25] and thermally cleavable anionic surfactants based on furan-maleimide Diels-Alder adducts [26] have been reported.

Recently, the syntheses and properties of gemini-type (dimeric) surfactants consisting of two hydrophobic alkyl chains, two hydrophilic groups and a linker in the same molecule have been extensively studied. The first report on dimeric cationics was published by Bunton *et al.* [27]. The term "gemini-surfactants" was coined for these dimeric surfactants by Menger *et al.* [28,29]. It was found that the surfactant properties of gemini-type surfactants, such as a low critical micelle concentration (cmc) and surface tension lowering, were superior to those of the corresponding linear-type (monomeric) surfactants [30-33]. These superior surfactant properties are due to the stronger intra- and intermolecular hydrophobic interactions between the two hydrophobic alkyl chains of gemini-type surfactants. Therefore, gemini-type surfactants can be regarded as green surfactants because they exhibit higher functionalities that lead to a reduction in their consumption. Biodegradable sugar-derived gemini-type nonionic surfactants have already been reported by Wilk and co-workers [34,35]. Also, Ono *et al.* reported a tartrate-derived bis(sodium sulfate)-type chemocleavable gemini-type surfactant containing ester linkages that exhibited good biodegradability [36]. Moreover, oligomeric surfactants composed of three or four hydrophobic and hydrophilic groups have been synthesized and characterized [37-39]. Their surfactant properties are superior to those of gemini-type surfactants. However, there are few reports on oligomeric surfactants because their syntheses are complicated.

1.3. Role of carbonate linkage in surfactant molecule: their current use

The carbonate linkage is hydrolyzed in aqueous media by lipase to produce two hydroxyl groups and a carbon dioxide, and no acidic or alkaline species are produced. It will be advantageous for surfactants that the carbonate linkages are more stable than ester linkages in aqueous media [40]; because they are usually used in aqueous media. It has also been reported that carbonate-type nonionic surfactants are readily biodegraded by environmental microbes [40]. However, they are synthesized using an acid chloride. The synthesis of next-generation surfactants using halogen-free and environmentally benign processes is needed from the viewpoint of green chemistry.

Figure 2 shows a simplified conceptual scheme of the synthesis, chemical recycling and biodegradation of novel green surfactants containing carbonate linkages. The designed surfactants are hydrolyzed in the environment to produce a hydrophobic 1-alkanol and a hydrophilic alcohol along with the evolution of carbon dioxide, which are further degraded by microbes as a part of their biodegradation or biorecycling process [40]. Also, the primary degradation products can be used for chemical recycling by lipase, which is an environmentally benign catalyst [41,42]. This sustainable chemical recycling may become an important issue for next-generation surfactants, particularly in industrial fields. A new strategy for molecular design involving ready chemical recyclability as well as biodegradability will be needed. Novel carbonate-type surfactants are produced by a two-step carbonate exchange reaction of 1-alkanol and a hydrophilic alcohol using diphenyl carbonate or dimethyl carbonate. Diphenyl carbonate is now industrially produced by the carbonate exchange reaction of dimethyl carbonate with phenol. Dimethyl carbonate can be produced from methanol and carbon dioxide by a green method [43-46].

In this paper, the design, synthesis and properties of novel cationic and nonionic surfactants containing carbonate linkages as biodegradable and chemically recyclable segments are described.

2. Synthesis and properties of novel green surfactants containing carbonate linkages

2.1. General methods

The equilibrium surface tensions of carbonate-type surfactants solutions were measured using an automatic digital Kyowa Precise Surface Tensiometer by the CBVP method (Kyowa Kagaku Co. Ltd., Tokyo, Japan). The measurement was carried out using the Wilhelmy vertical plate technique and a sandblasted glass plate. The test solutions were aged at 25 °C for at least 1 h before any measurements.

The antimicrobial activities of the surfactants were evaluated by the agar dilution method [47]. Gram-positive bacterial strains, *Staphylococcus aureus* KB210, *Bacillus subtilis* KB211 and *Micrococcus luteus* KB212, gram-negative bacterial strains, *Escherichia coli* KB213, *Salmonella typhimurium* KB20 and *Pseudomonas aeruginosa* KB115, six fungal strains, *Candida albicans* KF1, *Saccharomyces cerevisiae* KF25, *Trichophyton mentagrophytes* KF213, *Microsporium gypseum* KF64, *Penicillium chrysogenum* KF270 and *Aspergillus niger* KF103, were used. The bacteria and fungi were cultured by Nutrient agar and Sabouraud dextrose agar, respectively. The antimicrobial activity was expressed as the minimum inhibitory concentration (MIC). The MIC value shows the lowest concentration of a surfactant at which the tested microorganisms do not show visible growth.

The biodegradabilities of the carbonate-type surfactants were evaluated by the biochemical oxygen demand (BOD). The BOD was determined with a BOD Tester [VELP Scientifica s.r.l., Usmate (MI), Italy] using the oxygen consumption method according to the Modified MITI Test [48]. The BOD-biodegradation (BOD/ThOD) was calculated from the BOD values and the theoretical oxygen demand (ThOD).

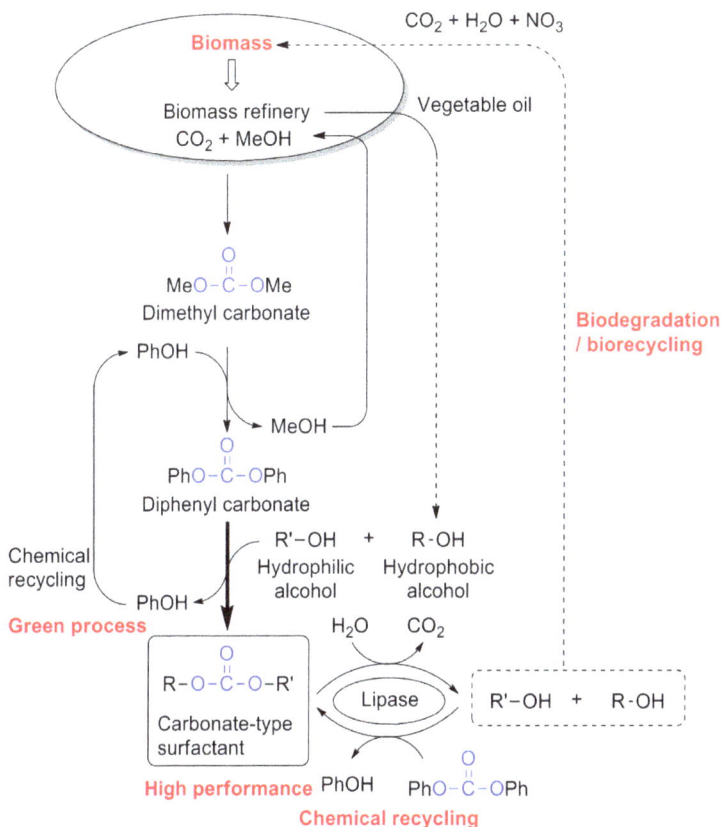

Figure 2. Simplified concepts of the synthesis, chemical recycling and biodegradation of novel green surfactants containing carbonate linkages.

2.2. Linear-type cationics containing carbonate linkage as green surfactants

In addition to good surfactant properties, it is known that cationic surfactants composed of quaternary-ammonium group and n-alkyl chain showed antimicrobial activities against a broad range of microorganisms such as gram-positive and gram-negative bacteria and fungi. However, they are generally highly resistant to biodegradation due to the lack of a primary degradation site in the molecule [49,50]. As a next-generation cationic surfactant, the biodegradabilities, chemical recyclabilities and higher functionalities that lead to a reduction in their consumption are needed as well as developing an environmentally benign synthetic route using renewable resources. Here, design and synthesis of linear-type cationics containing carbonate linkages (**SnX**) are described with respect to the effects of the introduction of carbonate linkages on biodegradation and chemical recycling (Figure 3) [51].

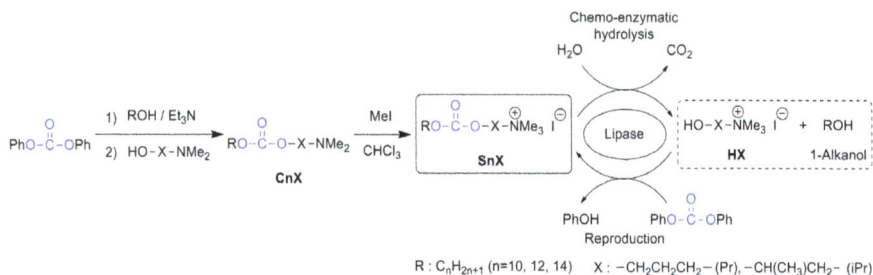

Figure 3. Synthesis and chemical recycling of novel linear-type cationics containing carbonate linkages (**SnX**) as biodegradable and chemically recyclable segments.

2.2.1. Synthesis of SnX

n-Alkyl-N,N-dimethylaminopropyl carbonate was first prepared by the reaction of diphenyl carbonate and 1-alkanol in the presence of Et_3N followed by the reaction of N,N-dimethylaminopropanol using one-pot, two-step reaction procedure. Then, the quaternarization of N,N-dimethylamino group of **CnX** readily occurred using CH_3I at room temperature (Figure 3). Purification was carried out by recrystallization from ethyl acetate to yield a series of quaternary-ammonium type cationics, **SnX**, in a total yield of 60-70%. The molecular structure was analyzed using ^{1}H-NMR, ^{13}C-NMR and elemental analysis.

2.2.2. Surfactant properties in aqueous solution

Carbonate-type cationics exhibited surfactant properties, such as surface tension lowering and micelle formation. From the surface tension $vs.$ concentration plots for the cationic surfactants in distilled water, their cmc and the surface tension at the cmc (γ_{cmc}) are determined and listed in Table 1. The surfactant properties of a typical cationics, N-dodecyl-N,N,N-trimethylammonium iodide (**S12**), were also measured under the same conditions. The cmc of carbonate-type cationics was depended on the hydrocarbon chain length. The shorter hydrocarbon chain length tended to show a higher cmc value. The cmc value of carbonate-type cationics containing an n-dodecyl chain (**S12X**) showed a lower value compared to that of **S12**. This result indicates that the propylene or isopropylene group between the quaternary-ammonium group and carbonate linkage behave as a hydrophobic group and contribute the hydrophobic interaction between the surfactant molecules. It was also found the cmc value of **S12iPr** was higher than that of **S12Pr**. This is due to the steric hindrance between surfactant molecules caused by the branched group of **S12iPr** [52].

2.2.3. Antimicrobial activities

The MICs of cationic surfactants are summarized in Table 2. The carbonate linkage of cationic surfactants may be hydrolyzed by bacterial enzyme to produce the corresponding 1-alkanol and quaternary-ammonium alcohol as shown in Figure 3. The antimicrobial activities of **HPr**

Cationics	cmc (mM)	γ_{cmc} (mN/m)
S10Pr	3.5	34
S12Pr	0.41	34
S14Pr	0.19	33
S12iPr	1.3	35
S12	5.4	35

Table 1. Surfactant properties of cationics in aqueous solution at 25 °C.

were also evaluated under the same conditions. Althogh the antimicrobial activities of **HPr** were low, **SnPr** showed high activities against some strains. Therefore the carbonate linkage of **SnPr** was not hydrolyzed under the tested conditions. It was also found that cationic surfactants containing both the carbonate linkage and *n*-dodecyl group showed higher antimicrobial activities, especially toward *Staphylococcus aureus* and *Escherichia coli*, when compared to those of the conventional cationics, **S12**. The **S12Pr** showed similar activities towards **S12iPr**. That is, no significant differences in the antimicrobial activities were observed by the branched methyl group.

Strain	MIC (µg/mL)					
	S10Pr	S12Pr	S14Pr	S12iPr	S12	HPr
S. aureus	25	2.5	2.5	2.5	10	>400
B. subtilis	50	2.5	2.5	5	10	>400
M. luteus	100	5	2.5	2.5	25	>400
E. coli	20	10	>400	10	25	>400
S. typhimurium	200	200	>400	200	100	>400
P. aeruginosa	100	100	100	400	50	>400
C. albicans	>400	>400	>400	400	200	>400
S. cerevisiae	400	400	400	400	100	>400
T. mentagrophytes	200	50	100	400	50	100
M. gypseum	25	10	10	5	25	200
P. chrysogenum	400	200	400	100	100	>400
A. niger	>400	>400	>400	>400	100	>400

Table 2. Antimicrobial activities of cationics and **SnPr**-derived **HPr**.

2.2.4. Biodegradabilities

The quick and complete biodegradation after use is one of the significant factors for the next-generation surfactants, because water-soluble surfactants are generally difficult to recover or recycle. The biodegradation of the carbonate-type cationics may first occur by hydrolysis due to environmental microbial enzymes as shown in Figure 3. If the quaternarized degradation products (**HX**), show higher biodegradabilities, the parent surfactants would be regarded as biodegradable.

Figure 4 shows the BOD-biodegradation (BOD/ThOD) for 28 days of cationic surfactants and their degradation products. **S10Pr** was readily biodegraded and showed a 60% BOD-biodegradability, which is the criterion for an acceptable biodegradation. The biodegradability of the carbonate-type cationics gradually decreased with the increasing hydrocarbon chain length. This is explained by the lower water solubility of cationics containing a longer hydrophobic alkyl chain. It was also found that **S12X**-derived 1-dodecanol (**DD**) was rapidly biodegraded by activated sludge. On the other hand, the biodegradability of **S12Pr**-derived **HPr** was better than that of **S12iPr**-derived **HiPr**: therefore, **S12Pr** showed a better biodegradability than **S12iPr**.

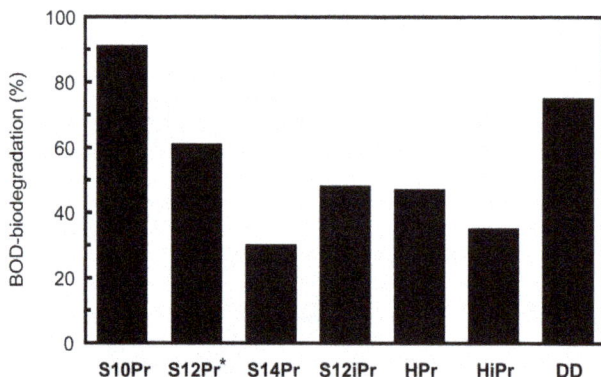

Figure 4. BOD-biodegradation of carbonate-type cationics and degradation intermediate (**HPr**, **HiPr** and **DD**) at 25 °C for 28 days (*45 days). Activated sludge: 30 ppm, carbonate-type cationics: ca. 40 ppm, **HPr** and **HiPr**: 57 ppm, **DD**: 38 ppm.

2.2.5. Enzymatic degradation and reproduction for chemical recycling

In the terms of green and sustainable chemistry, even water-soluble surfactants should be chemically recycled particularly in industrial fields. The **SnX** was hydrolyzed by lipase and accompanied by carbon dioxide evolution to produce the corresponding 1-alkanol and quaternary-ammonium alcohol, which could be converted into the initial cationics by the reaction with dialkyl carbonate. A lipase-catalyzed chemical recycling may become one of the green methods because lipase is a renewable catalyst with high catalytic activities [41,42].

The enzymatic degradation of **S12Pr** was carried out in toluene containing a small amount of water using immobilized lipase from *Candida antarctica* (CALB). The **S12Pr** was completely degraded at the carbonate linkage to produce **HPr** and **DD**. The **S12Pr** was regenerated in 69% yield by the lipase-catalyzed reaction of **HPr** and 1-dodecyl=phenyl=carbonate, which was prepared by the Et$_3$N-catalyzed reaction of **DD** and diphenyl carbonate (87% yield). The molecular structure of **S12Pr** was analyzed by ^{1}H-NMR. Based on these results, **S12Pr** showed a lipase-catalyzed chemical recyclability as shown in Figure 3.

2.3. Gemini-type cationics containing carbonate linkages as green surfactants

The syntheses and properties of gemini-type (dimeric) quaternary-ammonium cationics consisting of two hydrophobic alkyl chains and two quaternary-ammonium groups covalently attached through a linker moiety have been extensively studied by many researchers. Gemini-type cationics can be regarded as a next-generation green surfactant because they show higher functionalities that lead to a reduction in their consumption [32,33,53-55]. This saves carbon resources and production energies. However, there are few reports on the biodegradabilities of gemini-type cationics. It has been reported by Tehrani-Bagha *et al.* and Tatsumi *et al.* that gemini-type cationics containing ester linkages in the hydrophobic moiety were biodegraded by activated sludge [56-59]. However, ester linkages are generally labile to hydrolysis, particularly under alkaline conditions. More hydrolytically stable and biodegradable gemini-type cationics are, thus, needed. In this section, molecular design, synthesis and properties of gemini-type cationics containing carbonate linkages (**G12Pr** and **mG12Pr**) are described emphasizing the relations between carbonate linkage and biodegradation (Figure 5) [60,61].

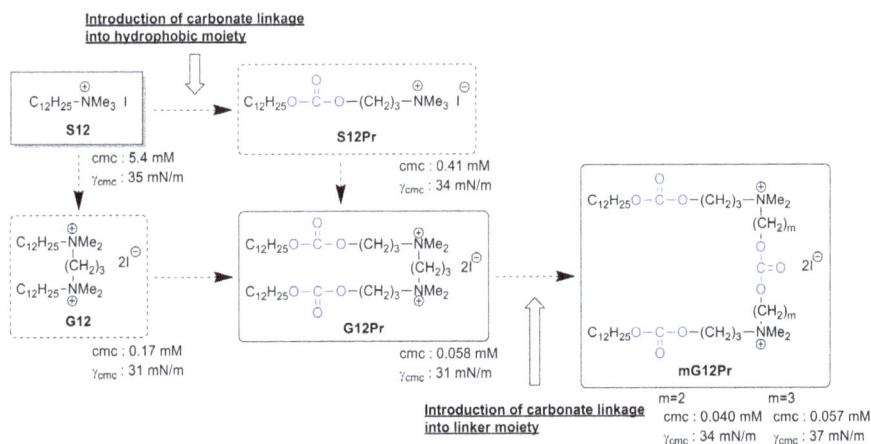

Figure 5. Design of biodegradable and chemically recyclable gemini-type cationics containing carbonate linkages.

2.3.1. Synthesis of gemini-type cationics containing carbonate linkages

Simultaneous gemini formation and quaternarization readily occurred at 80 °C for 1 day by the reaction with **C12Pr** and 1,3-diiodopropane in dry acetonitrile to produce **G12Pr** containing carbonate linkages in the hydrophobic moiety in 86% yield as shown in Figure 6. The **mG12Pr** containing carbonate linkages both in hydrophobic and linker moieties was prepared at 80 °C for 3 days by the reaction with **C12Pr** and α,ω-diiodide containing a carbonate linkage, which was prepared by the carbonate exchange reaction of diphenyl carbonate and iodoalkanol using K_2CO_3 (Figure 7). In a 1-day reaction, the yield of **2G12Pr** was only 20%. The maximum yield was 74% when the reaction time was 3 days. This could be due to the lower reactivity of the larger molecular size of di(iodoalkyl)carbonate.

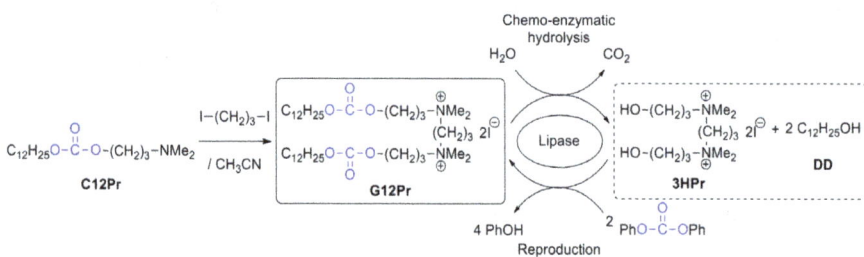

Figure 6. Synthesis and chemical recycling of **G12Pr** containing carbonate linkages in the hydrophobic moiety.

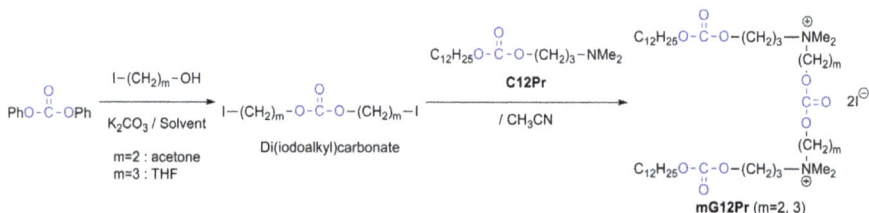

Figure 7. Synthesis of di(iodoalkyl)carbonate and **mG12Pr** containing carbonate linkages in both the linker and hydrophobic moieties.

2.3.2. Surfactant properties in aqueous solution

From the plots of surface tension *vs.* concentration of gemini-type cationics containing carbonate linkages in aqueous solution, the cmc and γ_{cmc} values are determined, and the results are shown in Figure 8. It was found that gemini-type cationics containing carbonate linkages showed lower cmc values compared to the corresponding linear-type cationics. These low cmc values of gemini-type surfactants were mainly caused by the simultaneous migration of the two alkyl chains, rather than one, from the aqueous phase to the micelle [32,33]. No significant differences in the cmc values of the tested gemini-type cationics containing both the carbonate linkages and n-dodecyl groups were observed according to the linker structure. The γ_{cmc} of **3G12Pr** was slightly higher than that of **2G12Pr**. This is due to the difference in the linker length between the quaternary-ammonium groups. That is, the **2G12Pr** having an ethoxycarbony-loxyethyl-type linker (m=2 in Figure 5) showed a lower γ_{cmc} value when compared to the corresponding **3G12Pr** having a propoxycarbonyloxypropyl-type linker (m=3). The lower γ_{cmc} of **2G12Pr** was due to the higher intra- and intermolecular hydrophobic interactions between the two hydrophobic alkyl chains of the gemini-type surfactant.

2.3.3. Enzymatic degradation and reproduction for chemical recycling

Gemini-type cationics containing carbonate linkages were hydrolyzed by lipase and accompanied by carbon dioxide evolution to produce the corresponding alcohol and quaternary-ammonium alkanol, which could be converted into the initial gemini-type cationics by the reaction with diphenyl carbonate. The enzymatic degradation of **G12Pr** was carried out in toluene containing a small amount of water using immobilized lipase CALB. The **G12Pr** was degraded at the carbonate linkage into the quaternary-ammonium alcohol (**3HPr**) and **DD** (92% yield). The **G12Pr** was regenerated by the reaction of **3HPr** and 1-dodecyl=phenyl=carbonate, which was prepared by the Et$_3$N-catalyzed reaction of **DD** and diphenyl carbonate (87% yield). That is, a mixture of **3HPr**, 1-dodecyl=phenyl=carbonate and immobilized lipase CALB was stirred in dry acetonitrile to obtain **G12Pr** in 30% yield (two-step). Based on these results, **G12Pr** showed chemical recyclability using lipase CALB (Figure 6).

2.3.4. Biodegradabilities

Figure 8 shows the BOD-biodegradation of gemini-type cationics and the **G12Pr**-derived degradation products, **3HPr** and **DD**. The conventional gemini-type cationics, **G12**, which had no hydrolytically cleavable moiety, showed no biodegradation by activated sludge. On the other hand, the **G12Pr** containing carbonate linkages showed higher biodegradability when compared to **G12**. It was found that the biodegradability was improved by the introduction of carbonate linkages into the hydrophobic moiety of the gemini-type cationics. However, the biodegradability of **G12Pr** was relatively low at around 25% after 28 days of incubation due to the relatively low biodegradability of **3HPr** as the primary biodegradation intermediate, i.e., 10% BOD-biodegradation after 28 days, as shown in Figure 8. On the other hand, **DD** as the degradation intermediate was readily biodegraded by activated sludge and its BOD-biode-gradability exceeded 60% after 28 days. Based on these results, the low biodegradability of **G12Pr** was due to the low biodegradability of the degradation intermediate containing two quaternary-ammonium groups, **3HPr**.

The biodegradability of gemini-type cationics containing carbonate linkages was strongly affected by the linker structure. Though the **2G12Pr** was quickly biodegraded, the **3G12Pr** showed relatively low biodegradability (Figure 8). This is due to the difference in the primary degradation products. In order to compare the hydrolytic degradability of **2G12Pr** and **3G12Pr**, an accelerated hydrolytic degradation test was carried out by dissolving the gemini-type cationics in distilled water and stirring at a higher temperature of 60 °C. The hydrolytic degradation of **mG12Pr** was analyzed by comparing the ^{1}H-NMR profiles of the reactants before and after the degradation, and the results are shown in Figure 9. Significant differences in the hydrolytic degradation were observed depending on whether the carbonate linkage was in the linker moiety or the hydrophobic moiety of the **2G12Pr** [Figure 9(a)]. Though the carbonate linkage in the linker moiety gradually hydrolyzed in water and only 18% remained after 9 h of reaction, 97% of the carbonate linkage in the hydrophobic moiety remained after 9 h of reaction. These results indicate that the carbonate linkage in the hydrophobic moiety was more stable against hydrolysis than the carbonate linkage in the linker moiety. Therefore, **2G12Pr** was readily hydrolyzed at the carbonate linkage in the linker moiety to produce the corresponding quaternary-ammonium alcohols having a similar molecular structure to the linear-type **S12Pr**, which exhibited a good biodegradability. On the other hand, both the carbonate linkages in the linker and hydrophobic moieties of **3G12Pr** were stable in water [Figure 9(b)]. It was found that the carbonate linkage in the linker moiety of **3G12Pr** was hydrolytically more stable than the carbonate linkage in the linker moiety of **2G12Pr**. Based on these results, **2G12Pr** was readily cleaved at the carbonate linkage in the linker moiety, indicating a higher degree of BOD-biodegradability than **3G12Pr**.

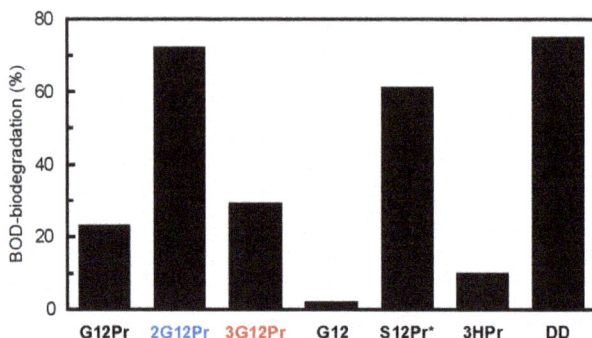

Figure 8. BOD-biodegradation of gemini-type cationics and **S12Pr** and **G12Pr**-derived degradation products, **3HPr** and **DD**, at 25 °C for 28 d (*45 d). Activated sludge: 30 ppm, cationics: ca. 40 ppm, **3HPr**: 52 ppm, **DD**: 38 ppm.

2.3.5. Antimicrobial activities

The gemini-type cationics were screened for their antimicrobial activities toward gram-positive and gram-negative bacterial strains and fungal strains based on the determination of their MICs. These results are shown in Table 3. It has been reported by some groups that cationic surfactants having multi-polar groups showed higher antimicrobial activities compared to the corresponding linear-type cationics because of the much higher charge density carried by multi-polar cationics [62,63]. However, **G12Pr** showed lower antimicrobial activities when compared to the corresponding linear-type **S12Pr**. The low antimicrobial activities of **G12Pr** could be due to the ready cleavability of the carbonate linkages of **G12Pr** by microbes forming surface-inactive molecules, **3HPr**, as shown in Figure 6. The **2G12Pr** containing

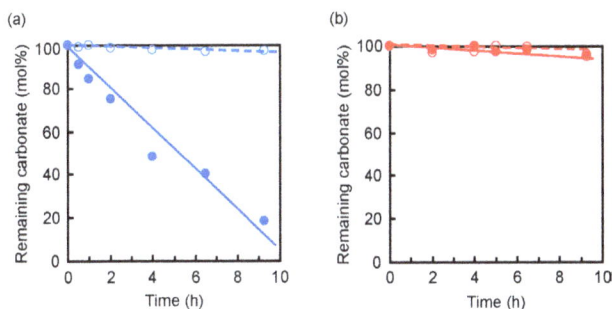

Figure 9. Time course of hydrolytic degradation of **mG12Pr** in distilled water at 60 °C. (a) Remaining carbonate in the hydrophobic moiety (○) and linker moiety (●) of **2G12Pr**. (b) Remaining carbonate in the hydrophobic moiety (○) and linker moiety (●) of **3G12Pr**. Concentration: 5 g/L.

carbonate linkages both in the hydrophobic and linker moieties showed higher antimicrobial activities than the **3G12Pr**. As discussed above, **2G12Pr** was readily hydrolyzed at the carbonate linkage in the linker moiety to produce the corresponding quaternary-ammonium alcohols having a similar molecular structure to the linear-type **S12Pr**, which exhibited strong antimicrobial activities. On the other hand, **3G12Pr** could be cleaved at the carbonate linkages in both the hydrophobic and linker moieties to produce the antimicrobially inactive alkyl chain-free fragments. Therefore the antimicrobial activities of **3G12Pr** were lower than those of **2G12Pr**.

Strain	MIC (μg/mL)					
	G12Pr	2G12Pr	3G12Pr	G12	S12Pr	3HPr
S. aureus	200	5	100	25	2.5	>400
B. subtilis	200	10	50	10	2.5	>400
M. luteus	400	25	100	25	5	>400
E. coli	400	200	200	100	10	>400
S. typhimurium	>400	400	>200	100	200	>400
P. aeruginosa	400	200	>200	100	100	>400
C. albicans	>400	400	>200	>400	>400	>400
S. cerevisiae	>400	400	>200	>400	400	>400
T. mentagrophytes	>400	100	>200	100	50	100
M. gypseum	400	25	100	50	10	200
P. chrysogenum	>400	400	>200	400	200	>400
A. niger	>400	>400	>200	400	>400	>400

Table 3. Antimicrobial activities of cationics and **G12Pr**-derived **3HPr**.

2.4. Chiral cationics containing carbonate linkages

The relationships between the physico-chemical properties and the stereochemistry of chiral surfactants derived from bio-based materials, such as sugars [64], amino acids [65,66], ascorbic acid [67] and succinic acid [68], have been studied by many researchers. Some of these showed better physico-chemical properties, such as surfactant properties and thermal properties, when compared to the corresponding stereochemically-mixed surfactants due to much stronger intermolecular interactions between the surfactant molecules. It is also known that the chirality of surfactants is a crucial issue in the interaction with a protein [69]. However, the syntheses of chiral surfactants are complicated, generating relatively low yields. Thus, a more facile synthetic route is required. Here, design, chemo-enzymatic synthesis, surface activities and biological properties of chiral cationics containing carbonate linkages [(R)-S12iPr and (S)-S12iPr] are described [70].

2.4.1. Optical resolution of 1-(N,N-dimethylamino)-2-propanol using a lipase and synthesis of chiral cationics

Optically active carbonate-type cationics were prepared using the (R)-alcohol or (S)-alcohol. The optical resolution of 1-(N,N-dimethylamino)-2-propanol using $_D$-tartaric or $_L$-tartaric acid has been reported by Chan et al. However, the yields of the optically active amino alcohols were relatively low at 7% [71]. Here, the optically active amino alcohol was prepared by the enantioselective transesterification of vinyl propionate and 1-(N,N-dimethylamino)-2-propanol using the lipase according to the method of Hull et al. as shown in Figure 10 [72].

First, the (R)-ester was prepared by the reaction of 1-(N,N-dimethylamino)-2-propanol (40 mmol) and vinyl propionate (20 mmol) using the lipase from *Burkholderia cepacia*, immobilized on diatomaceous (lipase PS-D). The (R)-ester was then hydrolyzed using sodium methoxide to obtain the (R)-alcohol. The two-step total yield of the (R)-alcohol was 43%. Next, the unreacted (S)-alcohol was collected by treatment with 1-(N,N-dimethylamino)-2-propanol (10 mmol) and vinyl propionate (10 mmol) in the presence of lipase PS-D. It is reported by Kazlauskas et al. that the (S)-secondary alcohol is the slower reacting enantiomer in lipase-catalyzed transesterifications [73]. The optical purity of the unreacted (S)-alcohol was 76% after 24 h of reaction. Thus, the recovered (S)-alcohol was further treated with vinyl propionate using lipase PS-D at room temperature for 24 h. Purification was carried out by silica gel column chromatography to obtain optically pure (S)-alcohol in 18% yield.

The intermediate was prepared by a two-step successive carbonate exchange reaction of diphenyl carbonate and 1-dodecanol in the presence of Et$_3$N followed by the reaction of the appropriate N,N-dimethylaminoalcohol as shown in Figure 3. The quaternarization of the N,N dimethylamino group readily occurred by methyl iodide at room temperature. No significant differences in the yields of chiral intermediate and cationics were observed due to stereochemistry. However, the total yields were 28% for (R)-S12iPr and 10% for (S)-S12iPr, respectively.

Lipase-catalyzed optical resolution

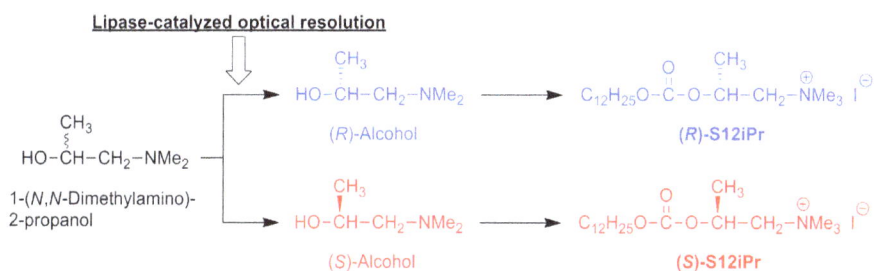

Figure 10. Preparation of chiral cationics by the lipase-catalyzed optical resolution of 1-(N,N-dimethylamino)-2-propanol.

2.4.2. Enantioselective hydrolysis of S12iPr using a lipase

To improve the yield of optically active cationics, the enantioselective hydrolysis of the **S12iPr** using lipase CALB was studied (Figure 11). Figure 12 shows the time course of the enzymatic hydrolysis of **(R)-S12iPr**, **(S)-S12iPr** and **S12iPr** in toluene containing a small amount of water. The enzymatic hydrolysis was strongly influenced by the stereochemistry of the cationics. Although the **(R)-S12iPr** was quickly hydrolyzed, the **(S)-S12iPr** was almost unchanged. The **S12iPr** was gradually hydrolyzed, and its hydrolysis rate exceeded 50% after 72 h. Therefore, it was hypothesized that the **(R)-S12iPr** would be preferentially hydrolyzed to produce the optically pure **(S)-S12iPr** as an unreacted material. Testing this hypothesis led to recovery the **(S)-S12iPr** in 27% yield (the theoretical yield was 50%) by the lipase-catalyzed hydrolysis of **S12iPr**.

Lipase-catalyzed optical resolution

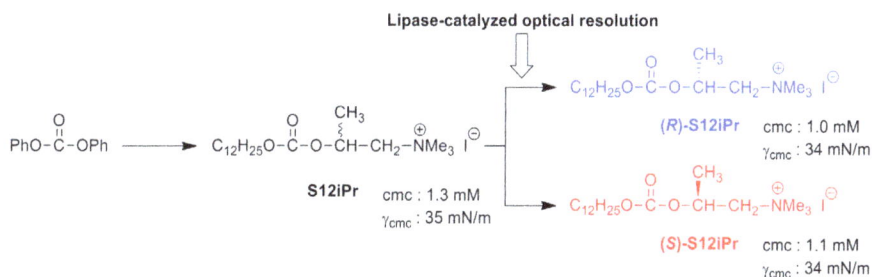

Figure 11. Preparation of chiral cationics by the lipase-catalyzed optical resolution of **S12iPr**.

2.4.3. Surfactant properties and antimicrobial activities

From the plots of surface tension *vs.* concentration curves of cationics in aqueous solution, the cmc and γ_{cmc} values are determined and the results are shown in Figure 11. No significant differences in the surfactant properties were observed by the stereochemistry of the tested

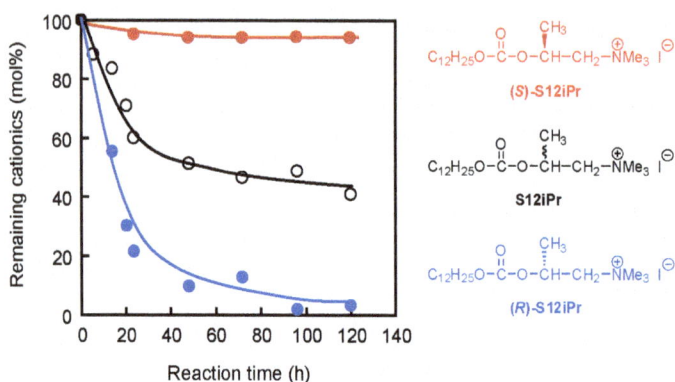

Figure 12. Time course of the enzymatic hydrolysis in toluene containing a small amount of water. Reaction conditions: cationics (10 mg) and 100 wt% immobilized lipase CALB (10 mg) were stirred in toluene (0.2 mL) and H_2O (10 µL) at 65°C.

cationics. These results demonstrated that the optical activity in the hydrophobic moiety did not affect the surfactant properties. The surfactant properties of enantiomeric or diastereoisomeric surfactants show similar tendencies, i.e., those containing the chiral center in the hydrophobic moiety have been found to have similar properties to those of the corresponding stereochemically mixed surfactants [74,75].

Also, the cationics were screened for their antimicrobial activities toward gram-positive and gram-negative bacterial and fungal strains based on the determination of their MICs. No significant differences in the antimicrobial activities were observed due to the stereochemistry of the cationics.

2.4.4. Biodegradabilities

Figure 13 shows the BOD-biodegradability of the cationics and their degradation products [(R)-HiPr, (S)-HiPr, HiPr and DD]. The biodegradability was strongly affected by the stereochemistry of optically active cationics. Although the (S)-S12iPr was quickly biodegraded, (R)-S12iPr was biodegraded to a relatively lower extent at around 30%. This is due to both the hydrolyzability at the carbonate linkage and the biodegradability of the primary degradation products. To compare the non-enzymatic hydrolyzability of the chiral cationics, an accelerated hydrolytic degradation test was carried out by dissolving cationics in distilled water and keeping them at 60 °C for 48 h. It was found that both (R)-S12iPr and (S)-S12iPr were remained 95%. That is, no significant differences in the hydrolysis were observed due to the stereochemistry of the cationics. On the other hand, the biodegradability of the primary degradation product was affected by the stereochemistry. Though the (S)-HiPr derived from (S)-S12iPr was biodegraded by activated sludge, the (R)-HiPr derived from (R)-S12iPr was not biodegraded. This suggests that the better biodegradability of (S)-S12iPr was due to the higher biodegradability of (S)-HiPr. The HiPr was biodegraded, but at a lower rate than (S)-HiPr.

Therefore, the biodegradability of **S12iPr** was lower than **(S)-S12iPr**, but higher than **(R)-S12iPr**. As discussed above, the hydrolysis of **(R)-S12iPr** by lipase CALB was much higher than that of **(S)-S12iPr**. This implies that the microbial hydrolysis in the biodegradation of the cationics by the activated sludge exhibited (S)-specificity rather than (R)-specificity of lipase CALB.

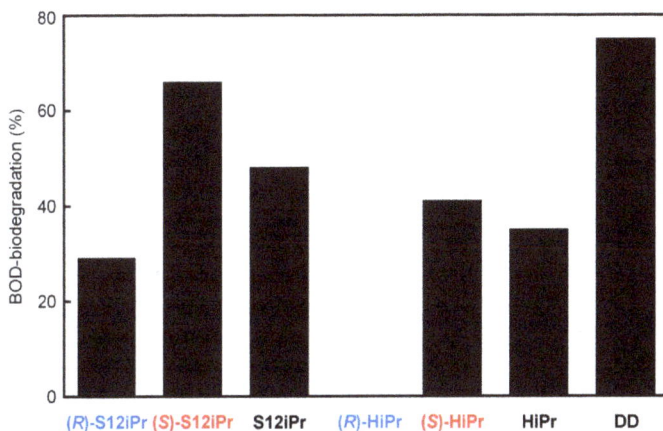

Figure 13. BOD-biodegradation of cationics and degradation intermediates at 25 °C for 28 d. Activated sludge: 30 ppm, cationics: ca. 40 ppm, **HiPr**: 57 ppm, **DD**: 38 ppm.

2.5. Green nonionics containing carbonate linkages

It has been reported that polyoxyethylene-type nonionic surfactant containing carbonate linkage was synthesized by the reaction of octanoyl chloroformate with tetra(ethylene glycol) [40]. More environmentally benign synthetic route without the use of halide is required in terms of green chemistry. Here, we describe synthesis of nonionics containing carbonate linkages (**mC-nEG**) using diphenyl carbonate as a green reagent in this section. We also report their properties and biodegradabilities [76].

2.5.1. Synthesis of mC-nEG

A series of **mC-nEG** was prepared by the two-step carbonate exchange reaction of diphenyl carbonate and 1-alkanol followed by the reaction of poly(ethylene glycol) as shown in Figure 14. The **12C-Ph** was prepared by the carbonate exchange reaction of diphenyl carbonate and **DD** in the presence of Et_3N at 70°C for 15 h (87% yield). Then, **12C-4EG** was produced in 70% yield by the reaction of **12C-Ph** with tetra(ethylene glycol) in the presence of K_2CO_3 in acetone at room temperature for 24 h. In a similar procedure, a series of **mC-nEG** was prepared in 66-80% yield. On the other hand, the yield of **12C-4EG** was around 50% by the lipase-catalyzed synthesis of **12-Ph** and tetra(ethylene glycol). The slightly lower yield from the lipase-catalyzed synthesis was due to the decarboxylation of the carbonate linkage of **12C-Ph** or **12C-4EG** during the reaction.

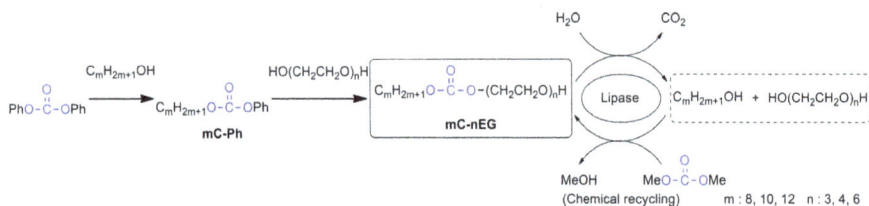

Figure 14. Synthesis and chemical recycling of **mC-nEG**.

2.5.2. Properties of mC-nEG

From the plots of surface tension *vs.* concentration curves of **mC-nEG**, their cmc and γ_{cmc} values were determined and listed in Table 4. Table 4 also shows the surfactant properties of tetra(oxyethylene)dodecanoate (**12Es-4EG**) and dodecyl tetra(oxyethylene)ether (**12Et-4EG**) as a reference. It was found that the cmc values for the ester-type **12Es-4EG** and ether-type **12Et-4EG** were roughly the same. These results imply that the carbonyl and methylene groups provide similar contributions to the surfactant properties. Różycka-Roszak *et al.* and Menger *et al.* reported that the carbonyl groups of ester-type cationics can act as the hydrophobic moiety [77,78]. Also, the cmc of the carbonate-type **12C-4EG** was lower than the cmc of the ether-type **12Et-4EG**. These results indicate that the oxycarbonyl part of the carbonate linkage provides the same hydrophobic contribution as a methylene group. There existed a relationship between the alkyl chain length and the cmc, such that the addition of an extra methylene group decreased the cmc of the nonionics by a factor of three [40]. Carbonate-type nonionics having a tetra(oxyethylene) chain had similar γ_{cmc} values. No significant differences in the γ_{cmc} values were observed among the linkages between the hydrophobic and hydrophilic moieties (*cf.* **12C-4EG**, **12Es-4EG** and **12Et-4EG**). Among the tested carbonate-type nonionics containing an *n*-dodecyl group, the shorter oxyethylene chain tended to have a lower γ_{cmc} value. The lowest γ_{cmc} of **12C-3EG** is due to the highest intermolecular hydrophobic interactions between the surfactant molecules.

Next, we carried out a hydrolytic degradation test by dissolving **12C-4EG** and **12Es-4EG** in distilled water at 30°C. The hydrolytic degradation of the nonionics was measured by size exclusion chromatography. The carbonate linkage of **12C-4EG** was stable in water, and 99% of the surfactant remained after 21 d. On the other hand, the ester linkage of **12Es-4EG** gradually hydrolyzed in water, and only 35% remained after 21 d. Based on these results, it was confirmed that the carbonate-type **12C-4EG** was more stable in water than the ester-type **12Es-4EG**.

Also, the synthesized carbonate-type nonionics containing an *n*-dodecyl group were readily biodegraded by activated sludge. Their BOD-biodegradability exceeded 60% after a 28-d

Surfactant	cmc (mM)	γ_{cmc} (mN/m)
8C-4EG	1.6	30
10C-4EG	0.17	30
12C-4EG	0.019	30
12C-3EG	0.015	29
12C-6EG	0.029	32
12Es-4EG	0.074	29
12Et-4EG	0.055	29

Table 4. Surfactant properties of carbonate, ester and ether-type nonionics in aqueous solution at 25°C

incubation, which is a criterion for acceptable biodegradation. No significant differences in the biodegradabilities of carbonate-type surfactants were observed by the oxyethylene chain length.

Furthermore, the enzymatic degradation of **12C-4EG** was carried out in toluene containing a small amount of water using immobilized lipase CALB. The **12C-4EG** was completely degraded at the carbonate linkage to produce tetra(ethylene glycol) and **DD**. The **12C-4EG** was reproduced by the two-step lipase-catalyzed carbonate exchange reaction of dimethyl carbonate and **DD** followed by the reaction of tetra(ethylene glycol) in two-step total yield of 24%. Based on these results, **12C-4EG** showed a lipase-catalyzed chemical recyclability as shown in Figure 14.

3. Conclusion

The design, synthesis and properties of green surfactants containing carbonate linkages were summarized in this paper. It was clarified that the carbonate linkages in surfactant molecules are effective as biodegradable and chemically recyclable segments. In particular, it was found that the biodegradability and antimicrobial activity are improved by the introduction of a carbonate linkage into the linker moiety of gemini-type cationics. Therefore, we propose a novel biodegradable and chemically recyclable gemini-type cationics having both good surfactant properties and high antimicrobial activities. It was confirmed that the biodegradability was strongly affected by the stereochemistry of cationics. To the best of our knowledge, this is the first report on the relationship between the stereochemistry of surfactants and biodegradability. These carbonate-type surfactants may contribute to the sustainable development of society. It is also important that the emulsification processes using biodegradable green surfactants are clarified from the viewpoint of industrial fields, such as the washing, hygiene, medicine and food industries. For example, we have demonstrated that oil-in-water emulsion droplets exhibit autonomous locomotion in an aqueous solution containing an ester-type biodegradable surfactant [79]. Since the natural environment is an open system, the elucidation of emulsification processes in such a non-equilibrium state lies on the frontier of surfactant research and development and is now needed for design the novel function of surfactants.

Acknowledgements

Immobilized lipase from *Candida antarctica* (lipase CALB, Novozym 435) was kindly supplied by Novozymes Japan Ltd. (Chiba, Japan). This work was supported by a Grant-in-Aid for General Scientific Research and JSPS Fellows 21·4882 from the Ministry of Education, Culture, Sports, Science, and Technology, Japan. This work was also supported by High-Tech Research Center Project for Private Universities: matching fund subsidy from MEXT, 2006-2011.

Author details

Taisuke Banno[1], Taro Toyota[1,2] and Shuichi Matsumura[3]

1 Department of Basic Science, Graduate School of Arts and Sciences, The University of Tokyo, Japan

2 Precursory Research of Embryonic Science and Technology (PRESTO), Japan Science and Technology Agency (JST), Japan

3 Department of Applied Chemistry, Faculty of Science and Technology, Keio University, Japan

References

[1] Anastas PT, Warner JC. Green chemistry: theory and practice. Oxford: Oxford University Press; 1998.

[2] Matsumura S. Enzyme-catalyzed synthesis and chemical recycling of polyesters. Macromolecular Bioscience 2002;2(3) 105-126.

[3] Tsarevsky NV, Matyjaszewski K. Green atom transfer radical polymerization: from process design to preparation of well-defined environmentally friendly polymeric materials. Chemical Reviews 2007;107(6) 2270-2299.

[4] Kobayashi S, Makino A. Enzymatic polymer synthesis: an opportunity for green polymer chemistry. Chemical Reviews 2009;109(11) 5288-5353.

[5] Akiyama R, Kobayashi S. "Microencapsulated" and related catalysts for organic chemistry and organic synthesis. Chemical Reviews 2009;109(2) 594-642.

[6] Li C-J, Chen L. Organic chemistry in water. Chemical Society Reviews 2006;35(1) 68-82.

[7] Bozell JJ, Petersen GR. Technology development for the production of biobased products from biorefinery carbohydrates—the US Department of Energy's "Top 10" revisited. Green Chemistry 2010;12(4) 539-554.

[8] Foley P, Kermanshahi pour A, Beach ES, Zimmerman JB. Derivation and synthesis of renewable surfactants. Chemical Society Review 2012;41(4) 1499-1518.

[9] von Rybinski W, Hill K. Alkyl polyglycosides—properties and applications of a new class of surfactants. Angewandte Chemie International Edition 1998;37(10) 1328-1345.

[10] Pérez L, Torres JL, Manresa A, Solans C, Infante MR. Synthesis, aggregation, and biological properties of a new class of gemini cationic amphiphilic compounds from arginine, bis(args). Langmuir 1996;12(22) 5296-5301.

[11] Morán C, Clapés P, Comelles F, Garcia T, Pérez L, Vinardell P, Mitjans M, Infante MR. Chemical structure/property relationship in single-chain arginine surfactants. Langmuir 2001;17(16) 5071-5075.

[12] Pegiadou S, Pérez L, Infante MR. Synthesis, characterization and surface properties of 1-N-L-tryptophan-glycerol-ether surfactants. Journal of Surfactants and Detergents 2000;3(4) 517-525.

[13] Seguer J, Infante MR, Allouch M, Mansuy L, Selve C, Vinardell P. Synthesis and evaluation of nonionic amphiphilic compounds form amino-acids – molecular mimics of lecithins. New Journal of Chemistry 1994;18(6) 765-774.

[14] Yoshimura T, Sakato A, Tsuchiya K, Ohkubo T, Sakai H, Abe M, Esumi K Adsorption and aggregation properties of amino acid-based N-alkyl cysteine monomeric and N,N'-dialkyl cystine gemini surfactants. Journal of Colloid Interface Science 2007;308(2) 466-473.

[15] Allouch M, Infante MR, Seguer J, Stebe MJ, Selve C. Nonionic amphiphilic compounds from aspartic and glutamic acids as structural mimics of lecithins. Journal of American Oil Chemists' Society 1996;73(1) 87-96.

[16] Okada Y, Banno T, Toshima K, Matsumura S. Synthesis and properties of polycarboxylate-type green surfactants with S- or N-linkages. Journal of Oleo Science 2009;58(10) 519–528.

[17] Kitamoto D, Morita T, Fukuoka T, Konishi M, Imura T. Self-assembling properties of glycolipid biosurfactants and their potential applications. Current Opinion in Colloid and Interface Science 2009;14(5) 315-328.

[18] Stjerndahl M, Ginkel CG, Holmberg K. Hydrolysis and biodegradation studies of surface-active esters. Journal of Surfactants and Detergents 2003;6(4) 319-324.

[19] Stjerndahl M, Holmberg K. Synthesis, stability, and biodegradability studies of a sur-
      face-active amide. Journal of Surfactants and Detergents 2005;8(4) 331-336.

[20] Tehrani-Bagha AR, Holmberg K. Cleavable surfactant. Current Opinion Colloid and
      Interface Science 2007;12(2) 81-91.

[21] Ono D, Masuyama A, Okahara M. Preparation of new acetal type cleavable surfac-
      tants from epichlorohydrin. The Journal of Organic Chemistry 1990;55(14) 4461-4464.

[22] Kida T, Yurugi K, Masuyama A, Nakatsuji Y, Ono D, Takeda T. Preparation and
      properties of new surfactants containing D-glucosamine as the building-block. Jour-
      nal of American Oil Chemists' Society 1995;72(7) 773-780.

[23] Kruger G, Boltersdorf D, Lewandowski H. Esterquats in Novel Surfactants. In: Holm-
      berg K. (ed.) New York: Marcel Dekker; 1998. p115.

[24] Epstein WW, Jones DS, Bruenger E, Rilling HC. The Synthesis of a photolabile deter-
      gent and its use in the isolation and characterization of protein. Analytical Biochem-
      istry 1982;119(2) 304-312.

[25] Masuyama A, Endo C, Takeda S, Nojima M, Ono D, Takeda T. Ozone-cleavable gem-
      ini surfactants. Their surface-active properties, ozonolysis, and biodegradability.
      Langmuir 2000;16(2) 368-373.

[26] McElhanon JR, Zifer T, Kline SR, Wheeler DR, Loy DA, Jamison GM, Long TM, Rahi-
      mian K, Simmons BA. Thermally cleavable surfactants based on furan-maleimide
      Diels-Alder adducts. Langmuir 2005;21(8) 3259-3266.

[27] Bunton CA, Robinson L, Schaak J, Stam MF. Catalysis of nucleophilic substitutions
      by micells of dicationic detergents. The Journal of Organic Chemistry 1971;36(16)
      2346-2350.

[28] Menger FM, Littau CA. Gemini surfactants: synthesis and properties. Journal of the
      American Chemical Society 1991;113(4) 1451-1452.

[29] Menger FM, Littau CA. Gemini surfactsnts: a new class of self-assembling molecules.
      Journal of the American Chemical Society 1993;115(22) 10083-10090.

[30] Rosen MJ, Tracy DJ. Gemini surfactants. Journal of Surfactants and Detergents
      1998;1(4) 547-554.

[31] Menger FM, Keiper JS. Gemini surfactants. Angewandte Chemie International Edi-
      tion 2000;39(11) 1906-1920.

[32] Zana R. Dimeric (gemini) surfactants: effect of the spacer group on the association
      behavior in aqueous solution. Journal of Colloid and Interface Science 2002;248(2)
      203-220.

[33] Zana R. Dimeric and oligomeric surfactants. Behavior at interfaces and in aqueous
      solution: a review. Advanced Colloid and Interface Science 2002;97(1-3) 205-253.

[34] Wilk KA, Syper L, Domagalska BW, Komorek U, Maliszewska I, Gancarz R. Aldona-mide-type gemini surfactants: synthesis, structural analysis, and biological proper-ties. Journal of Surfactants and Detergents 2002;5(3) 235-244.

[35] Laska U, Wilk KA, Maliszewska I, Syper L. Novel glucose-derived gemini surfac-tants with a 1,1'-ethylenebisurea spacer: preparation, thermotropic behavior, and bi-ological properties. Journal of Surfactants and Detergents;9(2) 115-124.

[36] Ono D, Yamamura S, Nakamura M, Takeda T. Preparation and properties of bis(so-dium sulfate) types of cleavable surfactants derived from diethyl tartrate. Journal of Oleo Science 2005;54(1) 51-57.

[37] Zhu Y, Masuyama A, Kirito Y, Okahara M, Rosen MJ. Preparation and properties of glycerol-based double- or triple-chain surfactants with two hydrophilic ionic group. Journal of American Oil Chemists' Society 1992;69(7) 626-632.

[38] Zana R, In M, Lévy H. Duportail G. Alkanediyl-α,ω-bis(dimethylalkylammonium bromide). 7. Fluorescence probing studies of micelle micropolarity and microviscosi-ty. Langmuir 1997;13(21) 5552-5557.

[39] Laschewsky A, Wattebled L, Arotcaréna M, Habib-Jiwan JL, Rakotoaly RH. Synthesis and properties of cationic oligomeric surfactants. Langmuir 2005;21(16) 7170-7179.

[40] Stjerndahl M, Holmberg K. Hydrolyzable nonionic surfactants: stability and physico-chemical properties of surfactants containing carbonate, ester, and amide bonds. Journal of Colloid and Interface Science 2005;291(2) 570-576.

[41] Kobayashi S, Litter H, Kaplan D. (ed.) Enzyme-catalyzed synthesis of polymers. Ad-vances in Polymer Science 194. Heidelberg: Springer; 2006.

[42] Varma IK, Albertsson A-C, Rajkhowa R, Srivastava RK. Enzyme catalyzed synthesis of polyesters. Progress in Polymer Science 2005;30(10) 949-981.

[43] Sakakura T, Choi JC, Saito Y, Sako T. Synthesis of dimethyl carbonate from carbon dioxide: catalysis and mechanism Polyhedron 2000;19(5) 573-576.

[44] Choi J, He L, Yasuda H, Sakakura T. Selective and high yield synthesis of dimethyl carbonate directly from carbon dioxide and methanol. Green Chemistry 2002;4(3) 230-234.

[45] Fujita S, Bhanage B, Ikushima Y, Arai M. Synthesis of dimethyl carbonate from car-bon dioxide and methanol in the presence of methyl iodide and base catalysts under mild conditions: effect of reaction conditions and reaction mechanism. Green Chem-istry 2001;3(2) 87-91.

[46] Hou Z, Han B, Liu Z, Jiang T, Yang G. Synthesis of dimethyl carbonate using CO2 and methanol: enhancing the conversion by controlling the phase behavior. Green Chemistry 2002;4(5) 467-471.

[47] Bristline Jr RG, Maurer EW, Smith FD, Linfield WM. Fatty acid amides and anilides, syntheses and antimicrobial properties. Journal of American Oil Chemists' Society 1980;57(2) 98-103.

[48] OECD guidelines for testing of chemicals, 301C, modified MITI test, Organization for Economic Cooperation and Development (OECD). Paris: 1981.

[49] Fernández P, Valls M, Bayona JM, Albalgés J. Occurrence of cationic surfactants and related products in urban coastal environments. Environmental Science and Technology 1991;25(3) 547-550.

[50] Games LM, King JE, Larson RJ. Fate and distribution of a quaternary ammonium surfactant, octadecyltrimethylammonium chloride (OTAC), in wastewater treatment. Environmental Science and Technology 1982;16(8) 483-488.

[51] Banno T, Toshima K, Kawada K, Matsumura S. Synthesis and properties of bioradable and chemically recyclable cationic surfactants containing carbonate linkages. Journal of Oleo Sciecnce 2007;56(9) 493-499.

[52] Patist A. Determining Critical Micelle Concentration. In: Holmberg K. (ed.) Handbook of Applied Surface and Colloid Chemistry. Chichester: John Wiley & Sons; 2002. p245.

[53] Zana R, Talmon Y. Dependence of aggregate morphology on structure of dimeric surfactants. Nature 1993;362(6417) 228-230.

[54] Devínsky F, Masárová L, Lacko I. Surface activity and micelle formation of some new bisquaternary ammonium salts. Journal of Colloid and Interface Science 1985;105(1) 235-239.

[55] Devínsky F, Lacko I, Bittererová F, Tomečková L. Relationship between structure, surface activity, and micelle formation of some new bisquaternary isosteres of 1,5-pentanediammonium dibromides. Journal of Colloid and Interface Science 1986;114(2) 314-322.

[56] Tehrani-Bagha AR, Oskarsson H, Ginkel CG, Holmberg K. Cationic ester-containing gemini surfactants: chemical hydrolysis and biodegradation. Journal of Colloid and Interface Science 2007;312(2) 444-452.

[57] Tehrani-Bagha AR, Holmberg K. Cationic ester-containing gemini surfactants: physical-chemical properties. Langmuir 2010;26(12) 9276-9282.

[58] Tatsumi T, Zhang W, Kida T, Nakatsuji Y, Ono D, Takeda T, Ikeda I. Novel hydrolyzable and biodegradable cationic gemini surfactants: 1,3-bis[(acyloxyalkyl) dimethylammoniol?-hydroxypropane dichloride. Journal of Surfactants and Detergents 2000;3(2) 167-172.

[59] Tatsumi T, Zhang W, Kida T, Nakatsuji Y, Ono D, Takeda T, Ikeda I. Novel hydrolyzable and biodegradable cationic gemini surfactants: bis(ester-ammonium) dichlor-

ide having a butenylene or a butynylene spacer. Journal of Surfactants and Detergents 2001;4(3) 279-285.

[60] Banno T, Toshima K, Kawada K, Matsumura S. Synthesis and properties of gemini-type cationic surfactants containing carbonate linkages in the linker moiety directed toward green and sustainable chemistry. Journal of Surfactants and Detergents 2009;12(3) 249-259.

[61] Banno T, Kawada K, Matsumura S. Creation of novel green and sustainable gemini-type cationics containing carbonate linkages. Journal of Surfactants and Detergents 2010;13(4) 387-398.

[62] Willemen HM, de Smet LCPM, Koudijs A, Stuart MCA, Heikamp-de Jong IGAM, Marcelis ATM, Sudhölter EJR. Micelle formation and antimicrobial activity of cholic acid derivatives with three permanent ionic head groups. Angewandte Chemie International Edition 2002;41(22) 4275-4277.

[63] Haldar J, Kondaiah P, Bhattacharya S. Synthesis and antimicrobial properties of novel hydrolyzable cationic amphiphiles. Incorporation of multiple head groups leads to impressive antimicrobial activity. Journal of Medicinal Chemistry 2005;48(11) 3823-3831.

[64] Boyd BJ, Krodkiewska I, Drummond CJ, Grieser F. Chiral glucose-derived surfactants: the effect of stereochemistry on thermotropic and lyotropic phase behavior. Langmuir 2002;18(3) 597-601.

[65] Harvey NG, Mirajovsky D, Rode PL, Verbiar R, Arnett EM. Molecular recognition in chiral monolayers of stearoylserine methyl ester. Journal of the American Chemical Society 1989;111(3) 1115-1122.

[66] Zhang YJ, Song Y, Zhao Y, Li TJ, Jiang L, Zhu D. Chiral discrimination in Langmuir monolayers of N-acyl glutamic acids inferred from π-A measurements and atomic force microscopy. Langmuir 2001;17(5) 1317-1320.

[67] Ambrosi M, Nostro PL, Fratini E, Giustini L, Ninham BW, Baglioni P. Effect of head-group chirality in nanoassemblies. Part 1. Self-assembly of D-isoascorbic acid derivatives in water. The Journal of Physical Chemistry B 2009;113(5) 1404-1412.

[68] Aisaka T, Oida T, Kawase T. A novel synthesis of succinic acid type gemini surfactant by the functional group interconversion of corynomicolic acid. Journal of Oleo Science 2007;56(12) 633-644.

[69] Faustino CMC, Calado ART, Garcia-Rio L. Gemini surfactant—protein interactions: effect of pH, temperature, and surfactant stereochemistry. Biomacromolecules 2009;10(9) 2508-2514.

[70] Banno T, Kawada K, Matsumura S. Chemo-enzymatic synthesis and properties of novel optically active cationics containing carbonate linkages. Journal of Oleo Science 2011;60(4) 185-195.

[71] Chan MML, Robinson JB. Stereoisomeric lactoyl-β-methylcholine iodides. interaction with cholinesterase and acetylcholinesterase. Journal of Medicinal Chemistry 1974;17(10) 1057-1060.

[72] Hull JD, Scheinmann F, Turner NJ. Synthesis of optically active methadones, LAAM and bufuralol by lipase-catalyzed acylations. Tetrahedron Asymmetry 2003;14(5) 567-576.

[73] Kazlauskas RJ, Weissfloch ANE. A structure-based rationalization of the enantiopreference of subtilisin toward secondary alcohols and isosteric primary amines. Journal of Molecular Catalysis B: Enzymatic 1997;3(1) 65-72.

[74] Larpent C, Chasseray X. Optically active surfactants -I- the first synthesis and properties of sodium bis[(S)-ethyl-2-hexyl] sulfosuccinates ("aerosol OT"). Tetrahedron 1992;48(19) 3903-3914.

[75] Moss RA, Sunshine WL. Micellar catalysis of ester hydrolysis. Influence of chirality and head group structure in "simple" surfactants. The Journal of Organic Chemistry 1974;39(8) 1083-1089.

[76] Banno T, Sato H, Tsuda T, Matsumura S. Synthesis and properties of green sustainable carbonate-type nonionics containing polyoxyethylene chain. Journal of Oleo Science 2011;60(3) 117-126.

[77] Różycka-Roszak B, Przestalski S, Witek S. Calorimetric studies of the micellization of some amphiphilic betaine ester derivatives. Journal of Colloid and Interface Science 1988;125(1) 80-85.

[78] Menger FM, Galloway AL. Contiguous versus segmented hydrophobicity in micellar systems. Journal of the American Chemical Society 2004;126(48) 15883-15889.

[79] Banno T, Kuroha R, Toyota T. pH-Sensitive self-propelled motion of oil droplets in the presence of cationic surfactants containing hydrolyzable ester linkages. Langmuir 2012;28(2) 1190-1195.

Chapter 7

Antimicrobial Modifications of Polymers

Vladimir Sedlarik

Additional information is available at the end of the chapter

1. Introduction

Polymers have been used for decades instead of metal, glass and wood in many applications due to their superior physicochemical properties, in addition to others, as well as for reasons of economy. For example, commercially available thermoplastic polymers, such as polyolefins, are hydrophobic and biologically inert. This has made them indispensable in the packaging industry, distribution of food stuffs and other perishable commodities [1]. Another example is agriculture, where plastics have largely replaced glass in the construction of green houses, in addition to which they have gained a unique position in the growing of soft fruit and vegetables over mulching films [2]. Increasing demands on polymer materials have led to their further development. In some applications, the polymer products also possess, besides the passive function (e.g. packaging or structural) an active function (e.g. protective and/or indicative). The polymeric materials with resistance to microbial colonization and pathogenic microorganism spreading (antimicrobial polymers) have been one of the examples of the active material functionality. The antimicrobial polymers are expected to protect against negative impact of the pathogenic microorganisms, which can seriously affect the society from the viewpoint of both health damages and unwanted economical loads connected with that.

This chapter is focused on antimicrobial modifications of polymer materials intended for medical devices production. Firstly, a brief introduction into the field of medical application of polymers is presented. Considering the fact that polymer medical devices are often connected with occurrence of nosocomial infections, the next part refers to this phenomenon and its causes. One of the possibilities of reducing of the infection occurrence is aimed at polymer modification. It is a key topic of the third part of the chapter. Finally, the methodology of the polymer materials antimicrobial properties determination is shown, together with references to the relevant standards.

2. Polymers in medicine

The opening paragraph highlights that polymers are now part and parcel of everyday life - in form of frequently utilized items (disposable packaging, textile fibers, and construction materials) as well as in specialized and complex fields such as electronics and pharmaceuticals. Polymers have found applications in medicine, too. They are commonly used to produce various medical devices, including implants, drug carriers, protective packaging materials, and healthcare items [3]. Such applications of polymers in the medicinal sphere are shown in Table 1.

Polymer	Medical device applications
Polyethylene	Orthopedic implants, containers, catheters, non-woven textiles
Polypropylene	Disposable items (e.g. syringes), non-woven textiles, membranes, sutures
Polyurethanes	Films, tubing, catheters
Polyvinylchloride	Catheters, tubing
Polyamides	Sutures, packaging, dental implants
Polyethylene terephthalate	Sutures, artificial vascular grafts
Polycarbonates	Containers, construction material
Poly(methyl methacrylate)	Membranes, implants, part of bone cement
Polydimethylsiloxane	Implants, catheters
Polytetrafluoroethylene	Artificial vascular grafts, catheters (albeit rarely)
Polyether ether ketone	Tubing
Polylactide	Resorbable implants

Table 1. Examples of synthetic polymer applications in medicine [4]

As can be seen in Table 1, polymer-based medical devices can consist of both synthetic non-biodegradable and biodegradable polymers, which are referred to as *bio-resorbable* in this instance. Another division of polymer biomaterials (i.e. materials used to produce medical devices) can be delineated according to polymer origin, i.e. synthetic and natural (e.g. cellulose, collagen, and derivatives) [5].

One might consider polymer an ideal biomaterial due to the benefits mentioned above. However, the use of polymers in medicine has a significant disadvantage. Polymer-based surfaces are often vulnerable to bacterial attack, which can give rise to serious complications in the form of so-called Nosocomial infections (see further sections) [6, 7]. Therefore, developing new polymer systems exhibiting antimicrobial activity has been the focus of extensive research, with numerous methods being applied to prevent and control the occurrence of these complications.

3. Bacterial colonization of polymer surfaces and its relation to nosocomial infections

Nosocomial infections have affected human beings for ages. There is evidence of surgical operations even taking place in ancient Egypt. However, the mortality rate of patients remained high due to subsequent complications. This was normal until the 1860s, when Dr. Joseph Lister introduced the practice of aseptic surgery, enabling surgeons to perform a greater variety of complex operations. Figure 1 shows a time scale of milestones in medical device development.

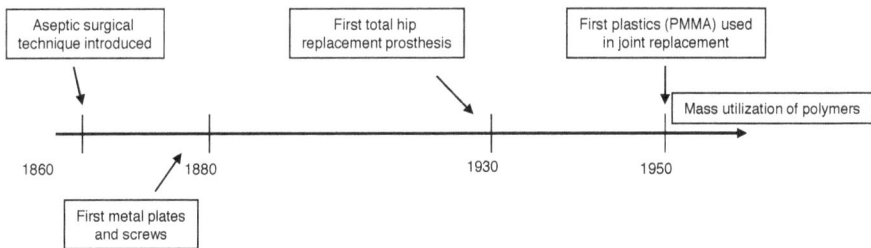

Figure 1. Notable milestones in relation to medical devices

Nosocomial infections (from Greek *nosos* = disease, *comeo* = care) are a consequence of bacteria colonizing the surface of a polymer-based medical device (see Figure 2).

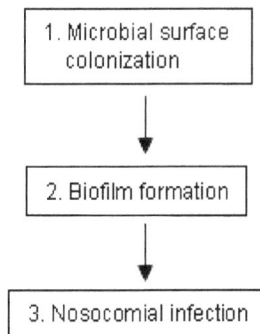

Figure 2. Consequences of microbial surface adhesion

A nosocomial infection (also known as secondary infections, hospital acquired infections) can be defined as a health complication that arises in the therapy of a patient in a medical facility. The infection can be both endogenous (e.g. mediated by surgery treatment) and exogenous (spread via an external environment) [8].

The repercussions of nosocomial infections are clear. Statistics reveal that the average incidence of secondary infections affects 8% of all hospitalized patients (for Great Britain it is 10%, Italy 6.7%, and Finland 8.7%). The United States observes the occurrence of nosocomial infection in over 1.7 million cases per year. Such health complications cause over 90,000 deaths a year. It is estimated that the annual cost of treatment of nosocomial infections brought about by mediated complications is between 4.5 and 11 billion US dollars [8, 9]. Nosocomial infections are often in the form of urinary infections (40%), wound infections (25%) and nosocomial pneumonia (20%).

The vast majority of the above-mentioned health complications are linked with the usage of medical devices. As already indicated in Figure 2, nosocomial infections are related to microbial colonisation and subsequent biofilm formation on the surfaces of polymer medical devices. Following text refers to the fundamentals of these phenomena.

3.1. Bacterial adhesion

Investigating the primary cause of bacterial adhesion on surfaces has proven a challenge to microbiologists for several decades. Initial works were, however, focused on microbial surface colonization in an aqueous environment. *Marshall* described the process of reversible and irreversible adhesion of the bacterial strain *Pseudomonas sp.* on a glass substrate in 1985 [10]. Further studies showed that the adhesion process can take from several seconds to a few minutes [11]. Extensive attention was paid to researching the extent of bacterial adhesion and colonization versus the physicochemical properties of surfaces. Here it is worth mentioning the example of polyethylene. The polyolefin is considered susceptible to bacterial colonization by various bacterial strains in a marine water environment, this being the cause of polymer detoriation [12]. However, a rise in unexplained infections in medical facilities brought this area to the attention of scientists. Reference is made later to various theories that exist to explain the primary cause of bacterial surface adhesion (for instance, thermodynamic theory [13]). Nevertheless, none of them has led to full clarification of this phenomenon [14].

One of the highest regarded approaches in bacteria adhesion process description was DVLO theory (Derjaguin, Landau, Verwey and Overbeek theory), which considers individual bacteria as colloid particles. These particles interact with surfaces on the basis of their charges [15]. Originally DVLO theory was introduced to explain stabilities in colloid systems in the 1940's. It formulates the stability of colloid systems as a measure of potential energy consisting of two components: (i) interactions of attraction based on Van der Walls forces; (ii) electrostatic repulsive forces. However, DVLO theory does not take into consideration a number of facts, and it cannot be accepted as satisfactory. Firstly, bacteria have an inhomogeneous surface. The surface of bacteria can be covered by an extracellular polysaccharide or peptide. Furthermore, appendages that protrude from the body of bacteria might be present (e.g. flagella) and significantly affect bacteria-surface interactions [16]. Conducting an experimentally confirmed process for irreversible adhesion also raises questions regarding the applicability of DVLO theory [10].

Generally, it is well accepted that bacterial adhesion on abiotic surfaces is mediated by non-specific interactions, such as electrostatic and Van der Waals forces, while adhesion to biotic surfaces is controlled by specific molecular mechanisms [17].

It was found experimentally that bacterial adhesion itself can be characterized by three parameters relating to [18-21]:

• bacteria (strain, bacterial growth, nutritional conditions, surface charge and energy)

• material surface (chemical composition, surface charge and energy, topology)

• surrounding environment (pH, temperature, presence of oxygen and other organic and inorganic compounds, hydrodynamic parameters)

3.2. Biofilm formation

In spite of the absence of a theory fully clarifying the initial phase of bacterial surface adhesion, the process of biofilm formation is described in relative detail. It can be divided into five stages as indicated in Figure 3. This also contains surface pictures taken by a camera connected to an optical microscope. Figure 4 shows a detailed surface picture (taken by a scanning electron microscope) of a polymer-based catheter colonized by the bacterial strain *Staphylococcus aureus*.

1. **Initial stage** – also called *the conditioning phase*. Initially, reversible bacteria-surface interactions are predominant. However, traces of extracellular substances remain on the surface after the detachment of bacteria. It is believed that such residua play an important role in the following stage of biofilm formation.

2. **Stage of irreversible attachment** – (primary adhesion); the predominant occurrence of stable bacterial attachment on the surface.

3. **Formation of micro-colonies** - (maturation phase 1); attached microorganisms reproduce and form colonies. The kinetics of cell division can, under ideal conditions, be expressed as "$m2^n$", where m is the number of colonies at the beginning and n represents the number of generations [16].

4. **Biofilm formation** – (maturation phase 2); adhered cells (microorganisms) produce extracellular compounds (often based on polysaccharides). Eventually the whole surface can be covered by extracellular secretion – the biofilm is formed. The environment of the biofilm is ideal for further reproduction of microorganisms. It was found that a lethal dose of antibiotics is more than one hundred times greater for bacteria present in the biofilm in comparison with their freely floating counterparts [22].

5. **Distribution stage** – the growing mass of biofilm leads to it rupturing. The bacteria present spread into the surrounding environment – i.e. the human body in the case of medical devices.

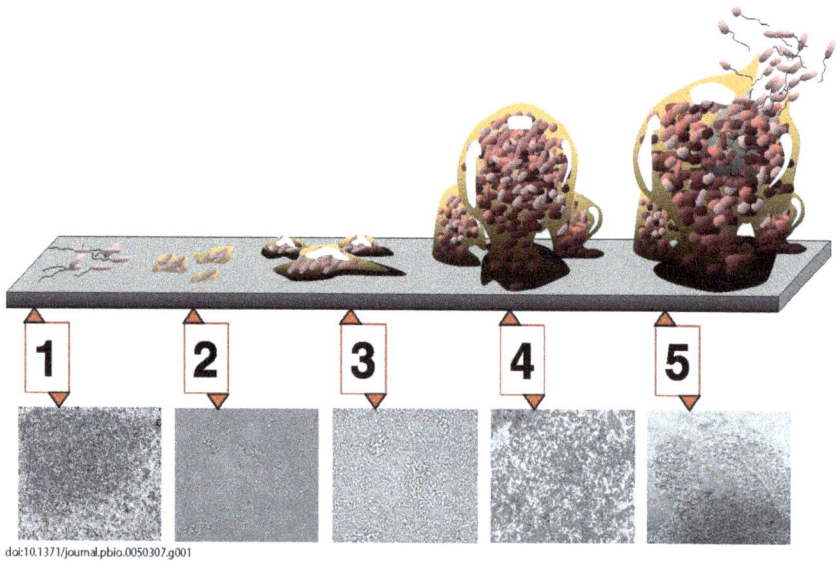

doi:10.1371/journal.pbio.0050307.g001

Figure 3. Five stages of biofilm formation (adopted from Monroe, D. doi:10.1371/journal.pbio.0050307 [22])

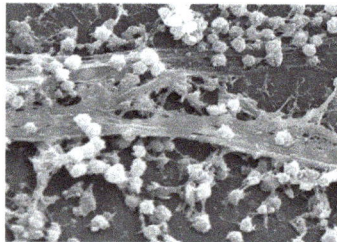

Figure 4. SEM picture of a plastic catheter colonized by *Staphylococcus aureus* (adopted from Monroe, D. doi: 10.1371/journal.pbio.0050307 [22])

Slight differences in describing the biofilm formation process can be found in the literature. However, these discrepancies are insignificant from the principle presented here. Semi empirical models explaining the kinetics of bacterial surface adsorption and desorption have been developed. These calculations are applied, for example, in the spheres of waste water treatment and biotechnology. However, such processes are too dissimilar in comparison with the colonization of medical devices (e.g. urinary catheters) to be mentioned in this work. Nevertheless, they can be found in the relevant literature [14, 16, 23, 24].

4. Antimicrobial modifications of polymers

An antimicrobial polymer system is a material intentionally modified (chemically or physically) to prevent its bacterial colonization. It usually consists of a polymer matrix and an antimicrobial agent. Antimicrobial polymers have been applied in medicine, food packaging as well as in the personal hygiene industry.

Strictly speaking there are four principles for antimicrobial modification of polymers.

Any selection of such a modification method should consider the following factors [16, 25, 26]:

- polymer properties (chemical and physical)

- intended use (humidity, temperature, pH factors, etc.)

- characteristics of the antimicrobial agent (toxicity, thermal stability, affinity with a certain component)

- technological factors (complexity, functionality, reproducibility)

- financial factors (financial burden versus added value, certification)

The majority of medical devices are made from a few known types of polymers (see Table 1). Beside these, a group of special polymers should be highlighted here as well. Special polymers are used as, for example, a hydrophilic coating layer (e.g. polyvinyl alcohol, polyvinylpyrrolidone), and bioactive substance carriers, etc.

a. Polymer modification without an antimicrobial compound

This technique originates from the basic assumption that modifying the surface properties of a material (surface free energy, polarity, topography) may result in diminishing bacterial adhesion during the initial stage of the biofilm formation process.

Modification can be performed either by applying wet chemistry through reaction with various chemical reagents, or by applying high energy electromagnetic radiation (e.g. by laser, ultraviolet radiation, gamma rays). The interaction of a polymer's surface with electromagnetic radiation causes surface activation (through the breakage of accessible polymer bonds), permitting subsequent chemical modification [27-29]. Another promising method is modifying polymer surfaces by ionized gas (plasma). This leads, naturally, to the selection of so-called *cold plasma* when the temperature of the treated material does not reach high figures in comparison with the ambient temperature. This method demands low pressure (0.1 – 100 Pa) and the presence of a working gas (usually N_2, O_2 or Ar, CF_4). Plasma modifications to polymer surfaces are characterized by their weak stability over time, as polymer surfaces tend to return to their original chemical state [30-40].

The extent of surface modification is a measure of free surface energy change, which can be simply detected by a contact angle technique, based on wettability by selected liquid determination. In the case of water, one measures the water contact angle (WCA).

The practical results of such surface modification are shown in Table 2.

In this experiment, the polymer surfaces were modified by various functional groups, which caused a dramatic change in WCA. These surfaces were subsequently tested for resistance to colonization by two bacterial strains (*Staphylococcus epidermidis* and *Deleya marina*). The results displayed in Table 2 represent the measure of bacterial colonization (100% stands for no effect in comparison with the unmodified surface). In this particular instance it is noticeable that a surface modification resulting in the WCA value 35° is optimal for reducing both strains. However, there is no universal surface modification technique that could prove applicable against all bacterial strains due to the rich diversity of microorganisms. Another disadvantage lies in the complexity of the surface modification methods.

Modification	WCA (°)	Colonization *Staphylococcus epidermidis* (%)	Colonization *Deleya marina* (%)
$-(CH_2)(CF_2)_7CF_3$	120	27	49
$-(CH_2)_{15}CH_3$	107	57	100
$-(CH_2)_{11}(OCH_2CH_2)_6$	35	0.3	0.3
$-(CH_2)_{15}COOH$	< 5	100	23

Table 2. Effect of polymer surface modification on the extent of bacterial colonization [25]

b. Method for direct deposition of an antimicrobial agent on a polymer surface

This represents the simplest technique, one widely used in medical practice. An antimicrobial agent is applied to the surface of a polymer-based medical device just prior to use. The antimicrobial agent is usually in the form of a solution or ointment. However, low efficiency, caused by fast resorption of the active component, is the disadvantage of this method [41].

c. Method for surface modification and chemical deposition of an antimicrobial agent

Here the antimicrobial agent is chemically deposited either on the surface of the polymer (after prior activation) or by way of a relevant mediator (often based on polyacrylic acid). The mediator is grafted on the polymer surface and it forms a polymer brush (see Figure 5). The ends of the mediator chains can be used to immobilize the antimicrobial component. The function is similar to the previous case – the antimicrobial agent is released from the surface and deactivates any potentially present bacteria. The advantage of this technique lies in its activity over a long period of time [39, 42, 43]. Nevertheless, the complexity of the modification process can prove limiting for this method. The issue of the mechanical stability of the thin surface layers should be also considered before opting for a modification technique.

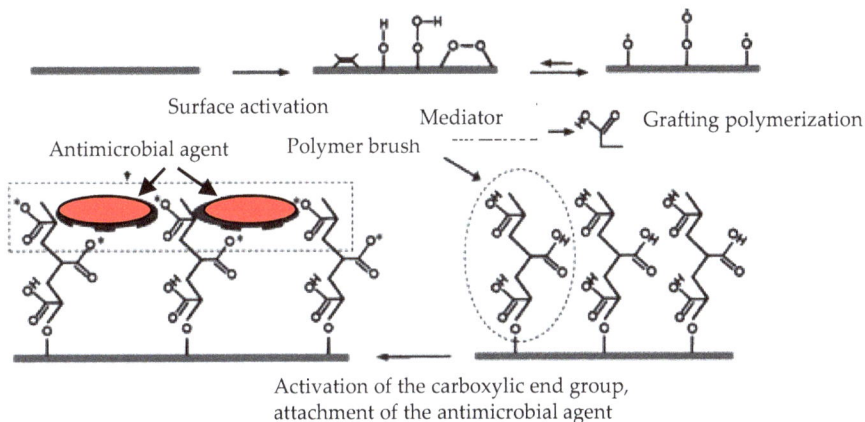

Figure 5. Schematic representation of antimicrobial agent immobilization on a polymer surface mediated by a polymer brush [43]

d. Method for bulk modification of a polymer with an antimicrobial agent

The last technique listed of antimicrobial polymer modification methods is based on directly incorporating the antimicrobial agent in a polymer matrix. This can be carried out in two ways. The first is suitable for preparing special coatings and films when mass production is not anticipated. The antimicrobial agent is introduced into a polymer solution that is cast subsequently. A dipping technique can be used in the case of coatings. The second way is applicable for large-scale production using thermoplastic polymer matrices. The antimicrobial agent is mixed with a polymer melt and processed by conventional techniques (extrusion, injection molding, blow molding, etc.).

Both methods for bulk modification of polymers are not technologically demanding. The antimicrobial additives behave in analogy with polymer fillers. The concentration of the antimicrobial agent in the polymer matrix does not usually exceed 20 vol. %.

A definite advantage of this method is the fact that, in most cases, the processing parameters as well as the technology necessary do not require significant modification. Moreover, with a lower additive content, the resulting mechanical properties of the modified materials are similar to those of the unmodified polymer matrix. However, the efficiency of antimicrobial modification (related to the amount of the incorporated agent) is low due to the restricted diffusibility of the antimicrobial agent molecules through the polymer matrix. It leads to most of the incorporated agent getting trapped in the polymer matrix, which cannot then become involved in the antimicrobial process. When designing a product, it is necessary to carefully select the antimicrobial agent (stability, efficiency, and economy), the processing techniques (bulk modification versus thin coating layers) and the parameters (temperature stability of an additive) [44, 45].

Nowadays, antimicrobial polymer additives are available commercially. They are designed for various types of polymer matrices and processing techniques. These additives are often based on organic compounds or some metals (Ag, Zn, Cu, etc.) [5, 46, 47]. However, it should be noted that only a marginal number of them are considered for medical use.

The pros and cons of the introduced method for antimicrobial polymer modification are summarized in Table 3.

Method	Principle	Pros	Cons
a)	Modification of polymer surface properties without an antimicrobial agent	Use of chemicals is avoided	Generally ineffective, technologically demanding
b)	Direct deposition of the antimicrobial agent on the polymer surface	Cheap, simple, fast	Low efficiency
c)	Chemical deposition of the antimicrobial agent on the polymer surface	Limited amount of the antimicrobial agent is required	Technologically demanding, expensive
d)	Direct incorporation of the antimicrobial agent in the polymer matrix	Simple, possible to prepare using conventional technology	Low efficiency, high concentration of antimicrobial agent is required, limitations as regards temperature stability in thermoplastic processing

Table 3. Summary of antimicrobial modification methods

5. Evaluation of antimicrobial properties

A methodology for evaluating antimicrobial activity is necessary from the viewpoint of material safety as well as to verify the efficiency of the antimicrobial modification process.

Historically, the fundamentals of antimicrobial testing methodology were developed for the purposes of the textile industry [48]. The standards for antimicrobial testing of textiles are well formulated as a consequence. A growing interest in polymer materials and optimizing their properties for more complex applications brought about the need to modify existing standards to make them applicable to polymers as well.

5.1. Testing the antimicrobial properties of polymer-based materials

The subheading reveals that testing antimicrobial properties involves observing the biological activity of microbes (mostly bacteria and mould). This means their ability to survive an effect of a given chemical compound at a certain concentration and for a certain time period. As regards terminology, resistance to bacteria is termed antibacterial, and to mould is termed antimicotic. The general word covering all microorganisms is antibacterial. In principle, methods for evaluating antimicrobial activity can be divided into two groups: (a) static methods and (b) dynamic methods [49].

a.   Static methods for antimicrobial testing

• Microscopy (optical, electron, fluorescence) – bacteria counting, morphology observation

• Survival of bacteria detection – observing growth inhibition, determining the number of colony forming units (CFU), special marking techniques

• Biofilm evaluation

The number of colony forming units (CFU) is usually related to a certain volume or area (in the case of plastic films), and it is the most commonly used parameter in an antimicrobial testing procedure. Other indicators (however less frequently used) are connected with observing microbial metabolic activity through detection with relevant enzymatic apparatus, products of substrate metabolization, or observing consumption of another component (e.g. oxygen). A special approach can be found in a test dedicated to evaluating bacterial adhesion on selected surfaces where a special procedure for sample preparation must be applied prior to microbiological investigation. The microorganisms obtained from isolated strains or, if originating in real conditions, can be used as an inoculum.

In the case of biofilm testing via static methods, a sample after cultivation under given conditions (selected in accordance with standard or real conditions) goes through specific procedures, which involves washing and mechanical removal of irreversibly attached cells. Some research papers state various intensity levels of mechanical removal (from classical mixing to ultrasound application). Finally, the amount of removed bacteria can be simply determined by dilution and the spread plate technique. The results of such an experiment are shown in Figure 6b.

Testing antimicrobial activity on agar plates (the Agar diffusion test or the Kirby-Bauer disk-diffusion method) provides semi-quantitative information on the diffusibility of the antimicrobial agent from the specimen, and its efficiency against the given microorganism. After a chosen period of incubation, the area of bacteria growth is observed (see Fig. 6a).

Other (not as experimentally demanding as cultivation procedure) techniques for bacterial growth/inhibition detection are based on the fact that the presence of bacteria in a medium affects its properties. One of them is turbidity, which is directly proportional to the number of CFU; i.e. the quantity of CFU present in a cultivation medium can be determined by a spectroscopic method against a calibration curve [50]. Unfortunately, this considers both living and dead cells, which can significantly influence the result. A relatively new method is an ATP assay, which permits relatively rapid determination of the CFU number [51].

1.   Dynamic methods for antimicrobial testing

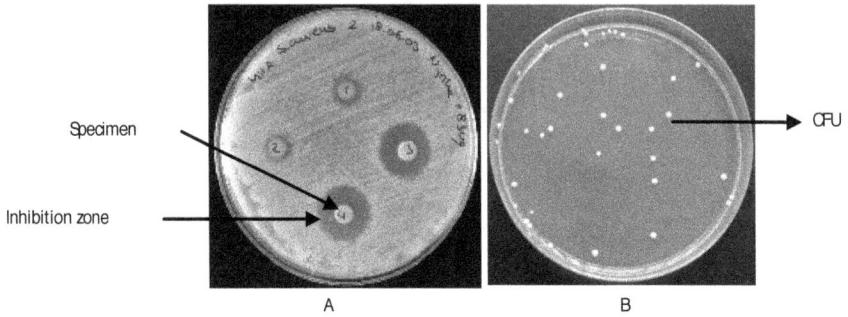

Figure 6. Outputs of antimicrobial testing (A) Agar diffusion test, (B) Dilution and spread plate technique test

This group of methods is based on measuring the flow characteristics of bacterial suspensions. The concentration of microorganisms in a given liquid medium is related to its viscosity. Dynamic methods can be divided, according the arrangement of the experiment, as follows [14]:

- Parallel plate flow system – the upper plate is made from glass while the lower part is made from the tested material. The flow of the liquid medium generates shear stress on the walls (τ_w), which can be calculated, providing values of pressure change (ΔP) and the dimensions of the channel (height (h) and width (L)) are known [52]:

$$\tau_w = \frac{\Delta P h}{2L} \tag{1}$$

- Measurement on a radial flow cell – the system consists of two concentric disks. The upper one of is made from transparent materials, while the lower one is made from the tested material. For a given flow (Q), the shear rate on the surface of the tested sample is inversely related to disk radius (r) and the gap between the discs (h) [53]:

$$S = \frac{3Q}{\pi\, r h^2} \tag{2}$$

- Measurement by rotating disc – this can be used in both the laminar and turbulent region of flow [54].

A schematic overview of antimicrobial testing methods is shown in Figure 7.

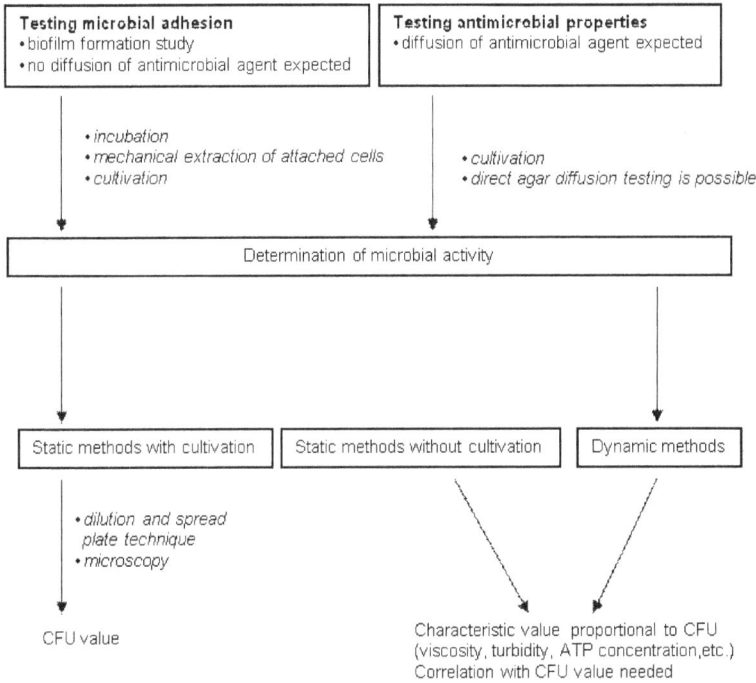

Figure 7. Schematic overview of the antimicrobial testing of polymer-based materials

5.2. Selected standards relating to the antimicrobial testing of materials

- ISO 22196:2011 "Measurement on the antibacterial activity of plastic and other non-porous surfaces"

- ASTM E2180-07 "Standard Test Method for Determining the Activity of Incorporated Antimicrobial Agent(s) In Polymeric or Hydrophobic Materials"

- ASTM E2149-10 "Standard Test Method for Determining the Antimicrobial Activity of Immobilized Antimicrobial Agents under Dynamic Contact Conditions"

- ASTM G21-09 "Standard Practice for Determining the Resistance of Synthetic Polymeric Materials to Fungi"

- ISO/CD 16256 "Clinical Laboratory Testing and In Vitro Diagnostic Systems – Reference Method for Testing the In Vitro Activity of Antimicrobial Agents Against Yeast Fungi Involved in Infectious Diseases"

- ISO 20645:2004 "Textile Fabrics -- Determination of Antibacterial Activity - Agar Diffusion Plate Test"

6. Conclusions

The bacterial colonization of surfaces, by subsequent biofilm formation on polymer-based medical devices, are the cause of serious health complications mediated by nosocomial infections. In addition, treating these infections gives rise to significant financial burden to health care systems. Applying antimicrobial polymers can noticeably reduce this negative phenomenon.

This text has provided an overview of the principles of bacterial adhesion and the biofilm formation process. General methods for antimicrobial modification of polymers have been discussed in accordance with current scientific works. Furthermore, techniques for antimicrobial properties testing have been introduced and summarized.

Research in the field of antimicrobial polymers is progressing rapidly due to intense demand for practical applications. Nevertheless, the overuse of antimicrobial agents should not occur, so as to avoid the occurrence of microbial resistance and environmental damage.

Acknowledgements

This work was financially supported by the Ministry of Education, Youth and Sports of the Czech Republic and Operational Program Research and Development for Innovations co-funded by the European Regional Development Fund (ERDF, project No. CZ. 1.05/2.1.00/03.0111).

Author details

Vladimir Sedlarik[1,2*]

1 Centre of Polymer Systems, University Institute, Tomas Bata University in Zlin, Zlin, Czech Republic

2 Polymer Centre, Faculty of Technology, Tomas Bata University in Zlin, Zlin, Czech Republic

References

[1] Scott, G. Green polymers. Polymer Degradation and Stability (2000). , 68-1.

[2] Scott, G, & Gilead, D. Degradable Polymers: Principles and Applications. Dordrecht: Kluwer Academic Publishers/Chapman and Hall; (1995).

[3] Zhang, W, Ji, J, Zhang, Y, Yan, Q, Kurmaev, E. Z, Moewes, A, Zhao, J, & Chu, P. K. Effect of NH_3, O_2 and N_2 Co-Implantation on Cu out Diffusion and Antimicrobial Properties of Copper Plasma-Implanted Polyethylene. Applied Surface Science (2007). , 2007, 253-8991.

[4] Buschow, K. H. J, Cahn, R. W, Flemings, R. C, Ilschnr, B, Kramer, E. J, & Mahajan, S. Encyclopedia of Materials- Science and Technology. Amsterdam: Elsevier; (2001).

[5] Galya, T. Antibacterial Modifications of Polymers by Using Metallic Compounds, Doctoral thesis, Zlin: Tomas Bata University in Zlin; (2009).

[6] Xu, X, Yang, Q, Wang, Y, Yu, H, Chen, X, & Jing, X. Biodegradable electrospun poly(L-lactide) fibres containing antibacterial silver nanoparticles. European Polymer Journal (2006). , 2006, 42-2081.

[7] Woo, G. L. Y, Mittelman, M. W, & Santere, J. P. Synthesis and Characterization of Novel Biodegradable and Antimicrobial Polymer. Biomaterials (2000). , 2000, 21-1235.

[8] Ducel, G, Fabry, J, & Nicolle, L. Prevention of Hospital Acquired Infections- Practical Guide. Document of World Health Organization WHO/CDS/EPH/2002.12; (2002).

[9] Pace, L. J, Rupp, M. E, & Finch, R. G. Biofilms, Infection, and Antimicrobial Therapy. Boca Raton: CRC Press Taylor&Francis Group; (2006).

[10] Marshall, K. C. Mechanisms of Bacterial Adhesion at Solid-Water Interfaces, published in Bacterial Adhesion, Savage G.C., Fletcher, M (eds) New York: Plenum Press; (1985).

[11] Wiencek, K. M, & Fletcher, M. Bacterial adhesion to hydroxyl-an Methyl-Terminated Alkanethiol Self-Assembled Monolayers. Journal of Bacteriology (1995). , 1995, 177-1959.

[12] Fletcher, M, & Loeb, G. I. Influence of Substratum Characteristics on the Attachment of a Marine Pseudomonad to Solid Surfaces. Applied and Environmental Microbiology (1979). , 1979, 37-67.

[13] Morra, M, & Cassinelli, C. Bacterial Adhesion to Polymer Surfaces: A Critical Review of Surface Thermodynamic Approaches. Journal of Biomaterial Science, Polymer Edition (1997). , 1997, 9-55.

[14] Katsikogianni, M, & Missirlis, Y. F. Concise Review of Mechanisms of Bacterial Adhesion to Biomaterials and of Techniques Estimating Bacteria Material Interactions. European Cells and Materials (2004). , 2004, 8-37.

[15] Verwey, E. J. W, & Overbeek, J. T. G. Theory of the Stability of Lyophobic Colloids. Amsterdam: Elsevier; (1948).

[16] Ratter, B. D, Hoffmann, A. S, Schoen, F. J, & Lemons, J. E. Biomaterials Science- An Introduction to Materials in Medicine. London: Elsevier Academic Press; (2004).

[17] Heilman, C, Schweitzer, O, Gerke, C, Vanittanakom, N, Mack, D, & Gotz, F. Molecular Basis of Intercellular Adhesion in the Biofilm-forming Staphylococcus epidermidis. Molecular Microbiology (1996). , 1996, 20-1083.

[18] An, Y. H, & Friedman, R. J. Cencise Review of Mechanisms of Bacterial Adhesion to Biomaterial Surfaces. Journal of Biomedical Materials Research (1998). , 1998, 43-338.

[19] Cordeno, J, Munuera, L, & Folgueira, M. D. The Influence of the Chemical Composition on Surface of the Implant Infection. Injury (1996). , 1996, 27-34.

[20] Balasz, D. J, Triandafillu, K, Chevolot, Y, Aronsson, B. O, Harms, H, Descouts, P, & Mathieu, H. J. Surface modification of PVC Endotracheal Tubes by Oxygen Glow Discharge to Reduce Bacterial Adhesion. Surface and Interface Analysis (2003). , 2003, 35-301.

[21] Vacheethasanee, K, Temenhoff, J. J, Higashi, J. M, Gary, A, Aanderson, J. M, Bayston, R, & Marchant, R. E. Bacterial Surface Properties of Clinically Isolated Staphylococcus epidermidis Strains Determine Adhesion onPolyethylene. Journal of Materials Research (1998). , 1998, 42-425.

[22] Monroe, D. Looking for Chinks in the Amour of Bacterial Biofilms. PLoS Biology (2007). , 2007, 5-2458.

[23] Anonymous Microbial Degradation of Materials and Methods of Protection, Published for the European Federation of Corrosion, The Institute of Materials, (1992). ISNB 0-901716-02-2

[24] Blashek, H. P, Wang, H. H, & Agle, M. E. Biofilms in the Food Environment. Oxford: IFT Press-Bllackwell Publishing; (2007).

[25] Wong, W. Y, & Brozino, J. D. Biomaterials. Boca Raton: CRC Press; (2007).

[26] Hin, T. S. Engineering of Materials for Biomedical Applications. New Jersey: World Scientific; (2004).

[27] Piccririllo, C, Perni, S, Cil-thomas, J, Prokopovich, P, Wilson, M, Pratten, J, & Parkin, I. P. Antimicrobial Activity of Methylene Blue Covalently Bound to a Modified Silicon Polymer Surface. Journal of Materials Science (2009). , 2009, 19-6167.

[28] Gozzelino, G, Dellaquila, A. G, & Romero, D. Hygienic Coatings by UV Curing of Diacrylic Oligomers with Added Triclosan. Journal of Coating Technology and Research (2010). , 2010, 7-167.

[29] Alvarez-Lorenzo, C, Bucio, E, Burillo, G, & Concheiro, A. Medical Devices Modified at The Surface by gamma-ray Grafting for Drug Loading and Delivery. Expert Opinion on Drug Delivery (2010). , 2010, 7 170.

[30] North, S. H, Lock, E. H, Coper, C. J, Franek, J. B, Taitt, C. R, & Walton, S. G. Plasma-Based Surface Modification of Polystyrene Microtiter Plates for Covalent Immobilization of Biomacromolecules. ACS Applied Materials & Interfaces (2010). , 2010, 2-2884.

[31] Asanovic, K, Mahailovic, T, Skundricic, P, & Simovic, L. Some Properties of Antimi-crobial Coated Knitted Textile Material Evaluation. Textile Research Journal (2010). , 2010, 80-1665.

[32] Cheruthazhekatt, S, Cermak, M, Slavicek, P, & Havel, J. Gas Plasmas and Plasma Modified Materials in Medicine. Journal of Applied Biomedicine (2010). , 2010, 8-55.

[33] Tseng, H. J, Hsu, S. H, Wu, M. W, Hsen, T. H, & Tu, P. C. Nylon Textiles Grafted with Chitosan by Open Air Plasma and their Antimicrobial Effect. Fibers and Poly-mers (2009). , 2009, 10-53.

[34] Jampala, S. N, Sarmadi, M, Somers, E. B, Wong, A. C. L, & Denes, F. S. Plasma- En-hanced Synthesis of Bactericidal Quartenary Ammonium Thin Layers on Stainless Steel and Cellulose Surfaces. Langmuir (2008). , 2008(24), 8583-8591.

[35] Gomathi, N, Surehkumar, A, Neogi, S, & Plasma-treated, R. F. Polymers for Biomedi-cam Applications. Current Science (2008). , 2008, 94-1478.

[36] Abdou, E. S, Elkholy, S. S, Elsabee, M. Z, & Mohamed, E. Inproved Antimicrobial Ac-tivity of Polypropylene and Cotton Nonwoven Fabrics by Surface Treatment and Modification with Chitosan. Journa of Applied Polymer Science (2008). , 2008, 108-2290.

[37] Zhang, W, Zhang, Y. H, Ji, J. H, Yan, Q, Huang, A. P, & Chul, P. K. Antimicrobial Polyethylene with Controlled Copper Release, Journal of Biomedical Materials Re-search A (2007). A , 838-844.

[38] Asandulesa, M, Topala, I, & Dimitrascu, N. Effect of Helium DBD Plasma Treatment on the Surface of Wood Samples. Holzforschung (2010). , 2010, 64-223.

[39] Asadinezhad, A, Novak, I, Lehocky, M, Sedlarik, V, Vesel, A, Junkar, I, Saha, P, & Chodak, I. An in Vitro Bacterial Adhesion Assesment of Surface-Modified Medical Grafe PVC. Colloids and Surfaces B- Biointerfaces (2010). , 2010, 77-245.

[40] Zhang, W, Chu, P. K, Zhang, Y, Fu, H, & Yan, R. K. Y. Q. Antibacterial properties of Plasma Modified and Triclosan or Bronopol Coated Polyethylene. Polymer (2006). , 2006, 47-931.

[41] Laporte, R. J. Hydrophilic Polymer Coatings for Medical Devices-Structure, Proper-ties, Development, Manufacture and Application. Boca Raton: CRC Press (1997).

[42] Asadinezhad, A, Novak, I, Lehocky, M, Bilek, F, Vesel, A, Junkar, I, Saha, P, & Popel-ka, A. Polysaccharides Coatings on Medical-Grade PVC: A Probe Into Surface Char-acteristics and the Extent of Bacterial Adhesion. Molecules (2010). , 2010, 15-1007.

[43] Asadineshad, A, Novak, I, Lehocky, M, Sedlarik, V, Vesel, A, Junkar, I, Saha, P, & Chodak, I. A Physicochemical Approach to render Antibacterial Surfaces on Plasma-Treated Medical-Grade PVC: Irgasan Coating. Plasma Processes and Polymers (2010). , 2010(7), 504-514.

[44] Ash, I, & Ash, M. Handbook of Preservatives. Endicott: Synapse Information Resources Inc.; (2009).

[45] Paulus, W. Directory of Herbicides for the Protection of Materials- A Handbook. Dordrecht: Springer; (2005).

[46] Radeshkumar, C, & Munstedt, H. Antimicrobial Polymers from Polypropylene/Silver Composites-Ag$^+$ Release Measured by Anode Stripping Voltametry. Reactive & Functional Polymers (2006). , 2006, 66-780.

[47] Radeshkumar, C, & Munstedt, H. Morphology and Mechanical Properties of Antimicrobial Polyamide/Silver Composites. Material Letters (2005). , 2005, 59-1945.

[48] Jakimiak, B, Rohm-rodowald, E, Stanieszewska, M, Cieslak, M, Malinowska, G, & Kaleta, A. Microbial Assessment of Efficiency of Antimicrobial Modified Textiles. Rocnik Panstwowego Zakladu Higieny (2006). , 2006, 57-177.

[49] An, Y. H, & Friedman, R. J. Laboratoory Methods for Studies of Bacterial Adhesion. Journal of Microbiological Methods (1997). , 1997, 30-141.

[50] Sutton, S. Determination of Inoculum for Microbial Testing. Microbiology Topics (2011). , 2011, 15-49.

[51] Baoshan, H, Xiaohong, L, Weiwei, Y, Aiyu, Z, Jinping, L, & Xinxia, C. Rapid detection of bacteria without cultivation with a portable bioluminescence sensor system. African Journal of Microbiology Research (2009). , 2009, 3-575.

[52] Bakker, D. P, Huijs, F. M, Devries, J, Klijnstra, J. W, Vusscher, H. H, & Van Der Mei, H. C. Bacterial deposition to fluoridated and non-fluoridated polyurethane coatings with different elastic modulus and surface tension in a parallel plate and a stagnation point flow chamber. Colloids and Surfaces B- Biointerfaces (2003). , 2003, 32-179.

[53] Dickinson, R. B, Nagel, J. A, Mcdevitt, D, Foster, T. J, Proctor, R. A, & Cooper, S. L. Quantitative comparison of clumping factor- and coagulase mediated *Staphylococcus aureus* adhesion to surface-bound fibrinogen under flow. Infection and Immunity (1995). , 1995, 63-3143.

[54] Dejong, P, Tegiffel, M. C, & Kiezebrink, E. A. Prediction of the adherence, growth and release of microorganisms in production chains. International Journal of Food Microbiology (2002). , 2002, 74-13.

Biodegradation: Microbial Behavior

Chapter 8

Biodegradation of Synthetic Detergents

Olusola Abayomi Ojo-Omoniyi

Additional information is available at the end of the chapter

1. Introduction

The synthetic detergent industry is a lucrative industry considering the world population and the need for washing and cleaning by every individual in the world. In USA detergent is often described as surfactant or syndet while in Europe the term 'tenside' (for tensio-active material) is fashionable. *The Comité International de Dérivés Tension Actifs* has agreed after extensive deliberation on the following definitions: A detergent is a product, with the formulation comprising essential constituents (surface-active agents) and subsidiary constituents (builders, boosters, fillers and auxilliaries) (that is, surface-active agents are chemicals made of hydrophilic and hydrophobic molecules, that is, amphiphyllic products). The first detergent to be utilized by man was soap, an innovation traditionally attributed to ancient Egyptian culture. However, when the word detergent is used today, it is assumed to refer to the synthetic detergents (tenside, syndets or surfactants), which have assumed increasing chemical and economic importance in the post-world war periods. The first synthetic detergents were developed by the Germans during the first world war period; they were generally referred to as Nekal. In the late nineteen-twenties and early thirties long-chained sulphonate-alcohols were sold as neutralized sodium salts. This metamorphosed to the long-chain alkyl aryl sulphonates sodium salts with benzene as the aromatic nucleus. The alkyl portion was from the kerosene fraction produced from petroleum industries. In the United Kingdom, Teepol a secondary Olefine sulphate from petrochemical sources was produced and still being produced in England and Western Europe to this day [1]. The worldwide manufacture of synthetic detergents has increased from the 1949 level of 30,000 tons to an estimated 1.5 to 2 million tons per year [2]. A significant increase in the material expectations of industrialized societies and the concomitant rise in popularity of the automatic washing machine has been the primary influence on the increased production levels demanded of the synthetic detergent industry. The failure of this increase in synthetic detergent manufacture and usage to cause a parallel increase in the detectable levels of waste detergent accumulating in various aquatic ecosystems is indirect evidence that the biodegradation of synthetic detergents occurs in nature. Moreover, due to its ease of manufacture and versatility, the alkyl benzene sulphonate (ABS) very quickly gained a foothold on the synthetic detergent market shortly after the Second World War more than half the detergents used throughout the world were based on the formulation.

Detergents are cleaning products derived from synthetic organic chemicals. The cheapness of detergent production from petrochemical sources with its ability to foam when used in acid or hard water gives it an advantage over soaps. Surfactants are the components mainly responsible for the cleaning action of detergents. In commercial detergents, the surfactant component is between 10 and 20% while the other components include bleach, filler, foam stabilizer, builders, perfume, soil-suspending agents, enzymes, dyes, optical brighteners and other materials designed to enhance the cleaning action of the surfactants[3].

A surfactant is formed when a strongly lipophyllic, hydrophobic group is bond together with a strongly hydrophilic group in the same molecule. Usually surfactants are disposed after use to sewage treatment plants (STPs). Biodegradation processes and adsorption on sludge particles remove these chemicals from wastewater to a greater or lesser extent, depending on the chemical structure of the surfactant molecule and on the operating conditions of the STP. After treatment, residual surfactants, refractory co-products and biodegradation products dissolved in STPs effluents or adsorbed on sludge are discharged into the environment. These chemicals through several transport mechanisms enter the hydro-geological cycle. Assessment of the environmental contamination levels of surfactants and related compounds is achieved through a wide range of laboratory biodegradation tests and eco-toxicological studies.

Linear alkylbenzene sulphonate (LAS) is a commonly used anionic surfactant in detergents and it is easily biodegraded than non-linear alkylbenzene sulphonate (ABS) even though, total biodegradation still requires several days. Commercial LAS also contain co-products called dialkyltetralinsulphonates (DATS) and iso-LAS. Over seventy major isomers of DATS have been detected in Commercial LAS [4]. Under aerobic conditions rate and pathway of LAS biodegradation have been studied; the primary biodegradation (biotransformation) begins with oxidation of the external methyl group (ω-oxidation) followed by stepwise shortening of the alkyl chain via oxidative cleavage of C_2 units (β-oxidation). The process leads to formation of Sulpho-phenyl carboxylic acids (SPACs) [5]. The second cycle (ultimate biodegradation or mineralization) involves opening of the aromatic ring and/or desulphonation of SPACs leading ultimately to CO_2, H_2O, inorganic salts and biomass.

It has however, been recognized that the molecular architecture of a synthetic detergent can influence its biodegradability potential. Linear alkylbenzene sulphonate (LAS) biodegradation rate and acute toxicity on aquatic life are both very much related to the chain length and phenyl position of the alkyl chain [7, 8]. The primary biodegradability of LAS has been established by the Methylene blue – Active substance (MBAS) method. Sulphonated aromatic compounds are xenobiotics and these compounds are produced in large amounts as linear alkylbenzene sulphonate (LAS). It was believed for a long time that desulphonation always occurs early in the aromatic sulphonate degradation pathway, however, it has been discovered that desulpho-nation occurs at later stages prior to the mineralization of the benzene ring. Commercial LAS is present in sewage at levels of 1- 20 mg/L and it is a mixture of many different isomers and homologues. Toluene sulphonate is an additive in some detergent formulations and as such it's released into the environment in large amounts [9, 10, 11-15].

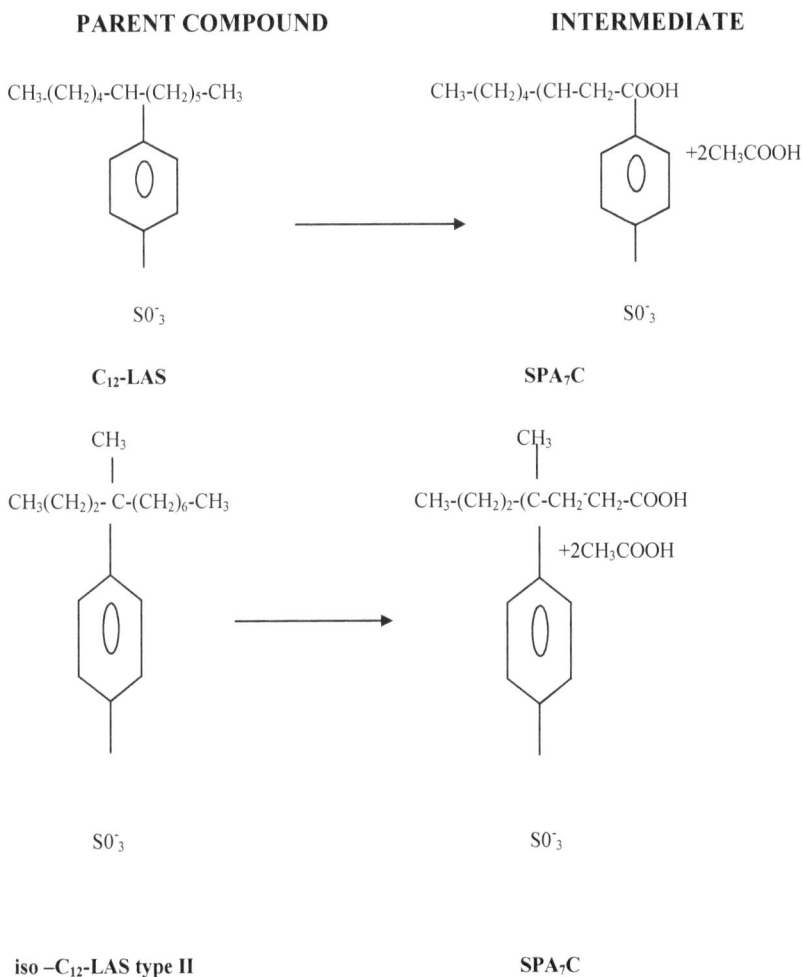

PARENT COMPOUND   **INTERMEDIATE**

$CH_3.(CH_2)_4-CH-(CH_2)_5-CH_3$ $CH_3-(CH_2)_4-(CH-CH_2-COOH$

$+2CH_3COOH$

SO_3^- SO_3^-

C_{12}-LAS   **SPA$_7$C**

CH_3 CH_3

$CH_3(CH_2)_2- C-(CH_2)_6-CH_3$ $CH_3-(CH_2)_2-(C-CH_2^-CH_2-COOH$

$+2CH_3COOH$

SO_3^- SO_3^-

iso –C_{12}-LAS type II   **SPA$_7$C**

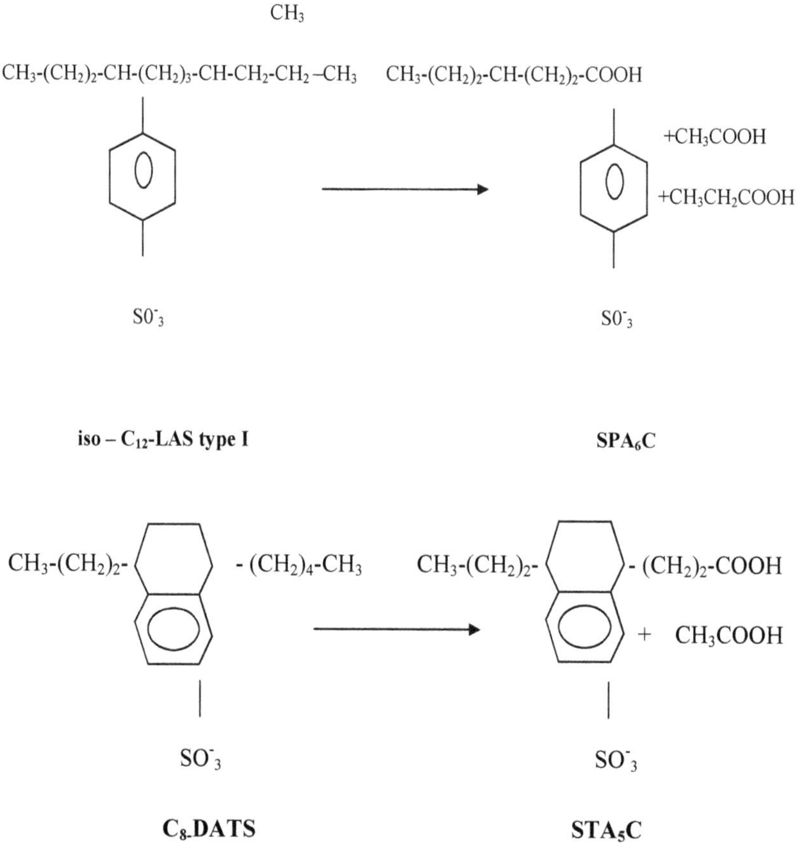

CH₃

CH₃-(CH₂)₂-CH-(CH₂)₃-CH-CH₂-CH₂–CH₃   CH₃-(CH₂)₂-CH-(CH₂)₂-COOH

+CH₃COOH

+CH₃CH₂COOH

SO⁻₃   SO⁻₃

iso – C₁₂-LAS type I   **SPA₆C**

CH₃-(CH₂)₂-   - (CH₂)₄-CH₃   CH₃-(CH₂)₂-   - (CH₂)₂-COOH

+   CH₃COOH

SO⁻₃   SO⁻₃

C₈-DATS   **STA₅C**

Figure 1. Examples of Structures and Acronyms of Linear Alkylbenzene Sulphonate (Surfactants), Co-Products and Related Catabolic Products

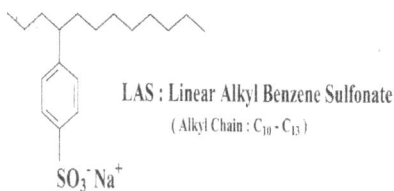

Source: [6]

Figure 2. Chemical Structure of LAS

2. Methods

The biodegradability of both domestic and industrial detergents were evaluated by collecting large numbers of representative untreated effluent samples (morning and evening) from many industrial and domestic sources, thereafter a composite wastewater sample can be generated. The determination of the physico-chemical properties of wastewater samples (morning and evening) is crucial to the outcome of the investigation and this can be done using the *Standard methods for the examination of water and wastewater* [17]. First, it is important to determine the heterotrophic microbial population in effluent samples using the Nutrient agar (bacterial) as well as Sabouraud dextrose agar (fungal). The Methylene-blue Active substance (MBAS) analysis is used to determine the percentage anionic matter (that is, surfactant component) in each of the commercial detergent product as well as in the wastewater samples. It is important to ascertain the presence of detergent components in the wastewater because this is the natural habitat of the 'detergent – degrader' microbial population. The strategic evaluation of the 'detergent – degrader' population would involve serial dilution technique, isolation of detergent-degraders on minimal salt medium supplemented with detergent at industry stipulated concentration(0.01% w/v) as well as the development of pure cultures of each microbial species that utilizes detergent components in wastewater environment.

The MBAS method is principally a titrimetric method similar to the common titration technique that identifies chemicals by neutralization process that involves Acid-Base reactions. When the end-point is reached a color change is observed. Thereafter, the concentration of the surfactant can be determined after the anionic substance is neutralized [3, 21, 22].

2.1. Aerobic heterotrophic microbial counts

The effluent samples collected from each sampling point at 0 -30cm depth were serially diluted and inoculated simultaneously onto Nutrient agar plates and Sabouraud dextrose agar in duplicates aseptically. The plates were incubated at ambient temperature for 24 - 48hrs and 7 days for fungi respectively. Thereafter, the colonies were counted [4,18,19]. The Micromorphology of both the fungal and bacterial isolates was then used as well as their biochemical reactions for their characterization and identification. This was done following standard and conventional methods [20]. *Thereafter, the schemes of Bergey's manual of systematic bacteriology and Smith's introduction to industrial mycology can be followed for the identification of bacterial and fungal isolates respectively.*

2.2. Microbial growth in wastewater spiked with detergents

The ability of native microorganisms to use detergent hydrocarbon components as sole source of carbon and energy for growth under laboratory simulated conditions suggested the biodegradation pattern in natural environment. Therefore, the composite effluent samples allowed to stand for minimum of 48hrs can be used to determine the microbial growth rate as well as the residual detergent components using Degradation – Time course graphical illustration. Wastewater samples were thereafter taken aseptically at intervals (Days; 0, 5, 10, 20, 25 and 30) to monitor the pH, total aerobic viable counts and spectrophotometry readings for growth changes. Serial dilution and spread plate methods were used to inoculate simultaneously Nutrient agar, Minimal medium and Sabouraud dextrose agar supplemented with test detergent at 0.01% (w/v), incubated at ambient temperature for 48hrs and 7 days respectively [21, 22]. This process is imperative to detect the actual utilization of detergent components for growth with time by the indigenous microbial population. Consequently, the pH of the simulated growth environment ought to change as biodegradation process proceeds.

2.3. Determination of LAS concentration using MBAS method

The titrimetric method (MBAS) is done in such a way that wastewater samples obtained from the biodegradation assay at intervals (Day 0, 5, 10, 20 and 30) were subjected to extraction procedures using organic solvent, that is, Chloroform (CCl_4).The chloroform- surfactant complex formed when the end-point is reached is indicated due to the presence of an indicator in the reaction medium. This process is repeated several times in order to extract all the detergent residues before its transfer to the Gas chromatogram(GC). The Gas chromatographic(GC)analysis detects the background LAS in wastewater as well as the residual detergent components from the biodegradation assay. Although, some other equipment such as High pressure liquefied chromatography (HPLC), spectrophotometer e.t.c can be used to monitor the ultimate biodegradation of LAS once these are calibrated with relevant standard aromatic hydrocarbon [19, 22, 23].

2.4. Gas chromatographic analysis

The gas chromatographic analyses of samples for selected time intervals Days 0, 10 and 20 were determined. The GC (Perkin Elmer Auto – System Gas Chromatography, USA) analysis of the total hydrocarbon was carried out using a GC equipped with flame ionization detector (FID). A 30m fused capillary column with internal diameter 0.25mm and 0.25m film thickness was used and the peak areas were analyzed with a SRI Model 203 Peak Simple Chromatography Data System. The column temperature was 60C for 2 minutes to 300°C programmed at 12°C/min. Nitrogen was used as carrier gas at 37 psi. Hydrogen and air flow rates were 9 psi and 13 psi respectively.

2.5. Molecular biology of detergent-degraders

2.5.1. Plasmid profile analysis

The plasmid profiles of the bacterial consortium metabolizing detergent components were evaluated purposely to use it as the basis of 'typing' for the detergent-degraders. The 'Mini-prep' plasmid isolation method was found to be inadequate because the plasmid DNA of isolated detergent-degrader consortium was very small hence they were not detectable with the 'Mini prep' method. However, the large-scale (Maxi-prep)plasmid DNA extraction method proved to be adequate for the size of plasmids encoded in this detergent-degrader microbial population (24, 25).

2.6. Curing of plasmid

This process becomes unavoidable since it is mandatory that plasmid encoded features would not be metabolically functional after curing process. The detergent-degrader bacterial population were thereafter inoculated on minimal salt medium supplemented with detergent, this is to evaluate the functionality of detergent utilization capacity of these isolates after going through curing process (25, 26).

2.7. Results

The findings of this study conducted in a tropical climate environment becomes very significant because previously this type of study were conducted in temperate climate environment. Although, similar results were obtained except for the fact that microbial detergent-degrader were observed to degrade and mineralize detergents at a slower rate than it occurred in temperate climate. This is suggestive of the consequential effect of the prevailing environmental factors in the tropics, sub-Saharan Africa. The mean physico-chemical parameter of the receiving wastewater body for the study was pre - determined following conventional and standard methods. The physico-chemical parameters (Table 1) measured revealed that the wastewater used for the study was within the mean alkaline pH range (10.54 -11.08), mesophilic temperature range (33.9 – 34.3°C), total hydrocarbon (THC) (13.6 – 15.0 mg/L) NH_4 –N content (178.7 – 193.5mg/L) and BOD (34.41 – 38.08mg/L). The physico – chemical properties of the wastewater is typical of that of a tropical environment. The organic matter component is evidently high and this is the reason for the high NH_4-N content. The oxidation of organic matter and hydrocarbon component being responsible for the high BOD.

The anionic matter (LAS) content of SDS (sodium dodecyl sulphate) was the highest of detergents used. Hence, it was used as test detergent in all GC analysis while Persil had the least LAS content (Fig. 3). Although, it should be noted that physico-chemical analysis of the composite wastewater used revealed that it was heavily polluted with organic matter; high BOD value, the COD falls short of Federal Environmental Protection Agency (FEPA), European Union (EU) and World Health Organization (WHO) standards. This was observed as one of the strong reasons for the slow rate of mineralization of xenobiotic components in the tropics.

PARAMETER	MORNING	EVENING	FEPA/WHO STANDARDS	EU STANDARDS
General appearance	Cloudy foaming	Foaming	NS	NS
Colour	Blue	Light green	NS	NS
Odour	Soapy smell	Soapy smell	NS	NS
pH(H$_2$0)	10.54	11.08	6 – 9	7.5 – 8.5
Conductivity @ 25 $^{\circ}$C	204 Usm^{-1}	185 Usm^{-1}	NS	340
Temperature	34.3 $^{\circ}$C	33.9 $^{\circ}$C	40 $^{\circ}$C	20 – 25 $^{\circ}$C
PO$_4$$^{3-}$	99.9mg/L	90.3mg/L	5mg/L	10 – 25mg/L
SO$_4$$^{2-}$	92.7mg/L	88.6mg/L	500mg/L	NS
NO$_3$$^{-1}$	26.29mg/L	22.86mg/L	20mg/L	20mg/L
Total suspended solid (TSS)	170mg/L	200mg/L	30mg/L	35mg/L
COD	57.51mg/L	52.01mg/L	200mg/L	<125mg/L
Specific gravity	1.009	1.022	NS	NS
NH$_4$ – N	193.5mg/L	178.7mg/L	NS	15mg/L
Cl^{-1}	36.18mg/L	37.95mg/L	600mg/L	600mg/L
Dissolved oxygen (DO)	9.05mg/L	9.45mg/L	"/>2mg/L	2mg/L
BOD	38.08mg/L	34.41mg/L	30mg/L	<25mg/L
Total hydrocarbon (THC)	15.0mg/L	13.6mg/L	10mg/L	<10mg/L
DO$_5$	36.04mg/L	32.67mg/L	"/>2mg/L	NS
Total dissolved solid (TDS)	NS	NS	NS	NS

NS = Not Specified

Table 1. Mean Physico-Chemical Properties of Composite Wastewater Sample

The NO$_3$ – N, SO$_4$$^{2-,}$ PO$_4$$^{3-}$, NH$_4$ – N and total hydrocarbon (THC) content of the composite wastewater used exceeds the WHO and EU limits which is suggestive of high organic chemical pollution and longer time required for mineralization to be effected (Table 1). The heterotrophic bacterial population was more than those of the fungal because the wastewater mean pH was alkaline. Hence, the microbial consortium involved in detergent degradation process had more of bacterial species than fungal. Alkalinophilic microbial consortium succeeded the acidophilic population that started the biodegradation process in an autogenic succession fashion. The alkaline pH as well as the mesophilic temperature range favored the acclimation of the indigenous microbial population to the test detergents, as well as enhanced the biodegradation efficiency of the microorganisms. This fluctuation in pH values over a 30-day period was suggestive of production of acidic and alkaline metabolites during detergent degradation as well as serves as the evidence of occurrence of chemical changes in the original formulation

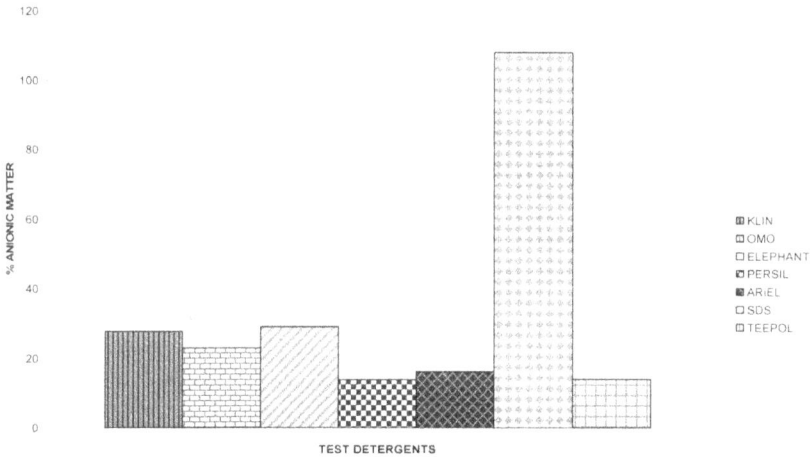

Figure 3. Determination of Anionic Matter in Test Detergents

of the test detergents (Fig. 4). The test detergents were coded as follows: AK 17 (Klin), AK 27 (OMO) AK 37 (Elephant), AK 47 (Peril), AK 57 (Ariel), AK 77 (Teepol) and AK 67 (SDS) which served as the standard detergent. *This is to facilitate unbiased analysis in the laboratory.*

This study revealed that though the test detergents contained similar formulation but the duration for the mineralization process was different from what has been previously reported, that is, 17-20 days in temperate climate. It has been reported that commercial LAS mixtures usually contain about 15% of co-products which may be responsible for differences in their rate of mineralization [28]. Major co-products of commercial mixtures of linear alkylbenzene sulphonate (LAS) surfactants are dialkyltetralinsulphonate (DATS) and methyl-branched isomers of LAS (iso-LAS). Unlike LAS, little and contrasting information on the fate of DATS and iso-LAS is available. The use of liquid chromatography/mass spectrometry (LC/ms) has confirmed that DATS were more resistant than iso-LAS to primary biodegradation. Biotrans-formation of both LAS-type compounds and DATS produced besides expected sulphophenyl alkyl monocarboxylated (SPAC) LAS and sulphotetralin alkyl carboxylated (STAC) DATS metabolites, significant amounts of sulphophenyl alkyl dicarboxylated and sulphotetralin alkyl dicarboxylated (SPADC and STADC) compounds which may be responsible for the measure of recalcitrancy found in synthetic detergents[28]. The mesophilic temperature range (25- 35°C) favored the activities of microbial detergent-degrader in tropical wastewater. The slow degradation of surfactants in natural environment may be as a result of unfavorable physico-chemical conditions (such as temperature, pH, redox potential, salinity, oxygen concentration) or the availability of other nutrients. The presence of optimal physico- chemical conditions will allow eventual evolution and growth of best-adapted microbial population to the detergent component. Although, industry specifications states that $C_{10} - C_{14}$ gives the best

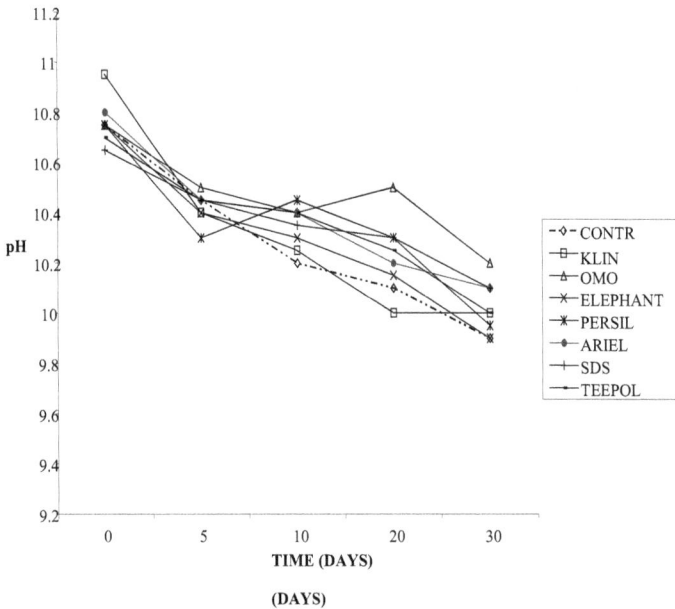

Figure 4. pH readings of primary biodegradation from field experiment (Shake-flask).

biodegradable detergent product. SDS was found to be the most rapidly biodegraded of all the test detergent products utilized for this study followed by Elephant. This is due to the fact that straight chain LAS are rapidly biodegraded than branched chain LAS. Although, SDS is a purer detergent of analytical grade often used in the laboratory with over 95% purity level while Elephant's purity level cannot be guaranteed up to 90% this was true for other commercial test detergents too. In comparisons, SDS relatively degraded faster than all the test detergents in the presence of the microbial consortium apart from the fact that it contains C_9 – C_{12}. Under 10 days, SDS (AK 67) was almost completely mineralized (except C_{17}), as at Day 20, ELEPHANT (AK 37) had components of C_{11}, C_{12} and C_{17} unmineralized (Fig. 5).Persil (AK 47) was quickly mineralized almost at the same time as SDS because of the level of purity and their molecular architecture (Fig 5).

The scientific evidence of microbial utilization of the synthetic detergent components for growth as well as source of energy is as shown in Fig.6. In less than 10 days of running the biodegradation assay, the bacterial detergent-degrader had reached the stationary phase on SDS whereas those growing on Elephant detergent got to stationary phase about 20th day. This was due to the differences in molecular architecture of the detergent being metabolized. The fungal population utilizing detergent as source of carbon and energy for growth dominated the culture vessel within the first 20 days prior to pH alteration from acidic to alkaline in an autogenic succession fashion (Fig.7). The pH of the intermediates formed also selects the microbial colonizers for detergents components, however , it should be noted that the test

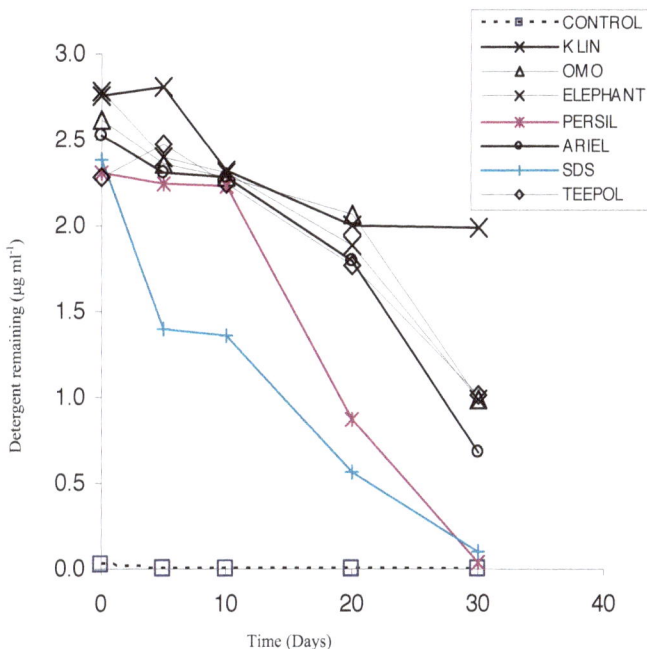

Figure 5. Biodegradation residues (shake-flask experiment).

detergents were from different manufacturers and that sometimes apart from industry stipulated additives for synthetic detergent products some manufacturers incorporates other chemicals as optical brighteners which sometimes are recalcitrant.

The GC profiles (Fig 8 and 9) represented the monitoring process for SDS degradation from the commencement (day 0) of the biodegradation assay as well as the detergent residues detected using the GC result for Day 10 (Fig. 9). Although, at Day 0, 9 peaks were detectable on sampling for SDS residues in the biodegradation assay (Fig. 8) but as at Day 10 only C17 residues peak was detectable. This showed that biodegradation was both progressive and successful. On further studies the SDS (AK 67) was completely mineralized by Day 20. This study was done for all the test deterrents and none of these test detergents were recalcitrant (19, 22). The present observation differs from the reported 17 to 20 days requirements for synthetic detergent mineralization in wastewater from temperate regions of the world. This present finding is peculiar to sub-Sahara Africa which has tropical climate. Moreover, at the end of the 30-day biodegradation assay, all test detergents were mineralized.

The second cycle (ultimate biodegradation or mineralization) involves opening of the aromatic ring and / or desulphonation of SPACs leading ultimately to CO_2, H_2O, inorganic salts and biomass formation [9, 28, 31]. It has been reported that the primary biodegradation of DATS

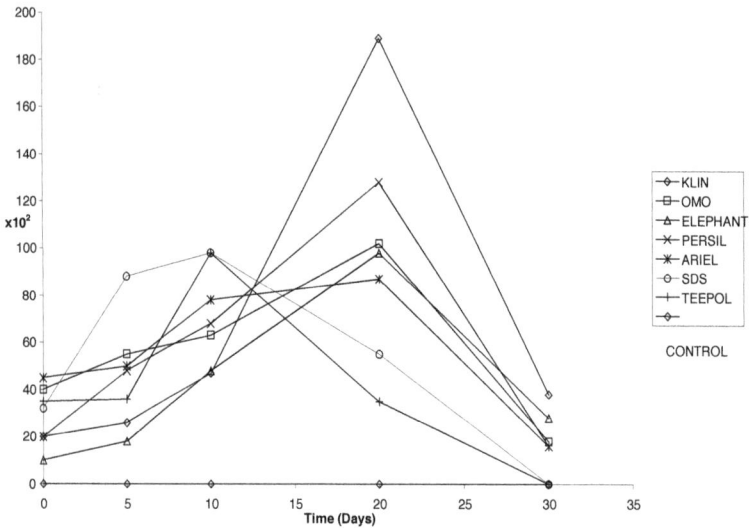

Figure 6. Mean aerobic bacteria detergent-degrader count (shake-flask experiment).

was completed by Day 17 [28]. Hence, both DATS and iso-LAS are not issues of concerns as regards their mineralization, both undergo primary biodegradation. The microbial populations of domestic and industrial activated sludge are very effective in the primary biodegradation of DATS and iso–LAS, but are not capable of mineralization of the related metabolites. However, these metabolites cannot be considered as refractory since under appropriate conditions they can be utilized as a sulphur source for bacterial growth and biomass accumulation [5, 15, 16].

These discoveries of co-products and LAS-type compounds explain the presence of some unusual peaks in the Gas chromatogram obtained in this study (Fig. 9). When the test synthetic detergents were subjected to ultimate biodegradation in laboratory simulated experiments, the native microorganisms metabolized the detergent components for growth and biomass accumulation as a result the gas chromatography was used at intervals to analyze the samples within a 30-day period. This is to monitor the transitory intermediates formed as well as to provide the convincing evidence for the mineralization of the detergent spiked into wastewater and nutrient broth respectively [18, 28].The Control was sterile water with no detergent addition as revealed by the GC analysis of the control flask (Fig. 10). Although, unusual peaks in GC profiles were detected by other researchers but it was sparsely reported [9]. Under aerobic conditions, rate and pathway of detergent biodegradation have been the objectives of many studies. Although, the biodegradation pathway of LAS has not been fully explained, there are many evidences showing that the first cycle of biotransformation (primary biodegradation) begins with

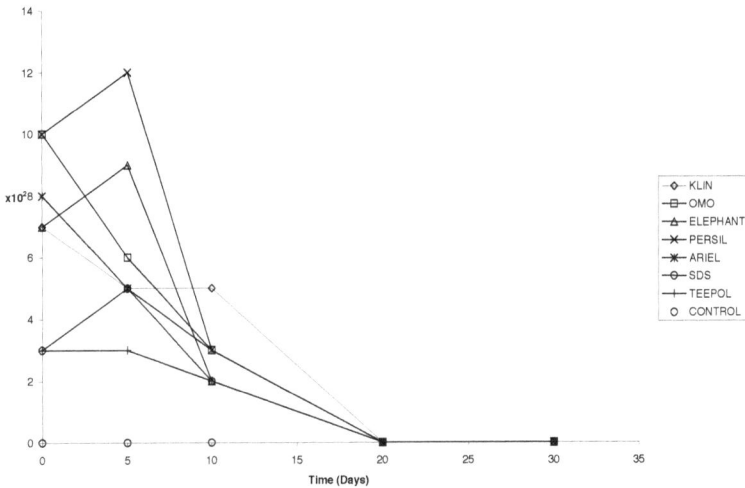

Figure 7. Mean fungal detergent-degrader count (shake-flask experiment).

oxidation of the external methyl groups (ψ- oxidation) followed by stepwise shortening of the alkyl chain via oxidative cleavage of C_2 units (β - oxidation) [28, 31].

The microbial isolates both from the shake-flask studies and sewage treatment plant (STP) utilizing the test detergents as C and energy sources include; *Enterococcus majodoratus, Klebsiella liquefasciens, Enterobacter liquefasciens, Klebsiella aerogenes, Escherichia coli, Enterobacter agglomerans, Staphylococcus albus, Pseudomonas aeruginosa, Myceliophthora thermophila, Geomyces sp., Alternaria alternata, Aspergillus flavus, A. oryzae* and *Trichoderma sp.* The other bacterial species for which their identity could not be ascertained were those that would require high through-put DNA technology to confirm their identity. The rate of multiplication of bacterial cells was highest *for Pseudomonas aeruginosa* followed by *Klebsiella aerogenes*, and *Enterobacter agglomerans*, then *E. coli* in this descending order. The bacterial consortium stabilized and reached the stationary growth phase at Day 10 for the microcosm experiment (Fig. 6). The fungal detergent utilizers were characterized following the methods of *Smith's Introduction to mycology (Smith, 1981)*. Comparative study of the mycelia features as well as microscopy was used to identify the fungal species. The fungal population could not tolerate the rise in pH value towards the alkaline pH range, hence their elimination. The fungal detergent-degrader population started decreasing after Day 10 (Fig.7). When *E. coli, Ps. aeruginosa, K. aerogenes, S. albus* and *K. liquefasciens* strains isolated from diseased-patients were inoculated onto detergent medium they utilized detergent component as C and energy source for growth. The production of detergent catabolic enzymes thus seems to be inductive by this observation.

Sample: AK67 Day 0
Comments: ALIPHATICS HYDROCARBON PROFILE FOR SAMPLE: AK67 Day 0

Temperature program:

Init temp	Hold	Ramp	Final temp
60.00	2.000	12.000	300.00

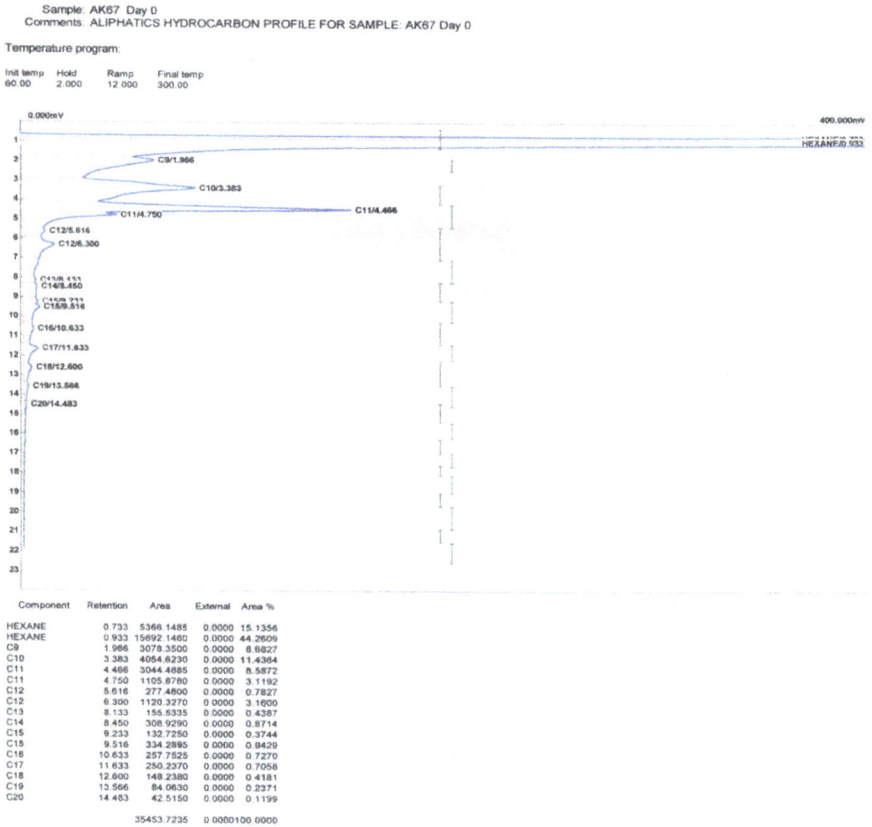

Component	Retention	Area	External	Area %
HEXANE	0.733	5366.1485	0.0000	15.1356
HEXANE	0.933	15692.1460	0.0000	44.2606
C9	1.966	3078.3500	0.0000	8.6827
C10	3.383	4054.6230	0.0000	11.4364
C11	4.466	3044.4885	0.0000	8.5872
C11	4.750	1105.8780	0.0000	3.1192
C12	5.616	277.4800	0.0000	0.7827
C12	6.300	1120.3270	0.0000	3.1600
C13	8.133	155.5335	0.0000	0.4387
C14	8.450	308.9290	0.0000	0.8714
C15	9.233	132.7250	0.0000	0.3744
C15	9.516	334.2895	0.0000	0.9429
C16	10.633	257.7525	0.0000	0.7270
C17	11.633	250.2370	0.0000	0.7058
C18	12.600	148.2380	0.0000	0.4181
C19	13.566	84.0630	0.0000	0.2371
C20	14.483	42.5150	0.0000	0.1199
		35453.7235	0.0000	100.0000

Figure 8. GC profile showing peaks of unmineralized detergent component (Day 0)

The GC profiles thus reveal that the actual process of detergent degradation truly occurs in natural environment (Fig. 8, 9 and10). The presence of residues represented by the unusual peaks was identifiable. This is due to activities of some detergent manufacturers who incorporated unwholesome chemicals in synthetic detergents to enhance detergency as well as aid optical brighteners in detergents. This unwholesome practice is common in sub-Sahara Africa as revealed in research works (19, 22) these inclusions are contrary to regulatory body specifications.

This observation lays to rest the issue of recalcitrancy of domestic and industrial detergents in natural environment. Desulphonation process is the rate limiting process and this has been shown to have taken place by microbial activities in natural environment.

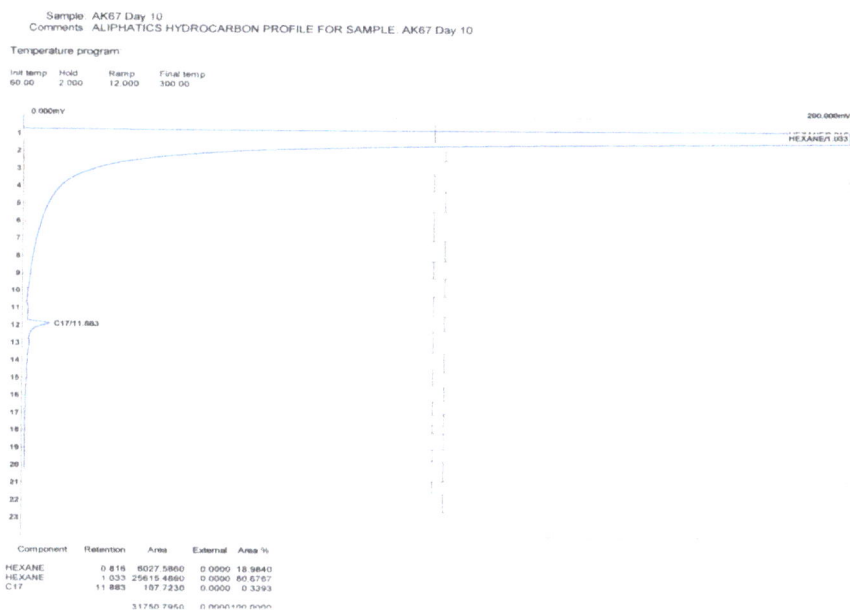

Figure 9. GC profile showing a single peak of detergent residues (C_{17}) Day 10

3. Molecular biology of detergent-degraders

The plasmid-profile of selected bacteria detergent-degraders was successfully detected with the Maxi-prep method. It was evident that the genetic information for detergent-hydrocarbon utilization was not plasmid - mediated since the cured isolates grew on detergent supplemented medium after plasmid was removed. Hence, the genes for detergent hydrocarbon utilization can be said to be resident within the genome of the selected bacterial population [19].

The ability to utilize detergent components as carbon and energy source for growth and biomass accumulation suggested some peculiarity in the genome of microorganisms with this trait. Since not all microorganisms have the ability, particularly acclimation to xenobiotic compounds. Plasmid DNA-coded character often plays significant role in bacteria adaptation to xenobiotics in the environment. However, the convincing evidence on the genetic linkage of the catabolic trait would be found using high through-put techniques such as the PCR and DNA sequencing to trace the nucleotide sequence encoding for detergent – catabolism which may be resident in the plasmid – DNA or the genomic DNA. The genetic diversity among xenobiotic-degrader has been confirmed by several researchers [32]. The recently developed high through-put technologies such as construction of metagenomic libraries and DNA array technology would authenticate as well as enhance the knowledge of the identity of detergent-

Sample: Control Day 0
Comments: ALIPHATICS HYDROCARBON PROFILE FOR SAMPLE : Control Day 0

Temperature program:

Init temp	Hold	Ramp	Final temp
60.00	2.000	12.000	300.00

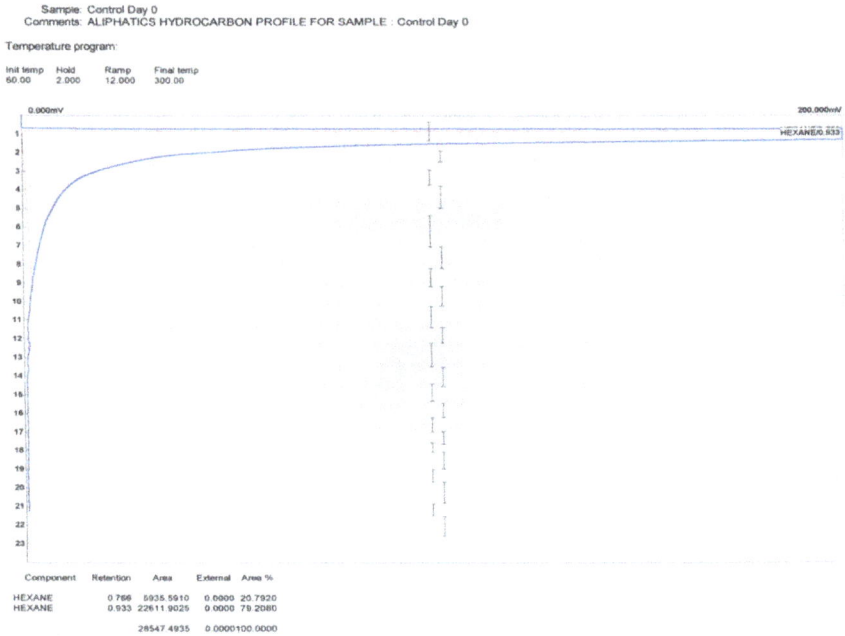

Component	Retention	Area	External	Area %
HEXANE	0.766	5935.5910	0.0000	20.7920
HEXANE	0.933	22611.9025	0.0000	79.2080
		28547.4935	0.0000	100.0000

Figure 10. GC profile of control experiment with no peak due to absence of detergent component (Day 0)

degrader genes with greater precision [32, 33]. These new technologies involve DNA hybrid-ization via construction of DNA probes which has been described as being a relatively rapid method of analysis compared to PCR [34]. Recent advances in molecular techniques, including high through-put approaches such as DNA microarrays and metagenomic libraries have opened up new perspectives towards new opportunities in pollution abatement and environ-ment management. [34]. Compared with traditional molecular techniques which are depend-ent on the isolation of pure cultures in the laboratory. DNA microarrays and metagenomic libraries allows the detection of many uncultivable and uncharacterized xenobiotic-degraders, thus enhancing the knowledge of the genetic diversity of environmentally relevant microor-ganisms.

Most countries are upgrading their effluent treatment plant to Membrane Bioreactor Technol-ogy (MBR) which improves the quality of domestic sewage and wastewater discharged without increasing the plant foot-print [29]. This MBR has a single line designed to handle effluent flow of 1000 fold/day more than that of conventional STPs. The up-graded process design increases the quality of discharged effluent to satisfy consent levels and achieve effluent of unrestricted irrigation re-use standard [27, 29]. The compact size and efficient operation of this advanced process allowed countries such as Japan, Breschia to achieve a higher quality discharged- effluent without expanding the plant's footprint [27].

The above strategic and holistic approach would enhance sustainable development as well as effluent quality when strictly adhered to. When a sub-lethal environment change is imposed on a mixed microbial population, two possible mechanisms of shift in the population may occur; (i) the predominance of the population may shift to those species whose activities are enhanced by the change; and (ii) those species of the population having the metabolic capability may acclimatize to the change.

The presently available information from this study suggested that microorganisms may eventually be able to deal with any kind of organic compound, particularly synthetic detergent provided that the compounds are intrinsically degradable. The surfactant component (LAS) supposedly recalcitrant is biodegradable under optimum environmental conditions. The following deductions can be made from this study;

- The biodegradation process was favored at the mesophilic temperature range 25 - 30^{0}C

- Alkaline pH was predominant in the field (STP) as well as in the microcosm studies.

- The pH of the wastewater ecosystem dictated the type of microbial heterotrophs that metabolized the detergent components.

- Although acidic pH selected the fungal detergent-degraders they were only able to tolerate the pH in the culture media spiked with detergent up to pH 7.0. This was responsible for the disappearance of fungal isolates after Day 10 while conducting the laboratory simulated detergent degradation.

- The NO_3-N in the composite wastewater was too high compared to newly legislated European Union environmental standard limits of 3mg/L for treated wastewater. The problem of eutrophication is inevitable and this would reduce the volume of oxygen (BOD) required for biodegradation process which may be responsible for the observed length of time taken to mineralize the test detergents in tropical environment as opposed to the 17 – 20-day often reported in temperate climates.

- Mineralization of LAS is true; hence its inclusion in synthetic detergent formulation poses no public health or aquatic life threats.

- Gas chromatographic analysis of biodegradation residues provided convincing evidence of the mineralization of the surfactant component of detergent products which was said to be recalcitrant.

- The result of this study showed the presence of C_{17} – C_{21} hydrocarbon component in detergent formulation, signifying some measures of inconsistency on the part of detergent manufacturers who might have incorporated other organic chemicals in detergent products to enhance their performance (that is, fluorescent optical brighteners) at the expense of public and aquatic life safety.

- Although aquatic toxicity of detergent products have been reported, this might be due to unwholesome additives incorporated (C_{17}–C_{21}) in the product which takes longer time than expected for it to be mineralized in STPs and open-river. These unwholesome additives include 4, 4,-bis (2-sulphostyryl) biphenyl and toluene sulphonate which are typical representative of the stilbene class of fluorescent optical brighteners.

• This study has shown that microorganisms possesses a range of genetic mechanisms allowing evolutionary changes in existing metabolic pathways, specialized enzyme systems and pathways for the microorganisms for degradation of xenobiotics such as LAS or synthetic detergents which have been found in geographically separated areas of the biosphere.

• Detergent – degraders were principally bacteria of Gram-negative group suggesting their ability to tolerate the surfactant than the Gram-positive bacteria.

• Plasmid profile analysis also revealed the presence of single plasmid in all the bacterial isolates but plasmid – DNA may not be absolutely responsible for the detergent–hydrocarbon utilization trait.

• Future studies with PCR and DNA sequence analysis would reveal the DNA' fingerprint' of each of the 'detergent-degraders' species. This would enhance the processes of surveillance for these organisms in similar ecosystems and the detection of new serotypes as well as assist in environmental impact assessment (EIA) study for sustainable development.

The question whether enzymes specialized for the degradation of xenobiotics in these bacteria evolved from more common isozymes only after the large-scale introduction of xenobiotic chemicals into the environment can now be answered with the recent advances in molecular techniques, including approaches such as DNA microarrays and metagenomic libraries.

This study has given insight into the molecular events and regulatory mechanisms responsible for evolution of detergent catabolic trait, and that the presence of a selective pressure of a specific substance or a toxic chemical (LAS) can spur microbes to metabolize such xenobiotics.

The current study summarized clearly evidences of the potential of microorganisms to evolve new and desired biochemical pathways themselves after surviving natural selection. The introduction of LAS (C_{10}-C_{14}) into detergent formulation as the principal surfactant component is thus environment friendly and would enhance sustainable development processes.

Acknowledgements

I am particularly indebted to Prof. B. A. Oso for his inestimable support, guidance and fatherly advice always. I also appreciate Mrs. Oluremi O. Adeyemo who typed my manuscripts. My sincere gratitude goes to all my Professors for their assistance when help was sought from them: O.O. Amund, G.C. Ukpokwasili, B.O. Elemo, N. A. Olasupo, O.A. Bamgboye, A.I. Sanni, O. E. Fagade and S. Nokoe. My sincere thanks to the entire staff of Nigerian Institute of Medical Research (NIMR) particularly, Drs. S. Smith, P. Agomo, Audu, and Niemogha. I am grateful to my spiritual coach, my wife and my daughter for their love and understanding.

Author details

Olusola Abayomi Ojo-Omoniyi*

Address all correspondence to: Prof.olusola2012@gmail.com solayom@yahoo.com

Department of Microbiology, Faculty of science, Lagos state university, Ojo. Lagos, Nigeria

References

[1] Davidsohn AS, Milwidsky BM. Synthetic Detergents (5th Ed.). International Textbook Company Ltd., Buckingham Palace Rd., London. 1972. Pp. 1 – 150.

[2] De Wolf W, Feijtel, T. Terrestrial risk assessment for linear alkylbenzene sulfonate (LAS) in sludge-amended soils. *Chemosphere* 1998; 36: 1319 – 1343.

[3] Okpokwasili GO, Nwabuzor CN. Primary biodegradation of anionic surfactants in laundry detergents. *Chemosphere* 1988; 17: 2175 – 2182.

[4] Trehy ML, Gledhill WE, Orth RG. Determination of linear alkylbenzene sulfonates and dialkyltetralinsulfonates in water and sediment by gas chromatography/mass spectrometry. *Analytical Chemistry* 1990; 62: 2581 – 2586.

[5] Cook AM. Sulfonated surfactants and related compounds: facets of their desulfonation by aerobic and anaerobic bacteria. *Tenside Surfactants Detergent* 1998; 35: 52 – 56.

[6] Cavalli L, Clerici R, Radici P, Valtorta L. Update on LAB / LAS. *Tenside Surfactants Detergents* 1999; 36: 254-258.

[7] Nomura Y, Ikebukuro K, Yokoyama K, Takeuchi T, Arikawa Y, Ohno S, Karube I. Application of a linear alkylbenzene sulfonate biosensor to river water monitoring. *Biosensors & Bioelectronics* 1998; 13: 1047 – 1053.

[8] Swisher RD. Biodegradation of ABS in relation to chemical structure. *Journal of Water Pollution Control Federation* 1963. 35: 877 – 892.

[9] Kertesz MA, Kolbener P, Stocinger H, Beil S, Cook AM. Desulfonation of Linear al-kylbenezene sulfonate Surfactants and related compounds by Bacteria. *Applied and Environmental Microbiology* 1994; 60(7): 2296 – 2303.

[10] Giger W, Alder AC, Brunner PH, Marcomini A, Sigrist H. Behaviour of LAS in sewage and sludge treatment and in sludge-treated soil. *Tenside surfactants Detergent* 1989; 26: 95 – 100.

[11] Poiger T, Field JA, Field TM, Giger W. Determination of detergent- derived fluorescent whitening agents in sewage sludges by liquid chromatography. *Analytical Methods and Instrumentation* 1993; 1: 104 – 113.

[12] Chung KT, Stevens SE, Jr. Degradation of azodyes by environmental microorganisms and helminths. *Environmental Toxicology and Chemistry* 1993; 12: 2121 – 2132.

[13] Feigel BJ, Knackmuss HJ. Syntrophic interactions during degradation of 4-aminobenzene sulfonic acid by a two species bacterial culture. *Archives of Microbiology* 1993; 159: 124 – 130.

[14] Schoberl P. Basic principles of LAS biodegradation. *Tenside Surfactants Detergent* 1989; 26(2): 86 – 94.

[15] Sigoillot JC, Nguyen MH. Complete oxidation of linear alkylbenzene sulfonate by bacterial communities selected from coastal seawater. *Applied and Environmental Microbiology* 1992; 58(4): 1308 – 1312.

[16] Zurrer D, Cook AM, Leisinger T. Microbial desulfonation of substituted naphthalene sulfonic acids and benzene sulfonic acids. *Applied and Environmental Microbiology* 1987; 53: 1459 – 1463.

[17] American Public Health Association. Standard Methods for the examination of water and wastewater, 18th Ed. American public Health Association (APHA), 1992. Washington, D.C.

[18] Larson RJ, Payne AG. Fate of the Benzene Ring of LAS in Natural waters. *Applied and Environmental Microbiology* 1981; 41(3): 626 – 627.

[19] OJO OA, OSO BA. Isolation of Plasmid – DNA from Synthetic detergent degraders in wastewater from a tropical environment. *African Journal of Microbiological Research* (AJMR) Nairobi, Kenya. 2009a; 3 (3): 123 – 127.

[20] Gerhardt P, Murray RGE, Costilow RN, Nester EW, Wood WA, Krieg N R, Phillips GB. Preservation In: *Manual of Methods of General Bacteriology. ASM Washington*, DC 1981. Pp. 208 – 210, 435.

[21] Okpokwasili GO, Olisa AO. River water biodegradability of surfactants in liquid detergent and shampoos. *Water Research* 1991; 25: 1425 – 1429.

[22]) OJO OA, OSO BA. Biodegradation of synthetic detergents in wastewater Southwest, Nigeria. *African Journal of Biotechnology (AJB)* Nairobi, Kenya. 2009; 8(6) : 1090 – 1109.

[23] Sullivan WT, Swisher RD. MBAS and LAS surfactants in the Illinois River. *Environmental Science and Technology* 1969; 3: 481 – 483.

[24] Zhou C, Yang Y, Yong AY. 'Mini-prep' in ten minutes *Biotechniques* 1990; 8(2): 172 – 173.

[25] Bhalakia N. Isolation and plasmid analysis of vancomycin-resistant *Staphylococcus aureus. Journal of Young Investigators* September, 2006. 15(4): 15 – 24.

[26] Ahrne S, Molin G, Stahl S. Plasmids in *Lactobacillus* strains isolated from meat and meat products. *Systematics and Applied Microbiology*, 1989; 11: 320 – 325.

[27] Water and Wastewater International (WWI). Sharing Risks and Rewards *alliance contracts profit desalination projects*. In: *MBR helps Breschia comply with EU regulations*. PennWell Publ. Ltd. (UK.) www.wwinternational.com 2005; 20(4): 23 – 25.

[28] Di Corcia A, Casassa F, Crescenzi C, Marcomini A, Samperi R. Linear alkylbenzene sulfonate chemical structure and bioavailability. Environmental Science and Technology 1999a. 33: 4112 – 4118.

[29] Water and Wastewater International (WWI). Reclamation sequences water from Australia under drought. In: Re-designed treatment System to improve industrial effluent quality. Pennwell Publ. Ltd. (UK.). www.wwinternational.com 2006; 20(9): 24.

[30] Maniatis T, Fritsch EJ, Sambrook EJ. Molecular Cloning: A laboratory manual. Cold Spring Harbor Lab. 1982; N.Y.

[31] Schleheck D, Lechner M, Schonenberger R, Suter MJF, Cook AM..Desulfonation and Degradation of the Disulfodiphenylether carboxylates from Linear alkylphenylether disulfonate surfactants. *Applied and Environmental Microbiology* 2003; 69(2): 938 – 944.

[32] Fessehaie A, De Boer SH,Le'vesque CA. An oligonucleotide array for the identification and differentiation of bacteria pathogenic on potato. *Phytopathology* 2003; 93: 262 – 269.

[33] Heiss G, Trachtmann N, Abe Y, Takeo M, Knackmuss H. Homologous *npdGI* genes in 2, 4-Dinitrophenol-and 4-Nitrophenol-Degrading *Rhodococcus spp*. *Applied Environmental Microbiology* 2003; 69(5): 2748 – 2754.

[34] Eyers L, George I, Schuler L, Stenuit B, Agathos SN, El- Fantroussi S. Environmental genomics: exploring the unmined richness of microbes to degrade xenobiotics. *Applied Microbiology and Biotechnology* 2004; 66: 123 – 130.

Chapter 9

Insecticide Resistence of Bumblebee Species

Marie Zarevúcka

Additional information is available at the end of the chapter

1. Introduction

Bumblebees are important pollinators of many crops and wild flowers and there are both conservation and economic reasons for taking action to assess the impact of pesticides on bumblebees. Pesticide risk assessments for honeybees are based on hazard ratios which rely on application rates and toxicity data and are unlikely to be appropriate for bumblebees. Bumblebees are active at different times and on different crop species and are, therefore, likely to have different exposure profiles. Unlike honeybees, deaths of bumblebees due to pesticides are unlikely to be reported, since the bees are not kept domestically and will die in small numbers. During the last decades side-effects of insecticides on bees have gained great attention due to their value as pollinators. In Europe insecticides are tested following the EPPO (European and Mediterranean Plant Protection Organization) guidelines to exclude any harm to honeybees *Apis mellifera*. Unfortunately there has been a decline in the abundance of bumblebees (*Bombus* sp., Apidae) in many countries, and it is possible that this is due in part to the use of certain pesticides.

Mechanisms of insecticide resistance found in insects may include three general categories. Modified behavioral mechanisms can let the insects avoid the exposure to toxic compounds. The second category is physiological mechanisms such as altered penetration, rapid excretion, lower rate transportation, or increase storage of insecticides by insects. The third category relies on biochemical mechanisms including the insensivity of target sites to insecticides and enhanced detoxification rate by several detoxifying mechanisms [1]. The representative detoxifying enzymes are general esterases and monooxygenases that catalyse the toxic compounds to be more water-soluble forms and the secondary metabolism is followed by conjugation reactions including those catalysed by glutathion S transferases.

Insecticide resistance is the development by some insects in a population of an ability to survive doses of a toxicant which would prove lethal to the majority of individuals in a normal population of the same species [2]. Insecticide resistance has serious consequences, such as outright control failure, increased application rates, decreased yields, environmental contamination and wildlife and natural enemy destruction [3].

2. Effect of insecticides

Insecticides affect insect behavior, such as reducing movements and affecting feeding levels. To date risks assessment studies conducting the side-effects of conventional insecticides are mostly limited to acute toxicity studies. Overall, when considering conventional insecticides it is remarkable that several litle of all compounds included (n=63) was considered as non-toxic (Figure 1)

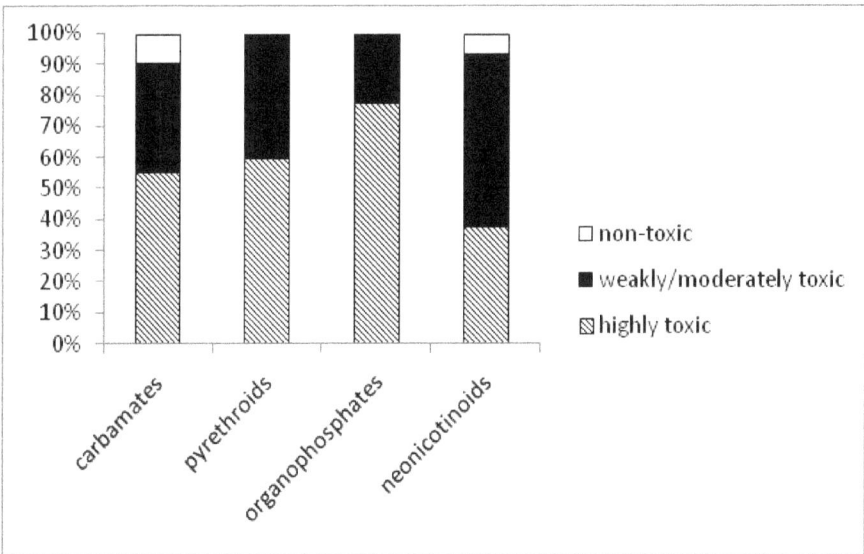

Figure 1. Overview of the toxicity of chemical insecticides towards bumblebees (*Bombus terrestris*). For each insecticide group the bars represent the percentage of compounds which are non-toxic, weakly/moderately toxic and toxic. The numbers of compounds considered per group are n=11 for carbamates, n=14 for pyrethroids, n=32 for organophosphates, n=6 for neonicotinoids [4].

Cases where insecticides have altered insect behavior have only been reported for imidacloprid (systemic neonicotinoid and insect neurotoxin). Perhaps imidacloprid has been the most investigated insecticide because for a long time it was considered to be relatively safe to non-target organisms, and thus was desired as a companion to biological control systems in integrated pest management systems. They may not only affect pest insects but also non-target organisms such as pollinators. Three different key aspects determining the risks of neonicotinoid concentrations for bee populations: (1) the environmental neonicotinoid residue levels in plants, bees and bee products in relation to pesticide application, (2) the reported side-effects with special attention for sublethal effects, and (3) the usefulness for the evaluation of neonicotinoids of an already existing risk assessment scheme for systemic compounds [5].

Imidacloprid probably migrate into nectar and pollen, then modify flower attractiveness, homing behavior, and colony development [6]. It was concluded that applying imidacloprid at the registered dose did not significantly affect the foraging and homing behavior of B. terestris and its colony development. The toxicity of imidacloprid to bees differs from most insecticides in that it is more toxic orally than by contact.

The imidacloprid was exposed adult worker bumble bees, Bombus terrestris, and honey bees, Apis mellifera L., to dietary in feeder syrup. Honey bees showed no response to dietary imidacloprid on any variable that was measured (feeding, locomotion and longevity) [7]. In contrast, bumble bees progressively developed over time a dose-dependent reduction in feeding rate but neither their locomotory activity nor longevity varied with diet. The honey bees are better pre-adapted than bumble bees to feed on nectars containing synthetic alkaloids, such as imidacloprid, by virtue of their ancestral adaptation to tropical nectars in which natural alkaloids are prevalent.

Neonicotinoid insecticides have been implicated in these declines because they occur at trace levels in the nectar and pollen of crop plants [8]. Colonies of the bumble bee Bombus terrestris in the laboratory were exposed to field-realistic levels of the neonicotinoid imidacloprid, then allowed them to develop naturally under field conditions. Treated colonies had a significantly reduced growth rate and suffered an 85% reduction in production of new queens compared with control colonies. The residual effects of insecticide treatments on colony vitality and behavior of the bumble bees Bombus impatiens foraging on turf containing white clover, Trifolium repens imidacloprid, was tested. The insecticide was applied as granules, followed by irrigation, or sprayed as a wettable powder, with or without irrigation [9] imidacloprid granules, and imidacloprid sprays applied with posttreatment irrigation, had no effect on colony vitality or workers' behavior, suggesting that such treatments pose little systemic or residual hazard to bumble bees. The direct contact toxicity of five technical grade insecticides-imidacloprid, clothianidin, deltamethrin, spinosad, and novaluron-was investigated on bees that may forage in canola: common eastern bumble bees (Bombus impatiens), alfalfa leafcutting bees (Megachile rotundata), and Osmia lignaria [10] Clothianidin and lesser extent imidacloprid were highly toxic to all three species, deltamethrin and spinosad were intermediate in toxicity, and novaluron was nontoxic. Bumble bees were generally more tolerant to the direct contact applications in comparison to leafcutting bees (Table 1).

The insecticide Teppeki can be selective about bumble and have a good compatibility with the activity of the apiaries [11]. This insecticide has the active ingredient flonicamid belonging to a new chemical class, called pyridinecarboxamides: the product has systemic effect and is known as having a long lasting efficacy against all important aphid species The flonicamid has a minor effect of interference with the activity of pollination by *B. terrestris*, compared to the standard used. Treatment with Teppeki has not given any acute effect on *B. terrestris*, nor any effect of interference in respect of its pollination activity.

insecticide	LC50 (% solution x 10^{-3})	Pesticide mode of activity
Clothianidin	0.39	act on the central nervous system of insects as an agonist of acetylcholine
Imidacloprid	3.22	central nervous system, acetylcholine agonist
Deltamethrin	6.90	due to irreversible damage to the nervous systém of insect
Spinosad	8.95[a]	binding sites on nicotinic acetylcholine receptors (nAChRs) of the insect nervous system, binding leads to disruption of acetylcholine neurotransmission
Novaluron	>100[a]	insect growth regulator that disrupts cuticle formation and prevents molting

[a]No mortality at the highest concentration tested (0.1% solution).

Table 1. Direct contact toxicity of technical grade insecticides after 48 h to bumble bees [10]

As has been already shown pesticide risk assessments are routinely carried out for honeybees, but the results of these are probably not directly applicable to bumblebees [12]. Pyrethroids are commonly applied to flowering oilseed rape in the early morning or evening, when bumblebees are often active. Laboratory and field-based bioassays appropriate to bumblebees have been developed in response to the growing use of bumblebees for the pollination of greenhouse crops, but these are not widely used and few toxicological data are available [12]. Almost all tests conducted so far have been on *B. terrestris*, and suggest that toxicity is similar to that found in honeybees. Tests with dimethoate and carbofuran suggest that these chemicals are selectively transported into the nectar where they can reach high concentrations [13]. When colonies are large it is likely that they can tolerate the loss of some of their workers. However, in the spring when queens are foraging, and subsequently when nests are small and contain just a few workers, mortality may have a more significant effect [14].

Despite risk assessments, widespread poisoning of honeybees has been reported [15]. Such effects are obvious in domestic hives where dead bees are ejected and form piles by the nest. It seems probable that pesticides would have similar effects on bumblebees but they are unlikely to be noticed in most situations. In Canada, the use of the insecticide fenitrothion in forests led to a decline in yield of nearby *Vaccinium* crops due to a reduction in abundance of bumblebee pollinators [16]. In the UK, bumblebee deaths have been reported following applications of dimethoate or alphacypermethrin to flowering oilseed rape, and of λ-cyhalothrin to field beans [12,14].

A growing appreciation of the damaging effects of broad-spectrum pesticides has led to the development of a new generation of more target-specific compounds. EU, US and Canadian law now demand that oral and acute toxicity tests are carried out on honeybees prior to the registration of any new pesticide [17]. However, there is no obligation to study sub-lethal effects on any bees, or to look at specific effects on bumblebees. Some of these substances cause no mortality in bumblebees if used appropriately [18,19]), but there is evidence that supplementary trials for non-lethal effects are necessary.

The spinosad is a commonly used insect neurotoxin which, based on studies of honeybees, has been deemed harmless to bees. However, it has recently been shown that bumblebee larvae fed with pollen containing this pesticide give rise to workers with reduced foraging efficiency [20]. Mommaerts et al. [21] screened eight chitin synthesis inhibitors currently registered as pesticides and found that although no lethal effect could be found on adults, the use of these pesticides has strong effects on colony growth and the development of larvae. Diflubenzuron and teflubenzuron were found to be the most harmful to bumblebees, greatly reducing reproductive output at concentrations far below the recommended field concentrations.

The effects of a naturally derived biopesticide, spinosad, was tested on bumble bee (*Bombus impatiens*) colony health, including adult mortality, brood development, weights of emerging bees and foraging efficiency of adults that underwent larval development during exposure to spinosad [20]. The colonies from an early stage was monitored, over a 10-week period, and fed spinosad to colonies in pollen at different concentrations during weeks The minimal negative effects to bumble bee colonies was detected at small concentrations of spinosad.

The toxicity of this Pyridalyl an insecticide of a novel chemical class (unclassified insecticides) to the pollinating insect *Bombus terrestris*, was evaluated using the body-dipping method or direct spray method [22]. No acute toxicity of this product was observed on these non-target insects. Moreover, the influence of pyridalyl to the nest of *Bombus terrestris* was evaluated using the direct spray to the inside of the nest. No apparent influence of this compound was observed by 21 days after treatment.

Gradish et al., determined the lethal and sub-lethal effects of four insecticides (imidacloprid, abamectin, metaflumizone and chlorantraniliprole) tested for use in greenhouse vegetable production to *B. impatiens* [23]. Imidacloprid, abamectin, and metaflumizone were harmful to worker bees following direct contact, while chlorantraniliprole was harmless. Worker bees fed imidacloprid-contaminated pollen had shortened life and were unable to produce brood. Worker bees consumed less pollen contaminated with abamectin. The new reduced risk insecticides metaflumizone and chlorantraniliprole are safe for greenhouse use in the presence of bumble bees.

The potential side effects of the novel insecticide spinetoram in comparison with spinosad on the bumblebee *Bombus terrestris* was determined by Besard et al. [24]. The potential lethal effects together with the ecologically relevant sublethal effects on aspects of bumblebee reproduction and foraging behaviour were evaluated. Overall, the results indicate that the use of spinetoram is safer for bumblebees by direct contact and oral exposure than the use of spinosad, and therefore it can be applied safely in combination with *B. terrestris*. The data provide strong evidence that neither spinosyn has a negative effect on the foraging behaviour of these beneficial insects.

Upon topical treatment, nitro-containing neonicotinoids (imidacloprid, clothianidin, thiamethoxam, nitenpyram and dinotefuran) were more toxic than the cyano-group containing ones (acetamiprid and thiacloprid) [25,26]. A similar high toxicity of imidacloprid and thiamethoxam was also found for the bumble bee *Bombus terrestris* [27]. The lower toxicity of the cyanogroup neonicotinoids can be attributed to their fast biotransformation [28-30] and the existence of different nAChR subtypes [31,32], however, found that bumble bees (*Bombus impatiens*) were more tolerant to clothianidin and imidacloprid than *Osmia lignaria* and *M. rotundata*.

3. The effect of insecticides on insect immunity

Individual immune defences in the bee parallel the innate immune systems of vertebrates [33], Figure 2. Insect immunity is basically composed of three parts: (1) the cuticle, which presents physical and chemical barriers to the outside world of microbes, (2) humoral responses, and (3) cellular responses. The primary defences are the cuticle, the spiracles and trachea and the alimentary canal including the intestinal epithelium and peritrophic membrane [34-36] If these are breached cellular immune defences include the cellular response represented by phagocytosis by haemocytes and melanisation.Insecticides affect the insect humoral and cellular immune responses. In the initial humoral response, pattern recognition proteins identify invading microbes and initiate the synthesis of various ofantimicrobial proteins [37], (Figure 2). Phagocytosis is typically accompanied by melanin production and melanization of nodules and capsules (Figure 1). Melanin production can occur morerapidly than the production of antimicrobial peptides, can lead to the formation of reactive oxidative species that can contribute to killing pathogens and are regulated through the phenoloxidase cascade [38], (Figure 2). The humoral and cellular responses are interdependent defensive forces involving detoxificationmechanisms that are also utilized by insects to prevent damage from environmental toxins. The humoral response is represented by secretion of antimicrobial peptides (AMPs, inducible antibioticpeptides such as apidaecins) [39,40] from within the fat body due to the activation of one or several intracellular signalling pathways (Toll, Imd and Jak-Stat) which degrade pathogens as well as the action of reactive oxygen and nitrogen species (Figure 2).

To date risks assessment studies conducting the side-effects of conventional insecticides are mostly limited to acute toxicity studies. A summary of the effects of the insecticides including the organochlorides, organophosphates, carbamates, pyrethroids, organophosphates and neonicotinoids on insect is given in Table 2.

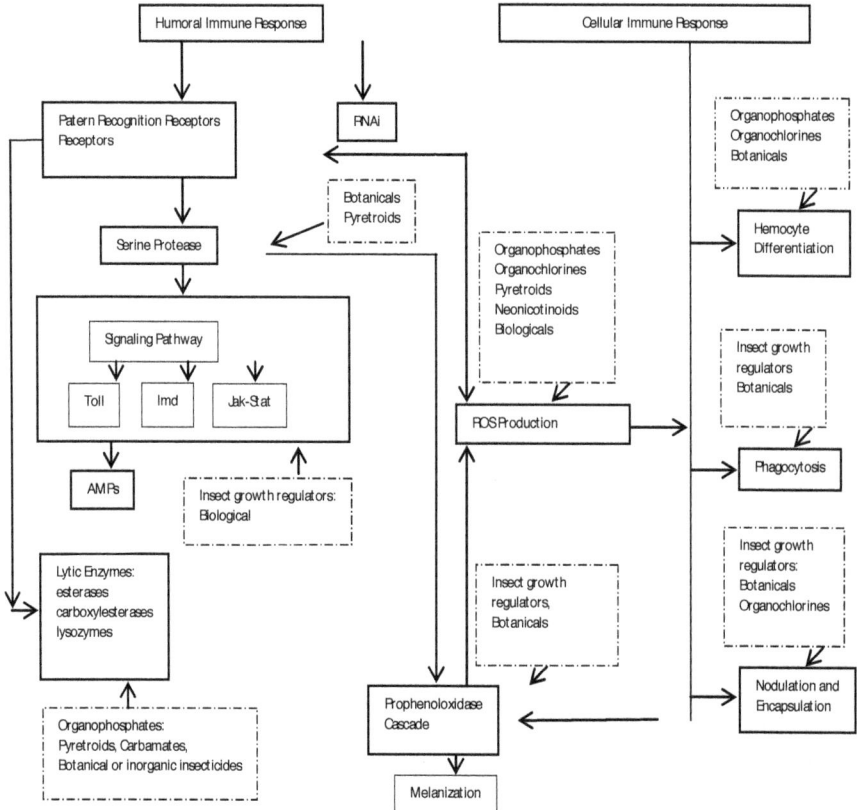

Figure 2. The effect of insecticides on insect immunity. Solid boxes and arrows represent a schematic of the insect immune system. Dashed boxes and arrows identify where pesticides have been documented to affect particular immune responses [33].

Insecticide class	Insecticide mode of activity
Organochlorides	- neurotoxin, disrupt the nervous system
Organophosphates	- neurotoxin, interfere with the enzymes acetylcholinesterase and other cholinesterases, disrupting nerve impulses, killing or disabling the insect
	- have an accumulative toxic effect to wildlife, so multiple exposures to the chemicals amplifies the toxicity
	- kill insects by inhibiting the enzyme cholinesterase, which is essential in the functioning of the nervous system
Carbamates	- have similar toxic mechanisms to organophosphates, but have a much shorter duration of action and are thus somewhat less toxic
	- are rapidly detoxified and eliminated from animal tissues
	- toxicity is thought to arise from a mechanism somewhat similar to that for the organophosphates.
Pyrethroids	- are nonpersistent sodium channel modulators, and are much less acutely toxic than organophosphates and carbamates.
Neonicotinoids	- broad spectrum systemic insecticides are nicotinic acetylcholine receptor agonists
	- have a rapid action (minutes-hours).
	- treated insects exhibit leg tremors, rapid wing motion, stylet withdrawal (aphids), disoriented movement, paralysis and death

Table 2. Chemical insecticides with effect on insect immunity

4. Detoxifying enzymes

Pesticide resistance in arthropods has been shown to evolve by two main mechanisms at the molecular level either (1) the increased production of metabolic enzymes (esterases, glutathione S-transferases and cytochrome P450 monooxygenases), which can break down or bind to the pesticide, or (2) structural changes (mutations) in the pesticide target protein, which render it less sensitive to the toxic effects of the pesticide [41]. The increased production of the metabolic enzymes is usually mediated via upregulation through mutations in *trans* and/or *cis*-acting regulatory loci or through amplification of the structural gene encoding the enzyme [42,43-47]. An alternative way of resistance to pesticides in some cases results from changes in the enzyme coding sequence, which confer an enhanced ability to metabolise the pesticide [48]. Enhanced production of esterases through gene amplification or upregulation has been implicated in resistance to organophosphates, carbamates and pyrethroids in a range of arthropod [42,49]. This can involve slow degradation of insecticide, which exhibit broad-spectrum resistance or metabolism of a limited range of insecticides containing a common ester bond [50]. Monooxygenases are mainly involved in the metabolism of pyrethroids and to a lesser extent, detoxification of organophosphates and carbamates [51]. Glutathione S-transferases (GSTs) on the other hand play a role in detoxification of a large range of xenobiotics such as to the organochlorine insecticide DDT. They also provide protection against oxidative stress and hormone biosynthesis [52]. The increased production of GSTs has been documented as a mechanism of resistance to organochlorines and pyrethroids and is most commonly mediated through upregulation [42]. The role of glutathione S-transferases in resistance to pyrethroids has been attributed to detoxification of lipid peroxidation products induced by pyrethroids rather than to direct metabolism of pyrethroid molecules [53]. P450s play a role in the metabolic resistence of arthropod to a range of different insecticide classes including the pyrethroids, organochlorines, neonicotinoids, organohosphates, carbamates and insect growth regulators [42].

Although acetylcholinesterase (AChE) is a specific target enzyme for organophosphate and carbamate poisoning (specific protein binding), other pesticides or their metabolites can inhibit AChE activity by unspecific protein binding (e.g., interaction with serine groups outside of the catalytic subunit). The latter interaction would also result in a reduction of AChE activity.

4.1. Glutathione-S-transferases

Glutathione-S-transferases (EC 2.5.1.18) constitute a family of enzymes that participe in 1) detoxification of xenobiotics, 2) protection from oxidative damage, and 3) intracellular transport of hormones, endogenous metabolites and exogenous chemicals. It is involved in insecticide resistance. Glutathione S-transferases (GSTs) are present in almost all animals and in most of them in multiple isoenzymic forms, constituting a significant intracellular mechanism of detoxification. The enzymes catalyse the conjugation of a large variety of compounds bearing an electrophilic site, with reduced glutathione (GSH) [54]. In insects, they represent a very interesting detoxification mechanism due to their involvement in tolerance to insecticides [55-58]. GSTs of different sub-families and classes are universally present in all forms of life, from invertebrates to vertebrates, plants and microorganisms. They have been identified and characterized in such diverse species as *Haliotis discus discus* [59], *Hyphantria cunea* [60], *Anopheles gambiae* [61], *Xenopus laevis* [62], and white leghorn chicks [63]. In addition, a number of GSTs have been purified and biochemically characterized [64-67]. Though plenty of studies have demonstrated the xenobiotic detoxification activity of GSTs, studies on immunomodulatory properties of GST members are scarce.

Increased levels of GSTs have been associated with resistence to a wide variety of insecticides, maily organophosphates, organochlorides and cyclodienes [68]. Other insecticide classes, such as pyrethroids and carbamates, are probably not detoxified by GSTs [69]. Induction of GST activity has been reported in many insects following treatment with insecticides [70,71]. The activity of insect GSTs has been found to be present in the midgut [72], fat body [73], hemolymph cells and other but more precise sites of GSTs need to be determined. The multiple isoenzymic forms of GST are distinguished by differences in structure and catalytic properties. The expression of isoenzymes depends on many internal and external factors. Age-dependent alteration of GST activities has been demonstrated in both vertebrates and invertebrates [74-76]. Alteration of GST expression induced by various substances in insets has also been reported (e.g.,food quality and administration of certain insecticides) [77-80]. Significant alterations in the isoenzymic profile, leading to the suggestion that the multiple GST isoenzymes present in insects are regulated independently, have also been reported as a result of stress, caused by a variety of factors [78,81]. The role of other enzymatic systems (such as esterases and monooxygenases) is emphasized by many investigators as even more important than GSTs in insecticide metabolism by insects [82].

Kim at all. [83] show the expression profiles of *GSTs* of the bumblebee *Bombus ignitus* in response to oxidative stress. The sigma-class GST from *B. ignitus* (*BiGSTS*) was identified. Comparative analysis indicates that the predicted amino acid sequence of BiGSTS shares a high identity with the sigma-class GSTs of hymenopteran insects such as *Apis mellifera*. Tissue distribution analyses showed the presence of BiGSTS in all tissues examined, including the fat body, midgut, muscle and epidermis. Under uniform conditions of H_2O_2 overload, the expression profile of GSTs and other antioxidant enzyme genes. These findings indicate that GSTs and other antioxidant enzyme genes in *B. ignitus* are differentially expressed in response to oxidative stress.

In line with the mammalian GST classification, insect GSTs have been classified into six classes: delta, epsilon, omega, theta, sigma, and zeta [84-86]. Due to the important role of GSTs, most studies of GSTs in insects have focused on their role in detoxifying exogenous compounds, in particular insecticides and plant allelochemicals [87-92]. In addition, insect GSTs have been studied for their role in mediating oxidative stress responses. *Drosophila melanogaster* sigma-class GST has been reported to play an important part in the detoxification of lipid peroxidation products, suggesting a protective role against oxidative stress [93]. In the mosquito *Anopheles gambiae*, the expression of epsilon GSTs increased in response to oxidative stress [94]. In the previous studies were investigated the antioxidant enzymes such as superoxide dismutase [95], and phospholipid-hydroperoxide glutathione peroxidase [96], and two [97] from the bumblebee *Bombus ignitus*. Furthermore, were have previously identified a delta-class GST (BiGSTD) from *B. ignitus* [98]

4.2. Esterases

The esterase family of enzymes hydrolyse ester bonds, which are present in a wide range of insecticides; therefore, these enzymes may be involved in resistance to the main chemicals employed in control programs. Acetylcholinesterase (AChE) is a specific target enzyme for organophosphate and carbamate poisoning (specific protein binding), other pesticides or their metabolites can inhibit AChE activity by unspecific protein binding (e.g., interaction with serine groups outside of the catalytic subunit). The latter interaction would also result in a reduction of AChE activity. AChE has been shown in aphid *Myzus persiace* [99], mosquitoes, *Culex quinquefasciatus* and *C. pipiens* [100] blowfly. In insects, the esterase enzyme patterns have shown high rates of intraspecific and interspecific variations [101]. Acetylcholinesterase enzyme a key enzyme in insect central nervous system, terminates nerve impulses by cata-lyzing the hydrolysis of the neurotransmitter acetylcholine. Resistance associated with modification of AChE makes it less sensitive to inhibition by organophosphates and carbamate insecticides.

Bombus hypocrita is one of the dominant bumblebees in China, and is widely used as one of the most crucial pollinators in greenhouse due to easy mass-rearing, strong population and effective pollinating performance [102]. However, it is often threatened by organophosphate and carbamate insecticides which are widely used in China, as these insecticides can inhibit the acetylcholinesterase activity in insects. The effect of six common organophosphate and carbamate insecticides on the AChE activity of *B. hypokrita* was evaluated. The inhibition of six insecticides to AChE from *B. hypocrita* showed a dosage effect. The inhibitory effects of six insecticides to AChE from high to low are in the following order: isoprocarb > chlorpyrifos > triazophos > bassa > propoxur > profenofos, and *B. hypocrita* is more sensitive to isoprocarb than to the other five insecticides.

4.3. Cytochrome P450 monooxygenases

Cytochrome P450 monooxygenases (P450s) are enzymes important in insects both for the detoxification of xenobiotics, such as pesticides and phytochemicals, and the biosynthesis of endogenous compounds, such as hormones and pheromones. The monooxygenases of insect have several functional roles, including growth, development, feeding and protection against xenobiotics, including resistance to pesticides and tolerance to plant toxins [103].

Monooxygenases are involved in the synthesis and degradation of insect hormones and pheromones. They catalyse many types of reactions including oxygenations, dehalogenations, dealkylations, deaminations, dehydrogenations and isomerations [104]. The enzyme reactions are based on activation of molecular oxygen within sertion of one of its atoms into the substrate and reduction of the other to form water [105]. The genome of every insect species may carry a hundred or so different P450 genes, all evolved from a common ancestral gene. The P450 family is thus one of the oldest and largest gene superfamilies. They do many things: P450 enzymes are found in the biosynthetic pathways of ecdysteroids and juvenile hormones, which are at the center stage of insect growth, development, and reproduction. P450 enzymes metabolize insecticides, resulting either in bioactivation or, more often, in detoxification, the latter process being enhanced in many strains with metabolic resistance to insecticides. Furthermore, P450 metabolism of certain plant chemicals is often the key to the adaptation of insect herbivores to their host plants. P450 enzymes play important roles in the synthesis or degradation of odorants, pheromones, or defensive chemicals. Insect P450 genes are expressed in many tissues; not surprisingly, the digestive tract and fat body are a rich source of P450–dependent metabolism of model substrates. Developmental regulation of P450 gene expression is well documented by biochemical means [106]. It is well established that many cases of metabolic resistance to insecticides are the result of elevated levels of P450 [92]. In vivo suppression or decrease in resistance by application of P450 inhibitors such as the synergist piperonyl butoxide is often used as a diagnostic for P450 involvement. In vitro demonstration of increased P450–dependent metabolism of the insecticide is done more infrequently. Because of the multiplicity of P450 genes, correlations between resistance and activity of P450 enzymes measured by model substrates, though indicative, are seldom demonstrative. The cytochrome P450-dependent monooxygenases are an extremely important metabolic system involved in the metabolism of xenobiotics and endogenous compounds. The number of P450s in a given insect species currently ranges from 48 in *Apis mellifera* to

164 in *Aedes aegypti*. There are never increasing number of studies that indicate P450s have a much broader role in insects. It is postulated that metabolism of xenobiotics might be the role for only a minority of the P450s in a given insect species. P450-mediated metabolism can result in detoxification of insecticides such as pyrethroids, or can be involved in both the bioactivation and detoxification of insecticides such as organophosphates.

Cytochrome P450 is a hemoprotein which acts as the terminal oxidase in monooxygenase systems. In eukaryotes, most P450s are found in the endoplasmic reticulum or mitochondria. Monooxygenases can oxidize widely diverse substrates and are capable of catalyzing a large array of reactions [107]. This is because each species contains numerous P450s and because of the broad substrate specificity of some isoforms. Insect monooxygenases can be detected in a wide range of tissues. Highest monooxygenase activities are usually associated with the midgut, fat bodies and Malpighian tubules [108], but again the expression of individual P450s can vary between these tissues [95]. In general, total P450 levels are undetectable in eggs, rise and fall in each larval instar, are undetectable in pupae and are expressed at high levels in adults [106]. The patterns of expression of individual P450s can vary within and/or between life stages [107-110].

5. Conclusion

Bumblebees are of crucial importance for the pollination of wild flowers and economical important crops in modern agri/horticulture. These pollinators as *Bombus terrestris*, *Bombus impatiens* and *Bombus ignitus* are also commercially reared for the pollination of agricultural and horticultural crops. Foraging worker bees are playing a key role as they are responsible for the amount of food brought to the colony. Therefore hazards of sublethal concentrations of pesticides on them are important. This lack of assessment, therefore, poses a serious problem and standardized laboratory tests for bumblebees are necessary to evaluate the impact of sublethal effects on their foraging behavior.

Over the last three decades, bumble bee populations have experienced global population declines due to environmental factors such as pathogens, pesticide exposure and habitat fragmentation. Given the importance of bumble bees as pollinators of crops and wild flowers, steps must be taken to prevent further declines.

Acknowledgements

Authors thank the research programme of the Institute of Organic Chemistry and Biochemistry RVO: 61388963 the Czech Science Foundation (grant No. 203/09/1446) and Technology Agency of the Czech Republic (grant No. TA01020969)

Author details

Marie Zarevúcka*

Address all correspondence to: zarevucka@uochb.cas.cz

Institute of Organic Chemistry and Biochemistry, AS CR, Flemingovo nám, Prague, Czech Republic

References

[1] Park B.S., Lee S.E. Proteomics in insecticide toxicology. Molecular Cellular Toxicology 2007:3(1) 11-18.

[2] Scott J.G., Wen Z. Cytochromes P450 of insects: the tip of the iceberg. Pest Management Science 2001:57(10) 958-967.

[3] Scott J.G., Insecticide resistance in insect. In Pimentel D.(ed.) Handbook of pest management in Agriculture, Vol.2, CRC Press, Boca Raton, FL, USA, 1991, p 663.

[4] Veerle M., Guy S. Side-Effects of Pesticides on the Pollinator Bombus: An Overview, Pesticides in the Modern World - Pests Control and Pesticides Exposure and Toxicity Assessment,M. Stoytcheva (Ed.), 2011 ISBN: 978-953-307-457-3, InTech, Available from: http://www.intechopen.com/books/pesticides-in-the-modern-world-pests-control-and-pesticides-exposure-andtoxicity-assessment/side-effects-of   pesticides-on-the-pollinator-bombus-an-overview.

[5] Blacquière T., Smagghe G., van Gestel C.A., Mommaerts V. Neonicotinoids in bees: a review on concentrations, side-effects and risk assessment. Ecotoxicology. 2012:21(4) 973-92.

[6] Tasei J.N., Ripault G., Rivault E. Hazards of imidacloprid seed coating to Bombus terrestris (Hymenoptera: Apidae) when applied to sunflower. Journal of Economic Entomology 2001:94(3) 623-7.

[7] Cresswell J.E., Page C.J., Uygun M.B., Holmbergh M., Li Y., Wheeler J.G., Laycock I., Pook C.J., de Ibarra N.H., Smirnoff N., Tyler C.R. Differential sensitivity of honey bees and bumble bees to a dietary insecticide (imidacloprid). Zoology 2012:115(6) 365– 371.

[8] Whitehorn P.R., O'Connor S., Wackers F.L., Goulson D. Neonicotinoid pesticide reduces bumble bee colony growth and queen production. Science. 2012:336(6079) 351-352.

[9] Gels J.A., Held D.W., Potter D.A. Hazards of insecticides to the bumble bees Bombus impatiens (Hymenoptera: Apidae) foraging on flowering white clover in turf. Journal of Economic Entomology 2002: 95(4) 722-8.

[10] Scott-Dupree C.D., Conroy L., Harris C.R. Impact of currently used or potentially useful insecticides for canola agroecosystems on Bombus impatiens (Hymenoptera: Apidae), Megachile rotundata (Hymentoptera: Megachilidae), and Osmia lignaria (Hymenoptera: Megachilidae). Journal of Economical Entomology 2009: 102(1) 177-82.

[11] Fanigliulo A., Filì V., Pacella R., Comes S., Crescenzi A. Teppeki, selective insecticide about Bombus terrestris. Communications in Agricultural and Applied Biological Sciences 2009:74(2) 407-10.

[12] Thompson H.M., Hunt L.V. 1999. Extrapolating from honeybees to bumblebees in pesticide risk assessment. Ecotoxicology 1999:8(3) 147-166.

[13] Davis A.R., Shuel R.W. Distribution of carbofuran and dimethoate in flowers and their secretion in nectar as related to nectary vascular supply. Canadian Journal of Botany 1988:66(7) 1248-1255.

[14] Thompson H.M. 2001. Assessing the exposure and toxicity of pesticides to bumble-bees (Bombus sp.). Apidologie 2001: 32(4) 305-321.

[15] Free J.B., Ferguson A.W. Foraging of bees on oil-seed rape (Brassica napus L.) in rela-tion to the flowering of the crop and pest control. Journal of Agricultural Sciences 1980: 94(1)151-154.

[16] Ernst W.R., Pearce P.A., Pollock T.L. 1989. Environmental effects of Fenitrothion use in forestry. Conservation and Protection, Environment Canada, Atlantic Region, Canada.

[17] Morandin L.A., Winston M.L., Franklin M.T., Abbott V.A. Lethal and sub-lethal ef-fects of spinosad on bumble bees (Bombus impatiens Cresson). Pest Management Science 2005: 61(7) 619-626.

[18] Franklin M.T., Winston M.L., Morandin L.A. Effects of clothianidin on Bombus impa-tiens (Hymenoptera: Apidae) colony health and foraging ability. Ecotoxicology 2004: 97(2) 369-373.

[19] Sechser B., Freuler J. The impact of thiomethoxam on bumble bee broods (Bombus terrestris L.) following drip application in covered tomato crops. Journal of Pest Sci-ence 2003:76(3)74-77.

[20] Morandin L.A., Winston M.L., Franklin M.T., Abbott V.A. Lethal and sub-lethal ef-fects of spinosad on bumble bees (Bombus impatiens Cresson). Pest Management Science 2005:61(7) 619-626.

[21] Mommaerts V., Sterk G., Smagghe G. Hazards and uptake of chitin synthesis inhibitors in bumblebees Bombus terrestris. Pest Management Science 2006:62(8) 752-758.

[22] Isayama S., Saito S., Kuroda K., Umeda K., Kasamatsu K. Pyridalyl, a novel insecticide: potency and insecticidal selectivity. Archives of Insect Biochemistry and Physiology 2005:58(4) 226-33.

[23] Gradish A.E., Scott-Dupree C.D., Shipp L., Harris C.R., Ferguson G. Effect of reduced risk pesticides for use in greenhouse vegetable production on Bombus impatiens (Hymenoptera: Apidae). Pest Management Science 2010:66(2) 142-6.

[24] Besard L., Mommaerts V., Abdu-Alla G., Smagghe G. Lethal and sublethal side-effect assessment supports a more benign profile of spinetoram compared with spinosad in the bumblebee Bombus terrestris. Pest Managemnet Science 2011:67(5) 541-7.

[25] Iwasa T., Motoyama N., Ambrose J.T., Roe M.R. Mechanism for the differential toxicity of neonicotinoid insecticides in the honey bee, Apis mellifera. Crop Protection 2004:23(5) 371–378.

[26] Laurino D., Porporato M., Patetta A., Manino A. Toxicity of neonicotinoid insecticides to honey bees laboratory tests. Bull Insectol 2011:64(1) 107–113.

[27] Mommaerts V., Reynders S., Boulet J. Besard L., Sterk G., Smagghe G. Risk assessment for side-effects of neonicotinoids against bumblebees with and without impairing foraging behaviour. Ecotoxicology 2010:19(1) 207–215.

[28] Suchail S., De Sousa G., Rahmani R., Belzunces L.P. In vivo distribution and metabolisation of C-14-imidacloprid in different compartments of Apis mellifera L. Pest Management Science 2004:60(11) 1056–1062.

[29] Suchail S., Debrauwer L., Belzunces L.P. Metabolism of imidacloprid in Apis mellifera. Pest Management Science 2004:60(3) 291–296.

[30] Brunet J.L., Badiou A., Belzunces L.P. In vivo metabolic fate of [C-14]-acetamiprid in six biological compartments of the honeybee, Apis mellifera L. Pest Management Science 2005:61(8) 742–748.

[31] Jones A.K., Raymond-Delpech V., Thany S.H., Gauthier M., Sattelle D.B. The nicotinic acetylcholine receptor gene family of the honey bee, Apis mellifera. Genome Research 2006:16(11) 1422–1430.

[32] Scott-Dupree C.D., Conroy L., Harris C.R. Impact of currently used or potentially useful insecticides for canola agroecosystems on Bombus impatiens (Hymenoptera: Apidae), Megachile rotundata (Hymenoptera: Megachilidae), and Osmia lignaria (Hymenoptera: Megachilidae). Journal of Economic Entomology 2009:102(1) 177–182.

[33] James R.R., Xu J. Mechanisms by which pesticides affect insect immunity. Journal of Invertebrate Pathology 2012:109(2) 175-182.

[34] Aronstein K.A., Murray K.D. Chalkbrood disease in honey bees. Journal of Inverte-
      brate Pathology 2010:103(Suppl 1) S20-S29.

[35] Glinski Z. Immuno-suppressive and immuno-toxic action of contaminated honey bee
      products on consumers. Medycina Weterynaryjna 2000: 56(10) 634-638.

[36] Glinski Z. Kauko L. Problems of immunosuppression and immunotoxicolgy in re-
      spect to the honeybee protection against microbial and parasitic nvaders. Apiacta
      2000:35(2) 65-76.

[37] Hetru C., Hoffmann D., Bulet P., 1998. Antimicrobial peptides from insects. In: Brey,
      P.T., Hultmark,D. (Eds.), Molecular Mechanisms of Immune Responses inInsects.
      Chapman and Hall, NY, pp. 40–66.

[38] Ragan E.J., An C., Jiang H., Kanost M.R. 2009. Roles of haemolymph proteins in anti-
      microbial defenses of Manduca sexta. In: Rolff, J., Reynolds, S.E. (Eds.), Insect Infec-
      tion and Immunity, Evolution, Ecology and Mechanisms. Oxford University Press,
      Oxford, pp. 34–48.

[39] Casteels P., Ampe C., Jacobs F., Vaeck M., Tempst P. Apidaecins, antibacterial pepti-
      des from honeybees. EMBO Journal 1989:8(8) 2387-2391.

[40] Li W.F., Ma G.X., Zhou X.X. 2006. Apidaecin-type peptides, Biodiversity, structure-
      function relationships and mode of action. Peptides 2006:27(9) 2350-2359.

[41] Bass C., Field L.M. Gene amplification and insecticide resistance Pest Management
      Science 2011: 67(8) 886–890.

[42] Li X.C., Schuler M.A., Berenbaum M.R. Molecular mechanisms of metabolic resist-
      ance to synthetic and natural xenobiotics. Annual Reviews Entomology 2007:52 231–
      253.

[43] Hemingway J., Karunaratne S.H. Mosquito carboxylesterases: a review of the molec-
      ular biology and biochemistry of a major insecticide resistance mechanism. Medical
      and Veterinary Entomology 1998:12(1) 1-12.

[44] Huang H.S., Hu N.T., Yao Y.E., Wu C.Y., Chiang S.W., Sun, C.N. Molecular cloning
      and heterologous expression of a glutathione-Stransferase involved in insecticide re-
      sistance from the diamondback moth Plutella xylostella. Insect Biochemistry and Mo-
      lecular Biology 1998:28(9) 651- 658.

[45] Daborn P.J., Yen J.L., Bogwitz M.R., Le Goff G., Feil E., Jeffers S., Tijet N., Perry T.,
      Heckel D., Batterham P., Feyereisen R., Wilson T.G., ffrench-Constant, R.H. A single
      P450 allele associated with insecticide resistance in Drosophila. Science
      2002:297(5590) 2253-2256.

[46] Nikou D., Ranson H., Hemingway J. An adult-specific CYP6 P450 gene is overex-
      pressed in a pyrethroid-resistant strain of the malaria vector, Anopheles gambiae.
      Gene 2003:318(1) 91-102.

[47] Festucci-Buselli R.A., Carvalho-Dias A.S., de Oliveira-Andrade M., Caixeta- Nunes C., Li, H.M., Stuart J.J., Muir, W., Scharf M.E., Pittendrigh B.R. Expression of Cyp6g1 and Cyp12d1 in DDT resistant and susceptible strains of Drosophila melanogaster. Insect Molecular Biology 2005:14(1) 69-77.

[48] Claudianos C., Russell R.J., Oakeshott J.G. The same amino acid substitution in or-thologous esterases confers organophosphate resistance on the house fly and a blow-fly. Insect Biochemistry and Molecular Biology 1999:29(8) 675–686.

[49] Hemingway J., Ranson H. Insecticide resistance in insect vectors of human disease. Annual Review of Entomology 2000:45 371-391.

[50] Hemingway J. The molecular basis of two contrasting metabolic mechanisms of in-secticide resistance. Insect Biochemistry and Molecular Biology 2000:30(11)1009-1015.

[51] Feyereisen R. Insect P450 enzymes. Annual review of Entomology 1999:44(1) 507-533.

[52] Enayati A.A., Ranson H., Hemingway J. Insect glutathione transferases and insecti-cide resistance. Insect Molecular Biology 2005:14(1) 3-8.

[53] Vontas J., Small G.J., Hemingway J. Glutathione S-transferases as antioxidant defence agents confer pyrethroid resistance in Nilaparvata lugens. Biochemical Journal 2001:357(1) 65–72.

[54] Mannervik B., lin P., Guthenberg C., Jensson H., Tahir M.K., Warholm M., Jornvall H.. Identification of three classes of cytosolic glutathione transferases common to several mammalian species: correlation between structural data and enzymatic prop-erties. Proceedings of the National Academy of Sciences USA 1985:82(21) 7202–7206.

[55] Motoyama N., Dauterman W.C., Glutathione S-transferases: their role in the metabo-lism of organophosphorus insecticides. Review of Biochemical Toxicology 1980:2(1) 49–69.

[56] Clark A.G., Dick G.L., Martindale S.M., Smith J.N. Glutathione S-transferases from the New Zealand grass grub, Costelytra zealandica: their isolation and characteriza-tion and the effect on their activity of endogenous factors. Insect Biochemistry 1985:15(1) 35–44.

[57] Fournier D., Bride J.M., Poirie M., Berge J.-B., Plapp Jr. F. Insect glutathione S-trans-ferases: biochemical characteristics of the major forms from houseflies susceptible and resistant to insecticides. Journal of Biological Chemistry 1992:267(3) 1840–1845.

[58] Kostaropoulos I., Papadopoulos A.I., Metaxakis A., Boukouvala E., Papadopoulou-Mourkidou E., 2001. Glutathione S-transferase in the defense against pyrethroids in insect. Insect Biochem Mol Biol 2001:31(4-5) 313–319.

[59] Wan Q., Whang I., Lee J. Molecular cloning and characterization of three sigma glu-tathione S-transferases from disk abalone (Haliotis discus discus). Comparative Bio-chemistry and Physiology B 2008:151(3) 257–267.

[60] Yamamoto K., Fujii H., Aso Y., Banno Y., Koga K. Expression and characterization of a sigma-class glutathione S-transferase of the fall webworm, Hyphantria cunea. Biosci Biotechnol Biochem 2007:71(9) 553–560.

[61] Ranson H., Rossiter L., Ortelli F., Jensen B., Wang X., Roth C.W., Collins F.H., Hemingway J., Identification of a novel class of insect glutathione Stransferases involved in resistance to DDT in the malaria vector Anopheles gambiae. Biochemical Journal 2001:359(Pt 2) 295–304.

[62] Carletti E., De Luca A., Urbani A., Sacchetta P., Di Ilio C. Sigma-class glutathione transferase from Xenopus laevis: molecular cloning, expression, and site-directed mutagenesis. Archives of Biochemistry and Biophysics 2003:419(2) 214–221.

[63] Hsieh C.H., Liu L.F., Tsai S.P., Tam M.F. Characterization and cloning of avianhepatic glutathione S-transferases. Biochemistry 1999:343(Pt 1) 87–93.

[64] Tomarev S.I., Zinovieva R.D., Guo K., Piatigorsky J. Squid glutathione Stransferase.Relationships with other glutathione S-transferases and S-crystallins of cephalopods. Journal of Biological Chemistry 1993:268(6) 4534–4542.

[65] Fitzpatrick P.J., Krag T.O., Hojrup P., Sheehan D. Characterization of a glutathione S-transferase and a related glutathione-binding protein from gill of the blue mussel, Mytilus edulis. Biochemical Journal 1995:305(Pt 1) 145–150.

[66] Hoarau P., Garello G., Gnassia-Barelli M., Romeo M., Girard J.P. Purification and partial characterization of seven glutathione S-transferase isoforms from the clam Ruditapes decussatus. European Journal of Biochemistry 2002:269(17) 4359–4366.

[67] Yang H.L., Zeng Q.Y., Nie L.J., Zhu S.G., Zhou X.W. Purification and characterization of a novel glutathione S-transferase from Atactodea striata. Biochemical and Biophysical Research Communications 2003:307(3) 626–631.

[68] Fournier D., Bride J.M., Poirie M., Berge J.B., Plapp F.W.Jr. Insect glutathione Stransferase: biochemical characteristics of the major forms from houseflies susceptible and resistant to insecticides. Journal of Biological Chemistry 1992:267(3) 1840-1845.

[69] Grant D.F., Matsumura F. Glutathione S-transferases 1 and 2 in susceptible and insecticide resistant Aedes aegypti. Pesticide Biochemistry and Physiology 1989:33(2) 132-143.

[70] lagadic L., Cuany A., Berge J.B. Echaubard M., Purification and partial characterisation of glutathione S-transferases from insecticide-resistant and lindane-induced susceptible Spodoptera littoralis (Boisd) larvae. Insect Biochemistry and Molecular Biology 1993:23(4) 467-474.

[71] Punzo F., Detoxification enzymes and the effects of temperature on the toxicity of pyrethrpoids to the fall armyworm, Spodoptera frugiperda (Lepidoptera: Noctudae). Comparative Biochemistry Physiology C 1993:105(1) 155-158.

[72] Tate L.G., Nakat S.S., Hodgson E. Comparison of detoxification activity in midgut and fat body during fifth instar development of the tobacco hornworm, Manduca sexta. Comparative Biochemistry and Physiology C 1982:72(1) 75–81.

[73] Chien C., Dauterman W.C. Studies on glutathione S-transferase in Helicoverpa. Insect Biochemistry 1991:21(8) 857–864.

[74] Gregus Z., Varga F., Schmelas A. Age-development and inducibility of hepatic glutathione S-transferase activities in mice, rats, rabbits and guinea-pigs. Comparative Biochemistry and Physiology 1985:C 80(1) 85–90.

[75] Hazelton G.A., Lang C.A. Glutathione S-transferase activities in the yellow-fever mosquito [Aedes aegypti (Louisville)] during growth and ageing. Biochemical Journal 1983:210(1) 281–287.

[76] Kostaropoulos I., Mantzari A.E., Papadopoulos A.I. Alterations of some glutathione S-transferase characteristics during the development of Tenebrio molitor (Insecta: Coleoptera). Insect Biochemistry and Molecular Biology 1996:26(8-9) 963–969.

[77] Hayaoka T., Gauterman W.C. Induction of glutathione S transferase by phenobarbital and pesticides in various housefly strains and its effect on toxicity. Pesticide Biochemistry and Physiology 1982:17(2) 113–119.

[78] Papadopoulos A.I., Stamkou E.I., Kostaropoulos I., Papadopoulou-Mourkidou E. Effect of organophosphate and pyrethroid insecticides on the expression of GSTs from Tenebrio molitor larvae. Pesticide Biochemistry and Physiology 1999:63(1) 26–33.

[79] Papadopoulos A.I., Boukouvala E., Kakaliouras G., Kostaropoulos I., Papadopoulou-Mourkidou E. Effect of organophosphate and pyrethroid insecticides on the expression of GSTs from Tenebrio molitor pupae. Pesticide Biochemistry and Physiology 2000:68(1) 26–33.

[80] Kostaropoulos I., Papadopoulos A.I., Metaxakis A., Boukouvala E., Papadopoulou-Mourkidou E., Glutathione S-transferase in the defense against pyrethroids in insect. Insect Biochemistry and Molecular Biology 2001:31(3-4) 313–319.

[81] Papadopoulos A.I., Polemitouc E., Laifi P., Yiangou A., Tananaki T. Glutathione S-transferase in the developmental stages of the insect Apis mellifera macedonica. Comparative Biochemistry and Physiology C 2004:C139(1) 87–92.

[82] Karunaratne S.H.P.P., Hemingway J. Different insecticides select multiple carboxyesterase isoenzymes and different resistance levels from a single population of Culex quinquefascians. Pestic Biochem Physiol 1996:54(1) 4-11.

[83] Kim BY, Hui WL, Lee KS, Wan H, Yoon HJ, Gui ZZ, Chen S, Jin BR. Molecular cloning and oxidative stress response of a sigma-class glutathione S-transferase of the bumblebee Bombus ignitus. Comparative Biochemistry and Physiology B 2011:158(1) 83-9.

[84] Chelvanayagam G., Parker M.W., Board P.G. Fly fishing for GSTs: a unified nomenclature for mammalian and insect glutathione transferases. Chemico-Biological Interactions 2001:133(1-3) 256–260.

[85] Ranson H., Claudianos C., Ortelli F., Abgrall C., Hemingway J., Sharkhova M.V., Unger M.F., Collins F.H., Feyereisen R. Evolution of supergene families associated with insecticide resistance. Science 2002:298 179–181.

[86] Claudianos C., Ranson H., Johnson R.M., Biswas S., Schuler M.A., Berenbaum M.R., Feyereisen R., Oakeshott J.G. A deficit of detoxification enzymes: pesticide sensitivity and environmental response in the honeybee. Insect Molecular Biology 2006:15(6) 615–636.

[87] Huang H.S., Hu N.T., Yao Y.E., Wu C.Y., Chiang S.W., Sun C.N. Molecular cloning and heterologous expression of a glutathione S-transferase involved in insecticide resistance from the diamondback moth, Plutella xylostella. Insect Biochemistry and Molecular Biology 1998:28(9) 651–658.

[88] Ranson H., Rossiter L., Ortelli F., Jensen B.,Wang X., Roth C.W., Collins F.H., Hemingway J., Identification of a novel class of insect glutathione S-transferases involved in resistance to DDT in themalaria vector Anopheles gambiae. Biochemical Journal 2001:359(2) 295–304.

[89] Ortelli F., Rossiter L.C., Vontas J., Ranson H., Hemingway J. Heterologous expression of four glutathione transferase genes genetically linked to a majorinsecticide-resistance locus from the malaria vector Anopheles gambiae. Biochemical Journal 2003:373(3) 957–963.

[90] Enayati A.A., Ranson H., Hemingway J. Insect glutathione transferases andinsecticide resistance. Insect Molecular Biology 2005:14(1) 3–8.

[91] Francis F., Vanhaelen N., Haubruge E. Glutathione S-transferases in theadaptation to plant secondary metabolites in the Myzus persicae aphid. Archives of Insect Biochemistry and Physiology 2005:58(3) 166–174.

[92] Lumjuan N., McCarroll L., Prapanthadara L., Hemingway J., Ranson H. Elevated activity of an epsilon class glutathione transferase confers DDT resistance in the dengue vector, Aedes aegypti. Insect Biochemistry and Molecular Biology 2005:35(8) 861–871.

[93] Singh S.P., Coronella J.A., Benes H., Cochrane B.J., Zimniak P. Catalytic function of Drosophila melanogaster glutathione S-transferase DmGSTS1-1 (GST-2) in conjugation of lipid peroxidation end products. European Journal of Biochemistry 2001:268(10) 2912–2923.

[94] Ding Y., Hawkes N., Meredith J., Eggleston P., Hemingway J., Ranson, H. Characterization of the promoters of epsilon glutathione transferases in themosquito Anopheles

gambiae and their response to oxidative stress. Biochemical Journal 2005:387(3) 879–888.

[95] Choi Y.S., Lee K.S., Yoon H.J., Kim I., Sohn H.D., Jin B.R. Bombus ignitus Cu, Zn superoxide dismutase (SOD1): cDNA cloning, gene structure, and up-regulation in response to paraquat, temperature stress, or lipopolysaccharide stimulation. Comparative Biochemistry and Physiology B 2006:144(2) 365–371.

[96] Hu Z.G., Lee K.S., Choo Y.M., Yoon H.J., Kim I., Wei Y.D., Gui Z.Z., Zhang G.Z., Sohn H.D., Jin B.R. Molecular characterization of a phospholipid-hydroxide glutathione peroxidase from the bumblebee Bombus ignitus. Comparative Biochemistry and Physiology B 2010:155(1) 54–61.

[97] Hu Z., Lee K.S., Choo Y.M., Yoon H.J., Lee S.M., Wei Y.D., Lee J.H., Sohn H.D., Jin B.R., Molecular cloning and characterization of 1-Cys and 2-Cys peroxiredoxins from the bumblebee Bombus ignitus. Comparative Biochemistry Physiology B 2010:155(3) 272–280.

[98] Park, J.H., Yoon, H.J., Gui, Z.Z., Jin, B.R., Sohn, H.D., 2009. Molecular cloning of a delta class glutathione S-transferase gene from Bombus ignitus. International Journal of Industrial Entomology 2009: 18(1) 31–35.

[99] Field L.M., Devonshire A.L.: Evidence that the E4 and FE4 esterase genes responsible for insecticide resistance in the aphid Myzus persicae (Sulzer) are part of a gene family. Biochemical Journal 1998:330(1) 169-173.

[100] Guillemaud T., Makate N., Raymond M., Hirst B., Callaghan A. Esterase gene amplification in Culex pipiens. Insect Molecular Biology 1997:6(4) 319-327.

[101] Nascimento A.P., Bicudo H.E.M.D. Esterase patterns and phylogenetic relationships of Drosophila species in the saltans subgroup (saltans group). Genetica 2002:114(3) 41- 51.

[102] Liao X.Li Luo ShuDong, Wu Xiang, Wu Jie Optimization of conditions for assaying activity of acetylcholinesterase in Bombus hypocrita (Hymenoptera: Apidae) and its sensitivity to six common, insecticides. Acta Entomologica Sinica 2011:54(12) 1361-1367.

[103] Scott J.G., Liu N., Wen Z.M. Insect cytochromes P450 diversity, insecticide resistance and tolerance to plant toxins. Comp Biochem Physiol C 1998:121(1-3) 147-155.

[104] Nebert D.W., Gonzalez F.J. P450 genes: structure, evolution and regulation. Annual Review of Biochemistry 1987:56 945-993.

[105] Guengerich P.F. Reactions and significance of cytochrome P-450enzymes. Journal of Biological Chemistry 1991;266(17) 10019-10022.

[106] Agosin M. 1985. Role of microsomal oxidations in insecticide degradation. In Kerkur G.A. (ed.) Comprehensive insect physiology, biochemistry and pharmacology. Vol. 12, pp 647-712 Pergamon, New York.

[107] Mansuy D. The great diversity of reactiins catalysed by cytochrome P450. Comparative Biochemical Physiology C 1998:121(1-3) 5-14.

[108] Hodgson E. The significance of cytochrome P450 in insects. Insect Biochemistry 1983:13(3) 237-246.

[109] Cohen M.B., Feyereisen R. A cluster of cytochrome P450 genes of the CYP6 family in the house fly. DNA and Cell Biology 1995:14(1) 73-82.

[110] Scott J.G., Sridhar P., Liu N. Adult specific expression and induction of cytochrome P450trp in house flies. Archives Insect Biochemistry and Physiology 1996:31(3) 313-323.

Chapter 10

Pesticide Biodegradation: Mechanisms, Genetics and Strategies to Enhance the Process

Ma. Laura Ortiz-Hernández,
Enrique Sánchez-Salinas,
Edgar Dantán-González and
María Luisa Castrejón-Godínez

Additional information is available at the end of the chapter

1. Introduction

As a result of human activities, currently a large number of pollutants and waste are eliminated to the environment. Worldwide, more than one billion pounds of toxins are released into the air and water. Approximately $6x10^6$ chemical compounds have been produced; annually 1,000 new products are synthetized and between 60,000 and 95,000 chemicals are commercially used [1]. Among these substances are chemical pesticides, which are used extensively in most areas of crop production in order to minimize pest infestations, to protect the crop yield losses and to avoid reducing the product quality.

The pesticides belong to a category of chemicals used worldwide as herbicides, insecticides, fungicides, rodenticides, molluscicides, nematicides, and plant growth regulators in order to control weeds, pests and diseases in crops as well as for health care of humans and animals. The positive aspect of application of pesticides renders enhanced crop/food productivity and drastic reduction of vector-borne diseases [2,3]. A pesticide is any substance or mixture of substances intended for preventing, destroying, repelling, or mitigating any pest (insects, mites, nematodes, weeds, rats, etc.), including insecticide, herbicide, fungicide, and various other substances used to control pests [4]. The definition of pesticide varies with times and countries. However, the essence of pesticide remains basically constant: it is a (mixed) substance that is poisonous and efficient to target organisms and is safe to non-target organisms and environments [5].

Pesticides are by no means a new invention. In fact, intentional pesticide use goes back thousand years when Sumerians, Greeks, and Romans killed pests using various compounds such as sulphur, mercury, arsenic, copper or plant extracts. However, results were frequently poor because of the primitive chemistry and the insufficient application methods. A rapid emergence in pesticide use began mainly after World War II with the introduction of DDT (dichlorodiphenyltrichloroethane), BHC (benzene hexachloride), aldrin, dieldrin, endrin, and 2,4-D (2,4-dichlorophenoxyacetic acid). These new chemicals were effective, easy to use, inexpensive, and thus enormously popular. However, under constant chemical pressure, some pests became genetically resistant to pesticides, non-target organisms were harmed, and pesticide residues often appeared in unexpected places [3].

Chemical pesticides can be classified in different ways, but one of the most used is according to their chemical composition, which allows to group pesticides in a uniform and scientific way and to establish a correlation between structure, activity, toxicity and degradation mechanisms, among others. Table 1 shows the most important pesticides according to their chemical composition. Some general characteristics of pesticides are shown in Table 2.

Group	Main composition
Organochlorine	Carbon atoms, chlorine, hydrogen and occasionally oxygen. They are nonpolar and lipophilic
Organophosphate	Possess central phosphorus atom in the molecule. In relation whit organochlorines, these compounds are more stable and less toxic in the environment. The organophosphate pesticides can be aliphatic, cyclic and heterocyclic.
Carbamates	Chemical structure based on a plant alkaloid *Physostigma venenosum*.
Pyrethroids	Compounds similar to the synthetic pyrethrins (alkaloids obtained from petals of *Chysanthemun cinerariefolium*).
Botanical origin	Products derived directly from plants. Not chemically synthesized.
Biological	Viruses, microorganisms or their metabolic products.
Copper	Inorganic compounds of copper.
Thiocarbamates	Differ from carbamates in their molecular structure, containing an-S-group in its composition.
Organotin	Presence of tin as a central atom of the molecule.
Organosulfur	They have a sulfur central atom in the molecule, very toxic to mites or insects.
Dinitrophenols	They are recognized by the presence of two nitro groups (NO_2) bonded to a phenol ring.
Urea derivatives	Compounds which include the urea bound to aromatic compounds.
Diverse composition	Triazines, talimides, carboxyamide, trichloroacetic and trichloropicolinic acids derivatives, guanidines and naphthoquinones.

Table 1. Classification of pesticides according to their chemical composition [6].

Pesticides	Characteristics	Examples
Organochlorines	Soluble in lipids, they accumulate in fatty tissue of animals, are transferred through the food chain; toxic to a variety of animals, long-term persistent.	DDT, aldrin, lindane, chlordane, mirex.
Organophosphates	Soluble in organic solvents but also in water. They infiltrate reaching groundwater, less persistent than chlorinated hydrocarbons; some affect the central nervous system. They are absorbed by plants, transferred to leaves and stems, which are the supply of leaf-eating insects or feed on wise.	Malathion, methyl parathion, diazinon
Carbamates	Carbamate acid derivatives; kill a limited spectrum of insects, but are highly toxic to vertebrates. Relatively low persistence	Sevin, carbaryl
Pyrethroids	Affect the nervous system; are less persistent than other pesticides; are the safest in terms of their use, some are used as household insecticides.	Pyrethrins
Biological	Only the *Bacillus thuringiensis* (Bt) and its subspecies are used with some frequency; are applied against forest pests and crops, particularly against butterflies. Also affect other caterpillars.	Dispel, foray, thuricide

Table 2. General characteristics of pesticides [7].

Worldwide approximately 9,000 species of insects and mites; 50,000 species of plant pathogens, and 8,000 species of weeds damage crops. Different pests such as insects and plants causing losses estimated in 14% and 13% respectively. Pesticides are indispensable in agricultural production. About one-third of the agricultural products are produced by using pesticides. Without pesticide application the loss of fruits, vegetables and cereals from pest injury would reach 78%, 54% and 32% respectively. Crop loss from pests declines to 35% to 42% when pesticides are used [8].

Over 1990s, the global pesticide sales remained relatively constant, between 270 to 300 billion dollars, of which 47% were herbicides, 79% were insecticides, 19% were fungicides/bactericides, and 5% the others. Over the period 2007 to 2008, herbicides ranked the first in three major categories of pesticides (insecticides, fungicides/bactericides, herbicides). Fungicides/bactericides increased rapidly and ranked the second. Europe is now the largest pesticide consumer in the world, followed by Asia. As for countries, China, the United States, France, Brazil and Japan are the largest pesticide producers, consumers or traders in the world. Most of the pesticides worldwide are used to fruit and vegetable crops. In the developed countries pesticides, mainly herbicides are mostly used to maize. Since the 1980s hundreds of thousands of pesticides have been developed, including various biopesticides [5].

The global agricultural sector is the primary user of pesticides, consuming over 4 million tons of pesticides annually. Pesticides have been extensively used for decades and have substantially increased the food production [9]. However a large amount of applied pesticides often never reach their intended target due to their degradation, volatilization and leaching, leading to serious ecological problems [9-10]. Under actual agricultural practices, different groups of pesticides are often simultaneously or consecutively applied interacting with each other [11].

Although pesticides are beneficial in controlling the proliferation of pests, their unregulated and indiscriminate applications for the application of pesticides can cause adverse effects to human health, to different life forms and to the ecosystems, which depend on the degree of sensitivity of organisms and toxicity of pesticides. The continued application of pesticides has increased its concentration in soils and waters, besides; they enter to the food chains. Dispersion mechanisms also have increased the level of environmental risk for the occupationally exposed population and the inhabitants of surrounding villages. Despite ban on application of some of the environmentally persistent and least biodegradable pesticides (like organochlorines), in many countries their use is ever on rise. Pesticides cause serious health hazards to living systems because of their rapid fat solubility and bioaccumulation in non-target organisms [2]. The main forms of pollution are direct applications to agricultural crops, accidental spills during transport and manufacturing, as well as waste from tanks where cattle are treated to ectoparasites control [4].

The effects of the impacts of pesticides can be analyzed from two different points of view: environmental and public health. The first occurs when pesticides are introduced to food chains, for example: a) producing a change in the decline of populations of phytoplankton and zooplankton (indicators of water pollution); b) producing carcinogenic, neurotoxic, and on fertility and viability (in invertebrates, fish, amphibians, insects and mammals) of their descendants; c) the presence of pesticides in the environment have caused the resistance of organisms considered as pests and disease vectors (for example malaria, dengue and Chagas disease), and instead other beneficial insect populations are diminished (like pollinators); d) alter biogeochemical cycles by decreasing the macro and microbiota, e) leaching of pesticides pollute water bodies, f) can be adsorbed pesticides when soil particles interact with positively or negatively charged, thus increasing their persistence in the environment (4-26 weeks). From the point of view of public health impact of pesticides is mainly acute intoxications (especially in occupationally exposed populations) or indirect exposure of the general population (through air, water and food contaminated with pesticide residues) [12].

In natural environments, pesticides or their degradation products may be further transformed or degraded by other microorganisms or eventually leading to complete degradation by the microbial consortium. However, persistent xenobiotics like pesticides and metabolic dead-end products will accumulate in the environment, become part of the soil humus, or enter the food chain leading to biomagnification (Figure 1).

The fate of pesticides in the environment is strongly related to the soil sorption processes that control not only their transfer but also their bioavailability [13]. Contamination of soil from pesticides as a result of their bulk handling at the farmyard or following application in the field or accidental release may lead occasionally to contamination of surface and ground water [14]. The behavior of pesticides in soils, the efficiency, persistence and potential as environmental contaminants, depend on their retention and degradation on soil constituents [15]. In soils, several parameters influence the rate of biodegradation processes: environmental factors such as moisture and temperature, physicochemical properties of the soil, presence of other

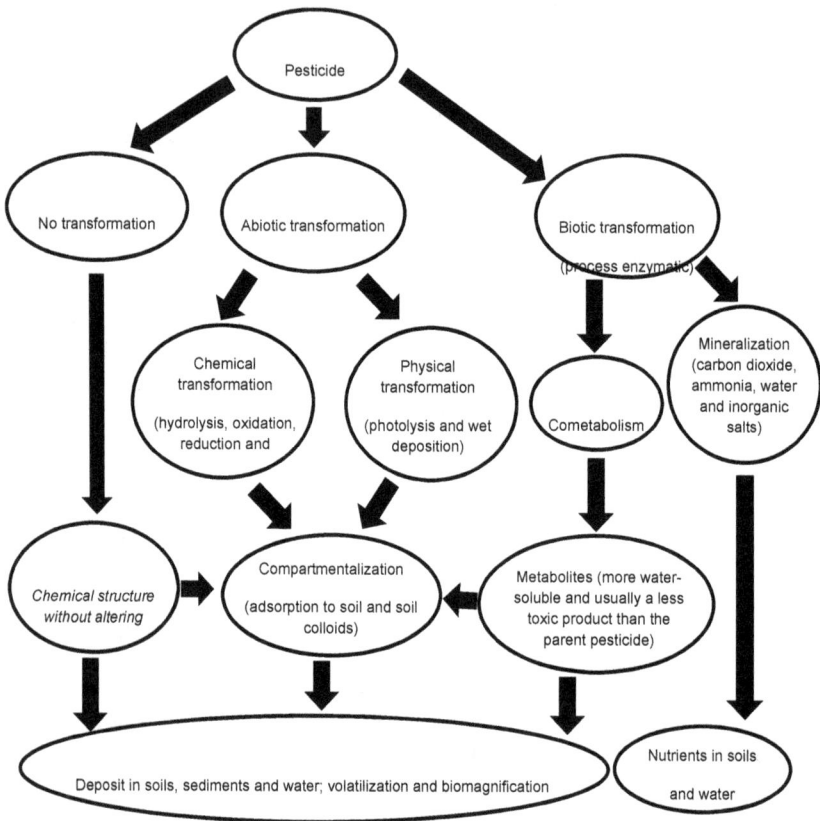

Figure 1. Fate of pesticides in the environment.

nitrogen sources or carbon, etc. can completely modify the microbial population and therefore the microbial activity [13].

On the other hand, liquid and solid wastes and obsolete products are stored or disposed in an inappropriate manner, which has favored the appearance of significant amounts of environmental liabilities, which in most cases are not reported to the appropriate authority. There are more than half a million tons of obsolete, unused, forbidden or outdated pesticides, in several developing and transitional countries, which endanger the environment and health of millions of people [16] In the absence of a clear obsolete pesticides management strategy, over the years, significant amounts of obsolete pesticides have been stockpiled in developing countries [17]. An obsolete pesticide may be recognized as one that is undesirable or impossible to use and has to be eliminated [17-20]. Because of their characteristics, obsolete pesticides are hazardous wastes that should be managed as such. Obsolete pesticides have accumulated in almost every developing country or economy in transition over the past several decades [17]. It is estimated that in Africa and Middle East there are more than 100,000 tons of these products, in Asia almost 200,000 and a similar quantity in East Europe and the old Soviet Union. Nowadays the FAO is elaborating the inventories of Latin America [19,21-22]. In Mexico, there is knowledge of the existence of obsolete pesticide products, both liquid and solid. A total of 551 records of obsolete pesticide products have been registered, distributed in 29 of 33 states of Mexico, achieving a total of 26,725.02 liters, 147,274 kg and 500 m^3 of highly polluted soils. In addition there are 28 reports of pesticide-contaminated sites in 15 states of the Mexican Republic [23]. Besides, some data indicate that the total of empty pesticide containers can be about 7,000 tons annually [24].

2. Strategies to reduce the impact to the environment and health

Due to the problems mentioned above, development of technologies that guarantee their elimination in a safe, efficient and economical way is important. In order to reduce the effects of pesticides on the environment and health, for remediation of contaminated sites and for the treatment of pesticide residues and/or obsolete pesticides, different methods have been developed. Among the existent technologies there are those that apply physical treatments, such as adsorption and percolator filters; chemical treatments such as the advanced oxidation which involve the use of powerful transient species, mainly the hydroxyl radical. Other technique used for the degradation of pesticides the heterogeneous photocatalysis with TiO$_2$ is a method for producing the radical mentioned [25]. A method currently used is high temperature incineration in special furnaces: pesticides are packaged in the places where they were abandoned, then transported to a country that has special facilities to dispose of hazardous wastes. FAO estimates that the cost of these operations varies between 3,000 and 4,000 USD/ton [6]. Other strategies that have been studied for the degradation of these compounds include the photodegradation [26]. However all these methods have several disadvantages such as the use of chemical catalysts, as titanium dioxide, and the use of expensive technology in the case of ozone. For some pesticides alkaline hydrolysis is used, such in the case of organophosphates, which must include a rigorous control of the conditions under which the experiments performed, such as maintenance of alkaline pH, as well as the presence of complexes formed with metal ions, which involves the formation of secondary pollutants.

These conventional physicochemical approaches are generally expensive and remediation process is often incomplete due to the conversion of the parent compound to metabolites which are more persistent and equally or more toxic to non-target organisms [14].

An alternative pesticides treatment with important global boom is bioremediation, which is conducted through the biodegradation of these chemical compounds. This technique relies on the ability of microorganisms to convert organic contaminants in simple and harmless compounds to the environment. Bioremediation overcomes the limitations of traditional methods for the disposal of hazardous compounds, so it has allowed the destruction of many organic contaminants at a reduced cost. Consequently, in the last years, bioremediation technology has progressed to an unknown virtual technology considered for the degradation of a wide range of pollutant compounds. Bioremediation can offer an efficient and cheap option for decontamination of polluted ecosystems and destruction of pesticides [14, 27-30]. As an efficient, economical and environmentally friendly technique, biodegradation has emerged as a potential alternative to the conventional techniques. However, the biodegradation process of many pesticides has not been fully investigated [31].

3. Microorganisms involved in the biodegradation of pesticides

Different biological systems, as microorganisms, have been used to biotransform pesticides. It has been reported that a fraction of the soil biota can quickly develop the ability to degrade certain pesticides, when they are continuously applied to the soil. These chemicals provide adequate carbon source and electron donors for certain soil microorganisms [32], establishing a way for the treatment of pesticide-contaminated sites [33-34].

Furthermore, the isolated microorganisms capable of degrading pesticides can be used for bioremediation of other chemical compounds to whom any microbial degradation system is known [14]. However, the transformation of such compounds depends not only on the presence of microorganisms with appropriate degrading enzymes, but also a wide range of environmental parameters [35]. Additionally, some physiological, ecological, biochemical and molecular aspects play an important role in the microbial transformation of pollutants [36-37].

There are different sources of microorganisms with the ability to degrade pesticides. Because pesticides are mainly applied to agricultural crops, soil is the medium that mostly gets these chemicals, besides pesticide industry's effluent, sewage sludge, activated sludge, wastewater, natural waters, sediments, areas surrounding the manufacture of pesticides, and even some live organisms. In general, microorganisms that have been identified as pesticide degraders have been isolated from a wide variety of sites contaminated with some kind of pesticide. At present, in different laboratories around the world there are collections of microorganisms characterized by their identification, growth and degradation of pesticides. The isolation and characterization of microorganisms that are able to degrade pesticides give the possibility to count with new tools to restore polluted environments or to treat wastes before the final disposition [16].

Microbial processes that eliminate organic environmental contamination are important. Progress in the biotechnology of biodegradation relies upon the underlying sciences of environmental microbiology and analytical geochemistry. Recent key discoveries advancing knowledge of biodegradation (in general) and the aromatic-hydrocarbon biodegradation (in particular) have relied upon characterization of microorganisms: pure-culture isolates, laboratory enrichment cultures, and in contaminated field sites. New analytical and molecular tools (ranging from sequencing the DNA of biodegrading microorganisms) have deepened our insights into the mechanisms (how), the occurrence (what), and the identity (who) of active players that effect biodegradation of organic environmental pollutants [38], (Figure 2).

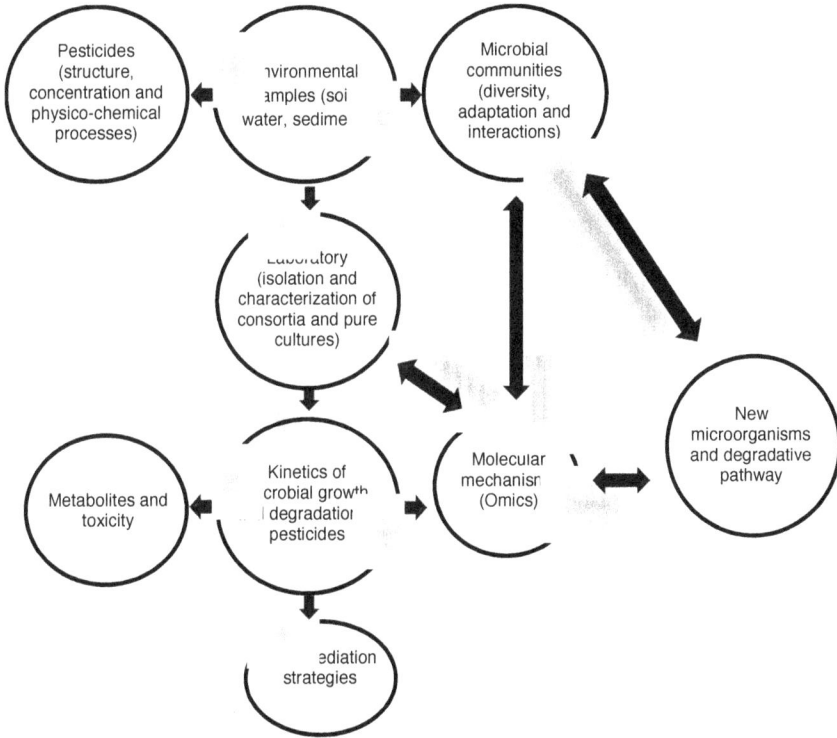

Figure 2. Representation of the relationships between pesticides, microbial communities, and the discovery of new biodegradation processes, Omics = high throughput-based characterization of biomolecules characteristic of bioprocesses; DNA, genomics; mRNA, transcriptomics; protein, proteomics; metabolites, metabolomics.

In the literature there are some examples of microbial pesticide degradation, among them, the following reports deserve mention:

According to [39], *Pseudomonas*, is the most efficient bacterial genus for the degradation of toxic compounds. The ability of these bacteria to degrade these compounds, is related to the contact time with the compound, the environmental conditions in which they develop and their physiological versatility. In other report [40], evaluated three *Pseudomonas* species for the biodegradation of the herbicide aroclor 1242, showing that these bacteria have a great ability to degrade it, according to their degradation percentage, 99.8, 89.4 and 98.4 respectively.

[41] isolated various fungi species from Algerian pesticide contaminated soils. Observing that the most frequent species isolated were *Aspergillus fumigatus, A. niger, A. terreus, Absidia and Rhizopus microsporus var corymberifera microsporis*. In this report, 53 of the isolated species were noted for their ability to degrade the herbicide metribuzin in liquid medium. It was demonstrated, at the same time, that the herbicide promoted the *Absidia* and *Fusarium* genera growth;

these genera were capable to eliminate the 50% of the compound after 5 days. Moreover, the species *Botrytis cinerea* could eliminate the linuron and metroburon herbicides almost completely, and other 31 isolated species also could eliminate metroburon. [42] reports the fungi *Trichoderma viridae* ability in the endosulfan and methyl parathion pesticides degradation.

Another experiments have been demonstrated the efficiency of the bacterium *Rhodococcus sp.* to degrade triazines to nitrate. [43] conducted a test to study the atrazine herbicide transformations resulting from microbial decomposition. After microbial action this compound was transformed into nitrite (30%), nitrous oxide (3.2%), ammonia (10%) and formaldehyde (27%).

Several bacterial genera are adapted to grow in pesticide contaminated soils. These microorganisms have enzymes involved in the hydrolysis of P-O, P-F, P-S and P-C bonds, which are found in a wide variety of organophosphorus pesticides [14]. Some bacteria isolated from the soil are capable of degrading pesticides as ethyl-parathion and methyl-parathion.

4. Biodegradation mechanisms

Biodegradation that involves the capabilities of microorganisms in the removal of pollutants is the most promising, relatively efficient and cost-effective technology. Biodegradation is a process that involves the complete rupture of an organic compound in its inorganic constituents. The microbial transformation may be driven by energy needs, or a need to detoxify the pollutants, or may be fortuitous in nature (cometabolism). Because of the ubiquitous nature of microorganisms, their numbers and large biomass relative to other living organisms in the earth, wider diversity and capabilities in their catalytic mechanisms [44], and their ability to function even in the absence of oxygen and other extreme conditions the search for pollutant-degrading microorganisms, understanding their genetics and biochemistry, and developing methods for their application in the field have become an important human endeavor [45].

As much as the diversity in sources and chemical complexities in organic pollutants exists, there is probably more diversity in microbial members and their capabilities to synthesize or degrade organic compounds [46-47].The microbial populations of soil or aquatic environments are composed of diverse, synergistic or antagonistic communities rather than a single strain. In natural environments, biodegradation involves transferring the substrates and products within a well-coordinated microbial community, a process referred to as metabolic cooperation [48]. Microorganisms have the ability to interact, both chemically and physically, with substances leading to structural changes or complete degradation of the target molecule. Among the microbial communities, bacteria, fungi, and actinomycetes are the main transformers and pesticide degraders [49]. Fungi generally biotransform pesticides and other xenobiotics by introducing minor structural changes to the molecule, rendering it nontoxic. The biotransformed pesticide is released into the environment, where it is susceptible to further degradation by bacteria [50].

Fungi and bacteria are considered as the extracellular enzyme-producing microorganisms for excellence. White rot fungi have been proposed as promising bioremediation agents, especially for compounds not readily degraded by bacteria. This ability arises from the production of extracellular enzymes that act on a broad array of organic compounds. Some of these extracellular enzymes are involved in lignin degradation, such as lignin peroxidase, manganese peroxidase, laccase and oxidases. Several bacterial that degrade pesticide have been isolated and the list is expanding rapidly. The three main enzyme families implicated in degradation are esterases, glutathione S-transferases (GSTs) and cytochrome P450 [51].

Enzymes are central to the biology of many pesticides [52]. Applying enzymes to transform or degrade pesticides is an innovative treatment technique for removal of these chemicals from polluted environments. Enzyme-catalyzed degradation of a pesticide may be more effective than existing chemical methods. Enzymes are central to the mode of action of many pesticides: some pesticides are activated *in situ* by enzymatic action, and many pesticides function by targeting particular enzymes with essential physiological roles. Enzymes are also involved in the degradation of pesticide compounds, both in the target organism, through intrinsic detoxification mechanisms and evolved metabolic resistance, and in the wider environment, via biodegradation by soil and water microorganisms [53]. [54] suggested that (i) the central metabolism of the global biodegradation networks involves transferases, isomerases, hydrolases and ligases, (ii) linear pathways converging on particular intermediates form a funnel topology, (iii) the novel reactions exist in the exterior part of the network, and (iv) the possible pathway between compounds and the central metabolism can be arrived at by considering all the required enzymes in a given organism and intermediate compounds [47].

For pesticides degradation, three are mainly enzyme systems involved: hydrolases, esterases (also hydrolases), the mixed function oxidases (MFO), these systems in the first metabolism stage, and the glutathione S-transferases (GST) system, in the second phase [55]. Several enzymes catalyze metabolic reactions including hydrolysis, oxidation, addition of an oxygen to a double bound, oxidation of an amino group (NH_2) to a nitro group, addition of a hydroxyl group to a benzene ring, dehalogenation, reduction of a nitro group (NO_2) to an amino group, replacement of a sulfur with an oxygen, metabolism of side chains, ring cleavage. The process of biodegradation depends on the metabolic potential of microorganisms to detoxify or transform the pollutant molecule, which is dependent on both accessibility and bioavailability [47].

Metabolism of pesticides may involve a three-phase process. In Phase I metabolism, the initial properties of a parent compound are transformed through oxidation, reduction, or hydrolysis to generally produce a more water-soluble and usually a less toxic product than the parent. The second phase involves conjugation of a pesticide or pesticide metabolite to a sugar or amino acid, which increases the water solubility and reduces toxicity compared with the parent pesticide. The third phase involves conversion of Phase II metabolites into secondary conjugates, which are also non-toxic. In these processes fungi and bacteria are involved producing intracellular or extra cellular enzymes including hydrolytic enzymes, peroxidases, oxygenases, etc [16, 56].

Due to the diversity of chemistries used in pesticides, the biochemistry of pesticide bioremediation requires a wide range of catalytic mechanisms, and therefore a wide range of enzyme classes. Information for some pesticide degrading enzymes could be founded in Table 3.

Enzyme	Organism	Pesticide
Oxidoreductases (Gox)	*Pseudomonas* sp. LBr	Glyphosate
	Agrobacterium strain T10	
Monooxygenases:		
ESd	*Mycobacterium* sp.	Endosulphan and Endosulphato
Ese	*Arthrobacter* sp.	Endosulphan, Aldrin, Malation, DDDT and Endosulphato
Cyp1A1/1ª2	Rats	Atrazine, Norflurazon and Isoproturon
Cyp76B1	*Helianthus tuberosus*	Linuron, Chlortoluron and Isoproturon
P450	*Pseudomonas putida*	Hexachlorobenzene and Pentachlorobenzene
Dioxygenases (TOD)	*Pseudomonas putida*	Herbicides Trifluralin
E3	*Lucilia cuprina*	Synthetic pyrethroids and insecticides phosphotriester
Phosphotriesterases:	Agrobacterium radiobacter	Insecticides phosphotriester
OPH/OpdA	Pseudomonas diminuta	
	Flavobacterium sp.	
Haloalkane Dehalogenases:	*Sphingobium* sp.	Hexachlorocyclohexane (β and δ isomers)
LinB	*Shingomonas* sp.	
AtzA	*Pseudomonas* sp. ADP	Herbicides chloro-s-trazina
TrzN	*Nocardioides* sp.	Herbicides chloro-s-trazina
LinA	*Sphingobium* sp.	Hexachlorocyclohexane (γ isomers)
	Shingomonas sp.	
TfdA	*Ralstonia eutropha*	2,4 - dichlorophenoxyacetic acid and pyridyl-oxyacetic
DMO	*Pseudomonas maltophilia*	Dicamba

Table 3. Relevant enzymes in the bioremediation of pesticides [52-53].

5. Generalities of the major enzymatic activities applied for pesticide biodegradation

5.1. Hydrolases

Hydrolases are a broad group of enzymes involved in pesticide biodegradation. Hydrolases catalyze the hydrolysis of several major biochemical classes of pesticide (esters, peptide bonds, carbon-halide bonds, ureas, thioesters, etc.) and generally operate in the absence of redox cofactors, making them ideal candidates for all of the current bioremediation strategies [53].

As an example of the catalytic activity of enzymes hydrolases, the degradation pathway of carbofuran, a pesticide the group of carbamates is presented (Figure 3). This pesticide can be transformed in the environment and different metabolites are generated and accumulated in potentially contaminated sites (soil, water and sediments, mainly). Different organisms isolated from contaminated sites that have been identified and characterized as transformers of carbofuran, resulting in different metabolites [57].

Among the hydrolases involved in the degradation of pesticides are including different types such as:

5.2. Phosphotriesterases (PTEs)

Among the most studied pesticide degrading enzymes, the PTEs are one of the most important groups [58]. These enzymes have been isolated from different microorganisms that hydrolyze and detoxify organophosphate pesticides (OPs). This reduces OP toxicity by decreasing the ability of OPs to inactivate AchE [14, 59-62]. The first isolated phosphotriesterase belongs to the *Pseudomonas diminuta* MG species; this enzyme shows a highly catalytic activity towards organophosphate pesticides. The phosphotriesterases are encoded by a gene called *opd* (organophosphate-degrading). *Flavobacterium* ATCC 27551 presents the *opd* gene encoding to a PTE [63]. The gene was cloned and sequenced by [64]. These enzymes specifically hydrolyze phosphoester bonds, such as P–O, P–F, P–NC, and P–S, and the hydrolysis mechanism involves a water molecule at the phosphorus center [65]. Different microbial enzymes with the capacity to hydrolyze MP have been identified, such as organophosphorus hydrolase (OPH; encoded by the *opd* gene), methyl-parathion hydrolase (MPH; encoded by the *mpd* gene), and hydrolysis of coroxon (HOCA; encoded by the *hocA* gene), which were isolated from *Flavobacterium* sp. [66], *Plesimonas* sp. strain M6 [67] and *Pseudomonas moteilli* [68], respectively.

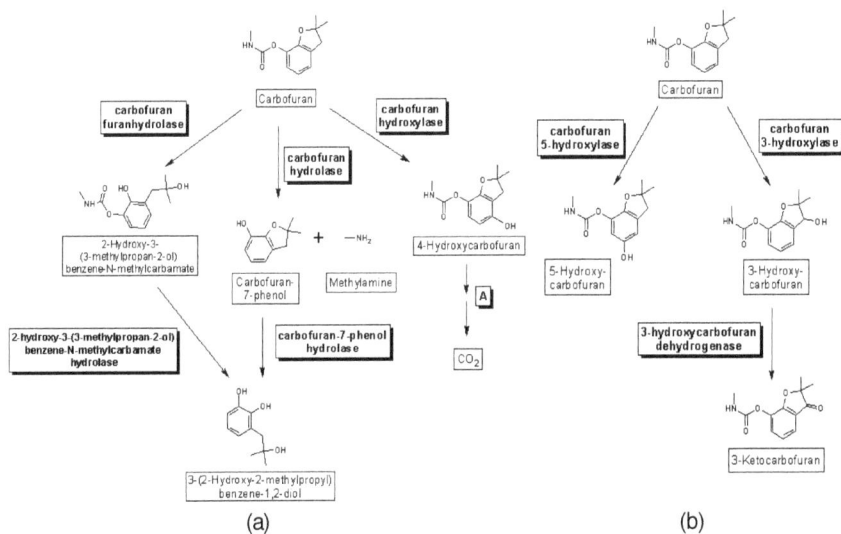

Figure 3. Degradation pathway of carbofuran. In a) several bacteria are involved in the hydrolysis of metabolites and b) fungal degradation of carbofuran may occur via hydroxylation at the three position and oxidation to 3-ketocarbofuran (University of Minnesota. Biocatalysis/Biodegradation Database, http://www.umbbd.ethz.ch/cbf/cbf_image_map1.html).

The phosphotriesterase enzyme is a homo-dimeric protein with a monomeric molecular weight of 36 Kda. As a first step in the PTE organophosphorous pesticide hydrolysis mechanism, the enzymatic active site removes a proton from water, activating this molecule, them, the activated water directly attacks the central phosphorus of the pesticide molecule producing an inversion in its configuration The oxygen is polarized by the active site, with the participation of a zinc atom [6, 69], (Figure 4). This enzyme has potential use for the cleaning of organophosphorus pesticides contaminated environments [65].

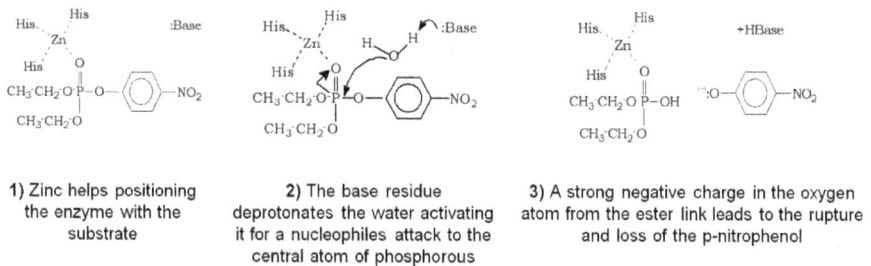

1) Zinc helps positioning the enzyme with the substrate

2) The base residue deprotonates the water activating it for a nucleophiles attack to the central atom of phosphorous

3) A strong negative charge in the oxygen atom from the ester link leads to the rupture and loss of the p-nitrophenol

Figure 4. Proposed mechanism for PTE activity. Zinc's active site functions in phosphate polarization, making phosphor more susceptible to the attack. 1) A base subtracts a proton from a water molecule with the subsequent attack of the hydroxyl to the central phosphorous. 2) The intermediary complex originates the products 3) p-nitrophenol and diethyl thiophosphate [6].

5.3. Esterases

Esterases are enzymes that catalyze hydrolysis reactions over carboxylic esters (carboxiesterases), amides (amidases), phosphate esters (phosphatases), etc. [70]. In the reaction catalyzed by esterases, hydrolysis of a wide range of ester substrates occurs in their alcohol and acid components as following:

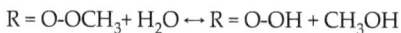

$$R = O\text{-}OCH_3 + H_2O \leftrightarrow R = O\text{-}OH + CH_3OH$$

Many insecticides (organophosphates, carbamates and pyrethroids) have associated a carboxylic ester, and the enzymes capable of hydrolyze such ester bond are known with the name of carboxylesterases.

At present, multiple classification nomenclature systems are used for these enzymes. According to the International Union of Biochemistry and Molecular Biology (IUBMB) nomenclature, carboxylesterases are located in the group of hydrolases (3), subgroup 1, and within it, in subtype 1 (Enzyme Commission 3.1.1.1, EC 3.1.1.1). Another common classification is the nomenclature divides the esterases into three groups according to the nature of their interactions with organophosphorus insecticides. Carboxylesterases belong, according to this classification, the group of ali-esterases and B-esterases. Esterases are a large family of enzymes in arthropods [71].

The esterases are a group of enzymes highly variable, which has been recognized as one of the most important in the metabolism of xenobiotics and its mechanism is associated with the mass production of multifunctional hydrolytic enzymes Organophosphate pesticides can be hydrolyzed by such enzymes [72-74]. There are different types of esterases and with very different distribution in tissues and organisms. The Carboxiesterases (type B esterases) are a group that hydrolyze, additionally to endogenous compounds, xenobiotics with ester, amide, thioester, phosphate esters (parathion, paraoxon) and acid anhydrides (DIPFP=DFP) in mammals.

Esterases A, contain a Cys residue in the active center and esterases B contain a Ser residue. In esterases A, the organophosphates interact with the functional group-SH forming a bond between P=S, which is easily hydrolyzed by H_2O. In the esterase B, organophosphates interaction with the SER-OH forming a P=O bond that is not hydrolyzed by H_2O. Organophosphates that bind to the esterase B stoichiometrically inhibit its enzymatic activity.

Esterases are a diverse group that protects the target site (acetylcholinesterase) by catalyzing the hydrolysis of insecticides, or acting as an alternative blank [75]. Esterases in general have a wide range of substrate specificities; they are capable of binding to phosphate triesters, esters, thioesters, amides and peptides [76].

5.4. Oxidoreductases

Oxidoreductases are a broad group of enzymes that catalyze the transfer of electrons from one molecule (the reductant or electron donor) to another (the oxidant, or electron acceptor). Many of these enzymes require additional cofactors, to act as either electron donors, electron acceptors or both. Oxidoreductases have been further sub classified into 22 subclasses (EC 1.1-1.21 and 1.97). Several of these have applications in bioremediation, albeit their need for cofactors complicates their use in some applications. There are enzymes that catalyze an oxidation/reduction reaction by including the molecular oxygen (O_2) as electron acceptor. In these reactions, oxygen is reduced to water (H_2O) or hydrogen peroxide (H_2O_2). The oxidases are a subclass of the oxidoreductases [53].

As an example of the many functions of these enzymes in the degradation of pesticides, as an example we present the ensodulfan degradation pathway. In this process not only oxidoreductase enzymes are involved, but different microorganisms and catalytic activities, in combination, can lead to complete mineralization of a pesticide (Figure 5). Endosulfan (*1,2,5,6,7,7-hexachloro-5-norbornene-2,3- dimethanolcyclic sulfite*) is an organochlorine insecticide of the cyclodiene family of pesticides. It is highly toxic and endocrine disruptor, and it is banned in European Union and several countries. Because it has been extensively applied directly to fields, it can be detected a considerable distance away from the original site of application. Contamination of drinking water and food, as well as detrimental effects to wildlife are important concerns [77]. The molecular structure has two stereochemical isomers α and β endosulfan. The end-use product of endosulfan is a mixture of two isomers, typically in a 2:1 ratio.

Microorganisms play a key role in removal of xenobiotics like endosulfan from the contaminated sites because of their dynamic, complex and complicated enzymatic systems which degrade these chemicals by eliminating their functional groups of the parent compound. This pesticide can undergo either oxidation or hydrolysis reactions. Several intensive studies on the degradation of endosulfan have been conducted showing the primary metabolites to normally be endosulfan sulfate and endosulfan diol (endodiol). Endosulfan sulphate will be present in the environment as a result of the use of endosulfan as an insecticide. If endosulfan sulphate is released to water, it is expected to absorb to the sediment and may bioconcentrate in aquatic organism. This metabolite has a similar toxicity as endosulfan and has a much longer half-life in the soil compared to endosulfan. Therefore, production of endosulfan sulfate by biological systems possesses an ecological hazard in that it contributes to long persistence of endosulfan in soil. Endodiol is much less toxic to fish and other organisms than the parent compound.

Thus, it is important to note that some microbial enzymes are specific to one isomer, or catalyze at different rates for each isomer [78]. For example, a *Mycobacterium tuberculosis* ESD enzyme degrades beta-endosulfan to the monoaldehyde and hydroxyether (depending on the reducing equivalent stoichiometry), but transforms alpha-endosulfan to the more toxic endosulfan sulfate. However, oxidation of endosulfan or endosulfan sulfate by the monooxygenase encoded by *ese* in *Arthrobacter* sp. KW yields endosulfan monoalcohol [79]. Both *ese* and *esd* proteins are part of the unique Two Component Flavin Dependent Monooxygenase Family, which require reduced flavin. They are conditionally expressed when no or very little sulfate or sulfite is available, and endosulfan is available to provide sulfur in these starved conditions.

Alternatively, hydrolysis of endosulfan in some bacteria (*Pseudomonas aeruginosa, Burkholderia cepaeia*) yields the less toxic metabolite endosulfan diol [80]. Endosulfan can spontaneously hydrolyze to the diol in alkaline conditions, so it is difficult to separate bacterial from abiotic hydrolysis. The diol can be converted to endosulfan ether or endosulfan hydroxyether and then endosulfan lactone [81]. Hydrolysis of endosulfan lactone yields endosulfan hydroxycarboxylate. These various branches of endosulfan degradation all result in desulfurization while leaving the chlorines intact, exhibiting the recalcitrance to bioremediation found in many organohalogen aromatics.

5.5. Mixed Function Oxidases (MFO)

In the reaction catalyzed by the MFO (EC 1.14.14.1), an atom of one molecule of oxygen is incorporated into the substrate, while the other is reduced to water. For this reason the MFO requires Nicotiamide-adenine dinucleotide phosphate (NADPH) and O_2 for its operation.

It is an enzyme system comprising two enzymes, cytochrome P450 and NADPH-cytochrome P450 reductase, both membrane proteins. They are also known as dependent cytochrome P-450 monooxygenases or P450 system. The genes encoding the different isozymes comprise a superfamily of over 200 genes grouped into 36 families based on their sequence similarity. Cytochrome P450 enzymes are active in the metabolism of wide variety of xenobiotics [82].

The cytochrome P450 family is a large, well characterized group of monooxygenase enzymes that have long been recognized for their potential in many industrial processes, particularly due to their ability to oxidize or hydroxylate substrates in an enantiospecific manner using molecular oxygen [83]. Many cytochrome P450 enzymes have a broad substrate range and have been shown to catalyse biochemically recalcitrant reactions such as the oxidation or hydroxylation of non-activated carbon atoms. These properties are ideal for the remediation of environmentally persistent pesticide residues. Over 200 subfamilies of P450 enzymes have been found across various prokaryotes and eukaryotes. All contain a catalytic iron-containing porphyrin group that absorbs at 450 nm upon binding of carbon monoxide. In common with many of the other oxidoreductases described before, P450 enzymes require a non-covalently bound cofactor to recycle their redox center (most frequently NAD(P)H is used), which limits their potential for pesticide bioremediation to strategies that employ live organisms.

In insects, MFOs are found in the endoplasmic reticulum and mitochondria, are involved in a large number of processes such as growth, development, reproduction, detoxification, etc. MFO are involved in the metabolism of both endogenous and exogenous substances, for this reason these compounds promote their induction Due to its high inspecificity, the MFOs metabolize a wide range of compounds such as organophosphates, carbamates, pyrethroids, DDT, inhibitors of the chitin synthesis, juvenile hormone mimics, etc. [84].

5.6. Glutathione S-Transferase (GST)

The GSTs (EC 2.5.1.18) are a group of enzymes that catalyze the conjugation of hydrophobic components with the tripeptide glutathione (Figure 6). In this reaction, the thiol group of glutathione reacts with an electrophilic place in the target compound to form a conjugate which can be metabolized or excreted, and they are involved in many cellular physiological activities, such as detoxification of endogenous and xenobiotic compounds, intracellular transport, biosynthesis of hormones and protection against oxidative stress [85].

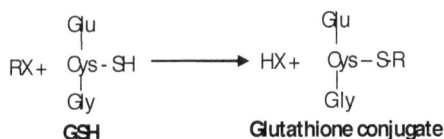

Figure 6. Representation of the conjugation reaction catalyzed by glutathione S-transferase (GST).

6. Genetics for pesticide degradation

In order to investigate genetic basis of pesticides biodegradation, several works with special emphasis on the role of catabolic genes and the application of recombinant DNA technology, had been reported. Pesticide-degrading genes of only a few microorganisms have been

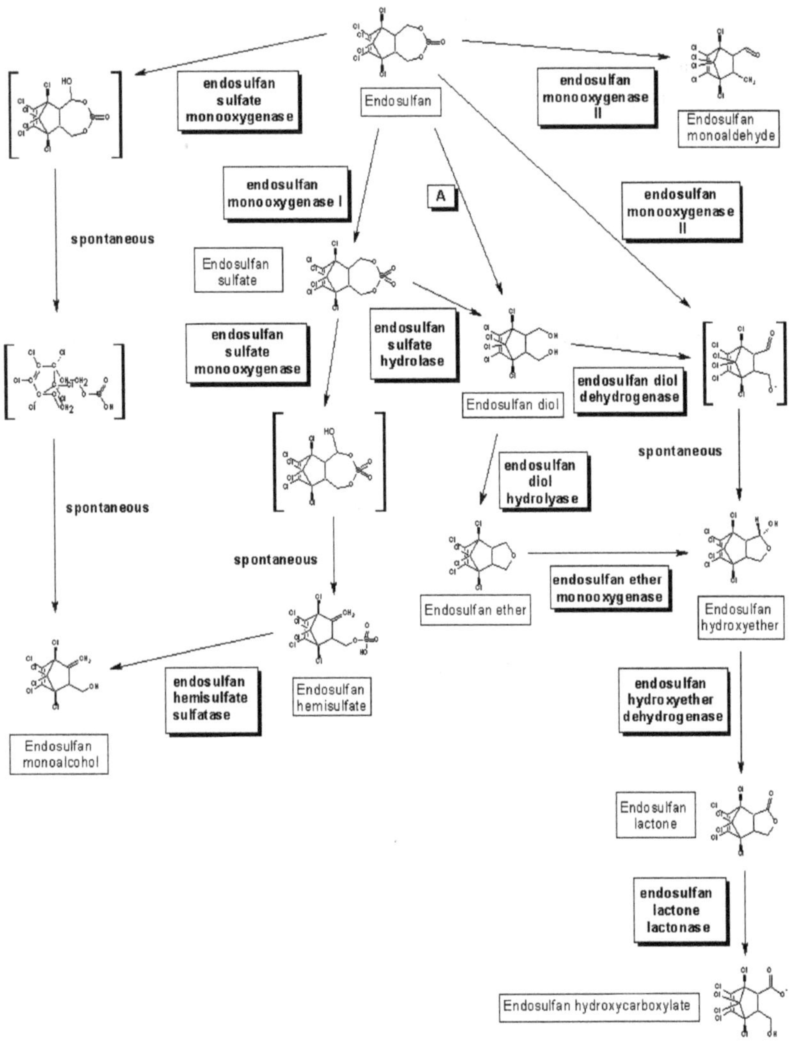

Figure 5. Degradation pathway of endosulfan (University of Minnesota. Biocatalysis/Biodegradation Database, http://www.umbbd.ethz.ch/end/end_map.html).

characterized. Most of genes responsible for catabolic degradation are located on the chromosomes, but in a few cases these genes are found in plasmids or transposons. The recent advances in metagenomics and whole genome sequencing have opened up new avenues for searching the novel pollutant degradative genes and their regulatory elements from both culturable and nonculturable microorganisms from the environment. Mobile genetic elements such as plasmids and transposons have been shown to encode enzymes responsible for the degradation of several pesticides. The isolation of pesticide degrading microorganisms and the characterization of genes encoding pesticide degradation enzymes, combined with new techniques for isolating and examining nucleic acids from soil microorganisms, will yield unique insights into the molecular events that lead to the development of enhanced pesticide degradation phenomenon.

An understanding of the genetic basis of the mechanisms of how microorganisms biodegrade pollutants and how they interact with the environment is important for successful implementation of the technology for in situ remediation [86].

Different microbial enzymes with the capacity to hydrolyze pesticides have been identified [57], such as **organophosphorus hydrolase** (OPH; encoded by the *opd* gene). This gene has been found in bacterial strains that can use organophosphate pesticides as carbon source; these have been isolated in different geographic regions. These plasmids show considerable genetic diversity, but the region containing the *opd* gene is highly conserved. **Methyl-parathion hydrolase** (MPH; encoded by the *mpd* gene), Are *Pseudaminobacter, Achrobacter, Brucella* and *Ochrobactrum* genes, they were identified by comparison with the gene *mpd* from *Pleisomonas* sp. M6 strain [87], the gene for the organophosphorus hydrolase has 996 nucleotides, a typical promoter sequence of the promoter TTGCAA N17 TATACT from *E. coli* [88].

In the various isolates of microorganisms capable of degrading pesticide, several genes have been described, in the table 4 shown the most studied.

7. Genetic engineering

Microorganisms respond differently to various kinds of stresses and gain fitness in the polluted environment. This process can be accelerated by applying genetic engineering techniques. The recombinant DNA and other molecular biological techniques have enabled (i) amplification, disruption, and/or modification of the targeted genes that encode the enzymes in the metabolic pathways, (ii) minimization of pathway bottlenecks, (iii) enhancement of redox and energy generation, and (iv) recruiting heterologous genes to give new characteristics [45,89]. Various genetic approaches have been developed and used to optimize the enzymes, metabolic pathways and organisms relevant for biodegradation [90]. New information on the metabolic routes and bottlenecks of degradation is still accumulating, requiring the need to reinforce the available molecular toolbox. Nevertheless, the introduced genes or enzymes, even in a single modified organism, need to be integrated within the regulatory and metabolic network for proper expression [89-91].

Detoxification of organophosphate pesticides was the first demonstrated by genetically engineered microorganisms and the genes encoding these hydrolases have been cloned and expressed in *P. pseudoalcaligenes, Escherichia coli, Streptomyces lividans, Yarrowia lipolytica* and *Pichia pastoris* [92-96].

Gene	Organism
Bacteria	
Opd	*Pseudomonas diminuta*
opaA	*Alteromonas* spp.
opdA	*A. radiobacter*
adpB	*Nocardia* sp.
pepA	*Escherichia coli*
hocA	*Pseudomonas monteilli*
pehA	*Burkholderia caryophilli*
Phn	*Bacillus cereus*
ophB	*Burkholderia* sp. JBA3.
ophC2	*Stenotrophomonas* sp. SMSP-1.
OpdB	*Lactobacillus brevis.*
Imh	*Arthrobacter* sp. scl-2.
Mpd	*Ochrobactrum* sp. Yw28, *Rhizobium radiobacter*
Oph	*Arthrobacter* sp
Mph	*Arthrobacter* sp. L1 (2006).
MpdB	*Burkholderia cepacia*
opdE	*Enterobacter* sp.
Fungi	
A-opd	*Aspargillus niger*
P-opd	*Penicillium lilacinum*

Table 4. Genes with the ability to degrade pesticides (Modified from 14).

Another strategy that has been used is phytoremediation, the use of plants to clean-up polluted soil and water resources is recognized as an economically cheaper, aesthetically pleasing, and environmentally friendly 'Green technology [97,98]. However, the limitation with plants is that they lack the catabolic pathways for complete degradation/mineralization of externally added organic compounds. The potential of plants to degrade organic pollutants can be further enhanced by engineering plants by introduction of efficient heterologous genes that are involved in the degradation of organic pollutants [98].

Unfortunately, the rates of hydrolysis several enzymes differ dramatically for members of the family of OP compounds, ranging from hydrolysis at the diffusion-controlled limit for paraoxon to several orders of magnitude slower for malathion, chlorpyrifos, and others pesticides [99]. Although site-directed mutagenesis has been used to improve the substrate specificity and stereoselectivity of OPH [99-100], the ability to deduce substitutions that are important for substrate specificity is still limited to the active-site residues.

Two interesting papers have shown that an biological solution for efficient decontamination might be to direct evolution. Directed evolution has recently been used to generate OPH variants with up to 25-fold improvements in hydrolysis of methyl parathion [101], a substrate that is hydrolyzed 30-fold less efficiently than paraoxon, and other report the directed evolution of OPH to improve the hydrolysis of a poorly hydrolyzable substrate, chlorpyrifos (1,200-fold less efficient than paraoxon). Up to 700-fold improvement was obtained, and the best variant hydrolyzes chlorpyrifos at a rate similar to that of the hydrolysis of paraoxon by wild-type OPH [102].

8. The application of genomics and functional genomics

8.1. Metagenomics

The complexity of microbial diversity results from multiple interacting parameters, which include pH, water content, soil structure, climatic variations and biotic activity. Current estimates indicate that more than 99% of the microorganisms present in many natural environments are not readily culturable and therefore not accessible for biotechnology or basic research [103]. During the last two decades, development of methods to isolate nucleic acids from environmental sources has opened a window to a previously unknown diversity of microorganisms. Analysis of nucleic acids directly extracted from environmental samples allows researchers to study natural microbial communities without the need for cultivation [103-104].

Each organism in an environment has a unique set of genes in its genome; the combined genomes of all the community members make up the "metagenome". Metagenome technology (metagenomics) has led to the accumulation of DNA sequences and these sequences are exploited for novel biotechnological applications [105,106]. Due to the overwhelming majority of non-culturable microbes in any environment, metagenome searches will always result in identification of hitherto unknown genes and proteins [105-106].

8.2. Functional genomics

In its broadest definition, functional genomics encompasses many traditional molecular genetics and biological approaches, such as the analysis of phenotypic changes resulting from mutagenesis and gene disruption [107]. Functional genomics has emerged recently as a new discipline employing major innovative technologies for genome-wide analysis supported by bioinformatics. These new techniques include proteomics for protein identification, characterization, expression, interactions and transcriptomic profiling by microarrays and metabolic engineering [107]. The application of proteomics in environmental bioremediation research provides a global view of the protein compositions of the microbial cells and offers a promising approach to address the molecular mechanisms of bioremediation. With the combination of proteomics, functional genomics provide an insight into global metabolic and regulatory networks that can enhance the understanding of gene functions.

The fundamental strategy in a functional genomics approach is to expand the scope of biological investigations from studying a single gene or protein to studying all the genes or proteins simultaneously in a systematic fashion. The classic approach to assess gene function is to identify which gene is required for a certain biological function at a given condition through gene disruption or complementation. With the combination of technologies, such as transcriptomics and proteomics complementing traditional genetic approaches, the detailed understanding of gene functions becomes feasible [105-107].

Metabolic engineering combines systematic analysis of metabolic and other pathways with molecular biological techniques to improve cellular properties by designing and implementing rational genetic modifications [108]. Understanding microbial physiology, will adapt to the host cells to support changes and become more efficient bioremediation processes, events that would be difficult to acquire during evolution [105].

With these new genomics tools scientists are in a better position to answer questions such as how oxygen stress, nutrient availability, or high contaminant concentrations along differing geochemical gradients or at transitional interfaces impact the organohalide respiring community structure and function. Ultimately, by tracking the overall microbial community structure and function in addition to key functional players, informed decisions can then be made regarding how to best manipulate the field conditions to achieve effective bioremediation of, e.g., pesticides.

9. Strategies to enhance the efficiency of pesticide degradation: Case cells immobilization

Cell immobilization has been employed for biological removal of pesticides because it confers the possibility of maintaining catalytic activity over long periods of time [109-111]. Whole-cell immobilization has been shown to have remarkable advantages over conventional biological systems using free cells, such as the possibility of employing a high cell density, the avoidance of cell washout, even at high dilution rates, easy separation of cells from the reaction system, repeated use of cells, and better protection of cells from harsh environments. Previous reports have suggested that this higher productivity results from cellular or genetic modifications induced by immobilization. There is evidence indicating that immobilized cells are much more tolerant to perturbations in the reaction environment and less susceptible to toxic substances, which makes immobilized cell systems particularly attractive for the treatment of toxic substances like pesticides [112]. In addition, the enhanced degradation capacity of immobilized cells is due primarily to the protection of the cells from inhibitory substances present in the environment. The degradation rates for repeated operations were observed to increase for successive batches, indicating that cells became better adapted to the reaction conditions over time [113].

There are two types of processes for cell immobilization: those based on physical retention (entrapment and inclusion membrane) and those based on chemical bonds, such as biofilm formation [114]. In cell immobilization methods may be used various materials or substrates inorganic (clays, silicates, glass and ceramics) and organic (cellulose, starch, dextran, agarose, alginate, chitin, collagen, keratin, etc.) [115]. Entrapment in polymeric gels natural has become the preferred technique for the immobilization of cells, however, immobilized cell on supports have been used more frequently in xenobiotics biodegradation as pesticides [116].

In order to degrade pesticides, is important to search for materials with favorable characteristics for the immobilization of cells, including aspects such physical structure, ease of sterilization, the possibility of using it repeatedly, but above all, the support must be cheap than allow in the future apply it for pesticide degradation. Table 5 describes the main methods of immobilization [115,117-120]. Thus, the methods can be grouped in two ways: the active that induce the capture of microorganisms in a matrix, and the passive that uses the tendency of microorganisms to attack surfaces either natural or synthetic, which form biofilms.

TYPE	METHOD	DESCRIPTION	MATERIAL / MATRIX
Chemical bond	Carrier union	Based on the union of biocatalyzers to insoluble carriers through covalent or ionic liknks, physical adsorption and biospecific union. The carrier materials must have enough mechanical strength, physical, chemical and biological stability. They must be economic and malleable but not toxic. This method does not apply with cells, for it is difficult to find the immobilization conditions.	• Water insoluble polysaccharides: dextran, cellulose, agarose • Proteins: albumin • Synthetic polymers: polystyrene deriver, ionic exchange resins • Organic materials: ceramics, magnetite and glass
	Cross-linked	It uses multifunctional reactive. Cells are linked to a matrix in such way that they form concentrate pellets	Dialdehydes, glutaraldehydes and diisocyanates are used.
Physical Retention	Entrapment	This method consists on the retention in inner cavities of a porous matrix	It uses porous matrix: alginate, agar, k-carrageenan, polyacrylamide, chitosan, collagen, polystyrene, cellulose triacetate, activated charcoal, porous ceramic and diatomaceous earth
	Inclusion in membrane	The enzymes or cells are surrounded by semipermeable membranes. This method allows multiple steps. Reactions to take place in reactors	Materials that form surfactant micelles

Table 5. Classification and description of methods of immobilization.

By the other hand, a biofilm can be defined as a coherent complex structure of microorganism organized in colony and cell products such as extracellular polymers (exopolymer), which either spontaneously or in forming dense granules, grow attached to a solid surface static (static biofilm) or in a suspension bracket [121,122]. The biofilm formation process is performed in several steps starting with the attack or recognition to the surface, followed by growth and utilization of various carbon and nitrogen sources for the formation of products with adhesive properties. In parallel a stratified organization dependent on oxygen gradients and other abiotic conditions takes place. This process is known as colonization. Then an intermediate period of maturation of the biofilm is carried out which varies depending on the presence of nutrients from the medium or friction with the surrounding water flow. Finally, a period of aging biofilm where a detachment of cells may occur and colonize other surfaces [123].

The hydrodynamic plays an important role in the development of biofilm as these organizations develop in a solid-liquid interface, where the flow rate passing through it influences the physical detachment of microorganisms. They possess a system of channels that allow the transport of nutrients and waste; this is vital when modify the environment that deprives microorganisms of molecules necessary for their development. Other biofilm characteristic is its resistance to host defenses and antimicrobial agents. While the microorganism are susceptible to different control factors, the colonies organized and included in a exopolymer form an impermeable layer where only the most superficial microorganisms are affected. Also when released biofilm cells, they can travel and to be deposited on new niche maintaining the same characteristics of a biofilm adhered to a surface. Microorganisms are communicated with each other. This is what has been called quorum sensing and involves regulation and expression of specific genes through signaling molecules that mediate intercellular communication [14, 124]. This characteristic is dependent on cell density; for example, biofilm with a high cell density, it induces expression of resistance genes that provide protection and survival [125]. Similarly, microorganisms can produce substances to promote the propagation of colonies and inhibit the growth of other leaving pathogens microorganisms in a more favorable position within the biofilm [126]. The supports may be of synthetic or natural origin (Table 6).

A material that has yielded good results in the degradation of mixtures of pesticides is the tezontle (in Nahuatl, tezt means rock and zontli means hair), that is a native volcanic rock of Morelos state (central Mexico). This rock is highly porous, provides a large contact surface, and can also be sterilized and reused. The presence of micropores allows the establishment of bacterial microcolonies. The immobilization method with this material is based on the colonization of the tezontle micropores through the formation of a biofilm. Subsequently, a current with the pesticides wastes is passed through to allow the contact with the immobilized microorganisms, so this way the biodegradation can be executed. This strategy has been really efficient and is a tool that can be used for the degradation of pesticides wastes [123]. In our work group, a bacterial consortium was immobilized in a biofilm on tezontle and exhibited a considerable capacity for the removal of a mixture of organophosphate pesticides, which are the pesticides widely used in agriculture and stockbreeding in Mexico. In addition, this material with immobilized cells was packaged in an up-flow reactor, which was obtained the greater viability of the bacteria as more efficient removal of pesticides [123].

Support	Xenobiotic	Microorganism	Efficiency	Reference
Glass beads	Coumaphos	*E. coli (transformed)*	Removal: 80%	[127]
Ceramic	Propachlor	*Pseudomonas GCH1*	Removal: 98% (39% absorption to the material)	[128]
Ca Alginate beads	Poliphenols	*Candida tropicalis* YMEC14	After 24 h of fermentation, removal of 69.2% and 55.3% of monophenols and polyphenols respectively,	[129]
Polyurethane Alginate Alginate poly vinyl alcohol	Phenol	*Pseudomonas spp.*	Removal time: 23 days (in suspension culture), 15 days (in polyurethane) and 7 days (in alginate)	[130]
Coffe beans	DDT Endosulfan	*P. aeruginosa* and *F. oryzihabitans,*	Removal: 68% DDT	[131]
Agave tequiliana Webber (blue)	Blue ácido113 Disperse Blue 3 Basic Green 4	*Trametes versicolor* *P. ostreatus* *Klebsiella sp.*	79.85% (9.98 mg/L/day) 62.39% (7.8 mg/L/day) 94.7% (11.84 mg/L/day) Mineralization	[132]
Ca Alginate beads	Coumaphos, diethylphosphate and chlorferon	*Escherichia coli (OPH)*	Degradation: 163 mg/g biomass/h (immobilized) 0.034 mg/g biomass/h (suspended)	[113]
Tezontle	2,4-D DDT	*Pseudomonas fluorescens*	Removal: 99%	[133]
Ca Alginate beads Tezontle	Methyl parathion/ tetrachlorvinphos	Bacterial consortia	Removal of 78% of methyl parathion Removal of 49 % of tetracholvinphos	[123]

Table 6. Different materials used as supports for immobilization of microorganisms in bioremediation.

Furthermore, there are several reports that indicate a variety of materials that provide the features necessary to immobilize microorganisms. For example, the use of various plant fibers as support for immobilizing bacterial consortium to degrade xenobiotics has important advantages. The use of the natural structural materials such as petiolar felt-sheath of palm for the cell entrapment has added another dimension to a variety of immobilization matrices. The advantages accruable from such biostructures are reusability, freedom from toxicity problems, mechanical strength for necessary support, and open spaces within the matrix for growing cells thus avoiding rupture and diffusion problems. These have suggested the need to search for other types of biomaterials from diverse plant sources that may be used for cell entrapment.

The loofa sponge (*Luffa cylindrica*) was used as carrier material for immobilizing various microorganisms for the purpose of either adsorption or degradation of various xenobiotics as shown in Table 7. This sponge have been used as natural support to immobilize various organisms such as *Chlorella sorokiniana, Porphyridium cruentrum, Penicillium cyrlopium, Funalia trogii* for nickel and cadmium II treatment, besides dyes and chlorinated substances. Loofa grows well in both tropical and subtropical climates and the sponges are produced in large quantities in Mexico where they are currently used for bathing and dish washing. They are light, cylindrical in shape and made up of an interconnecting void within an open network of matrix support materials. As a result of their random lattice of small cross sections coupled with very high porosity, their potentiality as carriers for cell immobilization is very high. The sponges are strong, chemically stable, and composed of interconnecting voids within an open network of fibers. Because of the random lattices of small cross sections of the sponges coupled with high porosity, the sponges are suitable for cell adhesion [134-136]. This sponge was used by our work group and we found methyl parathion removal efficiencies of 75%.

Xenobiotic	Immobilized microorganism	Efficiency	Reference
Nickel II, Chrome	*Chlorella sorokiniana:*	ND	[135].
Lead II ions, copper II and zinc II	*Phanerochaete chrysosporium*	Adsorption: 135.3, 102.8, 50.9 mg/g of Pb(II), Cu(II) y Zn(II) respectively.	[136]
Pb II, Hg II and Cd II ions	*Aspergillus terreus*	Adsorption: 247.2, 37.7 y 23.8 mg/g for Pb II, Hg II y Cd II respectively	[31]
Black 5 (RB5) reactive	*Funalia trogii*	ND	[134]
Blue 172 reactive	*Proteus vulgaris NCIM-2027*	Total discoloration at 37 ° C and pH 8.0 to 5-h in static incubation	[137]
Carbendazim and 2,4-diclorofexiacetic acid (2,4-D)	Bacterial consortium	Complete degradation to 5.5 and 1.5 days respectively	[138]
Carbendazim and 2,4-D	Bacterial consortium	Removal: 20 and, 50% respectively	[139]
Removal of organic matter and ammonium from wastewater	Aerobic bacteria	Chemical oxigen demand removal: 80% Nitrogen removal: 85.6%	[140]
Methylene blue, Crude oil Malachite green dye	--	Adsorption of 49 mg/g Adsorption of 4.6 g oil/g sorbent (in 24 hours.). Adsorption capacity of 29.4 mg / g	[141-143]

Table 7. Loofa use (*L. cylindrica*) as supports for immobilization of microorganisms in bioremediation. ND = Not detected.

10. Final considerations

For the biological degradation of pesticides, it is important to understand the molecular mechanisms involved in enzymatic catalysis, which will be possible to design new alternatives and/or efficient tools for the treatment of pesticide residues or for the bioremediation of contaminated sites. This information could be used in the future to treat pesticide residues in the field (such as waste resulting after washing pesticide containers), or the obsolete pesticides. Moreover, in implementing strategies to increase the efficiency of degradation, such as cell immobilization (bacteria or fungi), we may have tools to abate the existence of obsolete pesticides and waste generated, it will reduce the danger of pesticides on the environment and health.

Author details

Ma. Laura Ortiz-Hernández*, Enrique Sánchez-Salinas, Edgar Dantán-González and María Luisa Castrejón-Godínez

*Address all correspondence to: ortizhl@uaem.mx.

Biotechnology Research Center, Autonomous University of State of Morelos, Col. Chamilpa, Cuernavaca, Morelos, C.P., Mexico

References

[1] Shukla KP, Singh NK, Sharma S. Bioremediation: developments, current practices and perspectives. Genet. Eng. Biotechnol. J, 2010; 3: 1-20.

[2] Agrawal A, Pandey RS, Sharma B. Water Pollution with Special Reference to Pesticide Contamination in India. J. Water Resource and Protection 2010; 2: 432-448.

[3] Damalas CA. Understanding benefits and risks of pesticide use. Scientific Research and Essay. 2009; 4(10): 945-949.

[4] EPA. What is a Pesticide? http://www.epa.gov/opp00001/about/. (accessed 16 July 2012).

[5] Zhang W, Jiang F, Feng Ou J. Global pesticide consumption and pollution: with China as a focus. Proceedings of the International Academy of Ecology and Environmental Sciences, 2011; 1(2): 125-144

[6] Ortiz-Hernández ML. Biodegradación de plaguicidas organofosforados por nuevas bacterias aisladas del suelo. Thesis. Biotechnology PhD. Universidad Autónoma del Estado de Morelos. México. 2002.

[7] Badii M, Landeros J. Plaguicidas que afectan la salud humana y la sustentabilidad. CULCYT/ Toxicología De Plaguicidas. 2007: 4-19.

[8] Pimentel D. Environmental and economic costs of the application of pesticides primarily in the United States. In: Food, Energy, and Society, Third Edition Edited by. Pimentel M. and Pimentel D. CRC Press, 2007.

[9] Chen H, He X, Rong X, Chen W, Cai P, Liang W, Li S. Huang Q. Adsorption and biodegradation of carbaryl on montmorillonite, kaolinite and goethite. Applied Clay Science 2009; 46(1): 102–108.

[10] Chevillard A, Coussy HA, Guillard V, Gontard N, Gastaldi E. Investigating the biodegradation pattern of an ecofriendly pesticide delivery system based on wheat gluten and organically modified montmorillonites. *Polymer Degradation and Stability* 2012; 97(10): 2060–2068.

[11] Myresiotis CK, Vryzas Z, Papadopoulou-Mourkidou E. Biodegradation of soil-applied pesticides by selected strains of plant growth-promoting rhizobacteria (PGPR) and their effects on bacterial growth. *Biodegradation* 2012; 23: 297–310.

[12] Sanchez SE, Ortiz HL. Riesgos y estrategias en el uso de plaguicidas. INVENTIO 2011; 7(14): 21-27.

[13] Besse-Hoggan P, Alekseeva T, Sancelme M, Delort A, Forano C. Atrazine biodegradation modulated by clays and clay/humic acid complexes. *Environmental Pollution* 2009; 157(10): 2837–2844.

[14] Singh BK, Walker A. Microbial degradation of organophosphorus compounds. *FEMS Microbiol. Rev.* 2006; 30 (3): 428–471.

[15] Worrall F, Fernandez-Perez M, Johnson A, Flores-Cesperedes F, Gonzalez-Pradas E. Limitations on the role of incorporated organic matter in reducing pesticide leaching. *J Contain. Hydrol.* 2001. 49: 241-262.

[16] Ortiz-Hernández ML, Sánchez-Salinas E, Olvera-Velona A, Folch-Mallol JL. Pesticides in the Environment: Impacts and its Biodegradation as a Strategy for Residues Treatment, Pesticides - Formulations, Effects, Fate, Margarita Stoytcheva (Ed.), InTech, DOI: 10.5772/13534. Available from: http://www.intechopen.com/books/pesticides-formulations-effects-fate/pesticides-in-the-environment-impacts-and-its-biodegradation-as-a-strategy-for-residues-treatment. 2011.

[17] Dasgupta S, Meisner C, Wheeler D. Stockpiles of obsolete pesticides and cleanup priorities: A methodology and application for Tunisia. *Journal of Environmental Management.* 2010; 91(4): 824-830.

[18] Martinez J. *Practical Guideline on Environmentally Sound Management of Obsolete Pesticides in the Latin America and Caribbean Countries.* Basel Convention Coordinating Centre for Latin America and the Caribbean, Montevideo, Uruguay. 2004.

[19]   Karstensen KH, Nguyen KK, Le BT, Pham HV, Nguyen DT, Doan TT, Nguyen HH, Tao MQ, Luong DH, Doan HT. Environmentally sound destruction of obsolete pesticides in developing countries using cement kilns. *Environmental Science & Policy*. 2006; 9(6): 577-586.

[20]   Shah BP, Devkota B. Obsolete Pesticides: Their Environmental and Human Health Hazards. *The Journal of Agriculture and Environment*. 2009; 10: 51-56.

[21]   Farrera PR. *Acerca de los plaguicidas y su uso en la agricultura*. Revista Digital Ceniap Hoy, 2004; 6.

[22]   Ortiz-Hernández ML, Sánchez-Salinas E. Biodegradation of the organophosphate pesticide tetrachlorvinphos by bacteria isolated from agricultural soils in México. Rev. Int. Contam. Ambient. 2010; 26(1): 27-38.

[23]   Giner de los Ríos DCF. *Estudio: Precisión del Inventario de Plaguicidas Obsoletos y Sitios Contaminados con Éstos*. Donative TF-053710. Actividades de habilitación para ayudar a México a cumplir con el Convenio de Estocolmo sobre Contaminantes Orgánicos Persistentes. 2007.

[24]   Albert LA. Panorama de los plaguicidas en México. *Revista de Toxicología en Línea*. 2005.

[25]   Ferrusquía Ch, Roa G, García M, Amaya A, Pavón T. Evaluación de la degradación de metil paratión en solución usando fotocatálisis heterogénea. *Rev. Latinoamericana de Recursos Naturales* 2008; 4 (2): 285-290.

[26]   Torres-Duarte C, Roman R, Tinoco R, Vazquez-Duhalt R. Halogenated pesticide transformation by a laccase mediator system. *Chemosphere* 2009; 77 (5): 687–692.

[27]   Vidali M. Bioremediation. An overview. *Pure Appl. Chem.* 2001; 73(7): 1163–1172.

[28]   Singleton I. Microbial metabolism of xenobiotics: fundamental and applied research. Journal of Chemical Technology and Biotechnology, 2004. DOI:10.1002/jctb. 280590104.

[29]   Blackburn JW, Hafker WR. The impact of biochemistry, bioavilability, and bioactivity on the selection of bioremediation technologies. Trends in Biotechnology. 1993. 11 (8): 328-333.

[30]   Dua M, Singh A, Sethunathan N, Johri AK. Biotechnology and bioremediation: successes and limitations. Appl Microbiol Biotechnol. 2002; 59:143–152. DOI 10.1007/s00253-002-1024-6.

[31]   Sun,W, Chen Y, Liu L, Tang J, Chen J, Liu P. Conidia immobilization of T-DNA inserted *Trichoderma atroviride* mutant AMT-28 with dichlorvos degradation ability and exploration of biodegradation mechanism. *Bioresource Technology* 2010; 101 (23): 9197-9203.

[32] Galli C. *Degradación por medios bacterianos de compuestos químicos tóxicos*. Comisión Técnica Asesora en: Ambiente y desarrollo sostenible, Buenos Aires, Argentina. 2002.

[33] Qiu X, Zhong Q, Li M, Bai W, Li, B. Biodegradation of p-nitrophenol by methyl parathion-degrading *Ochrobactrum* sp. B2. International Biodeterioration and Biodegradation 2007; 59: 297-301.

[34] Araya M, Lakhi A. Response to consecutive nematicide applications using the same product in mussa AAAcv. Grande naine originated from in vitro propagative material and cultivated in virgin soil. Nematologia Brasileira 2004; 28(1): 55-61.

[35] Aislabie J, Lloyd-Jones G. A review of bacterial-degradation of pesticides. *Australian Journal of Soil Research* 1995; 33(6); 925-942.

[36] Iranzo M, Sain-Pardo I, Boluda R, Sanchez J, Mormeneo S. The use of microorganisms in environmental remediation. *Annals Microbiol* 2001; 51: 135-143.

[37] Vischetti C, Casucci C, Perucci P. Relationship between changes of soil microbial biomass content and imazamox and benfluralin degradation. Springer Verlag, 2002.

[38] Jeon CO, Madsen EL. *In situ* microbial metabolism of aromatic-hydrocarbon environmental pollutants. Current Opinion in Biotechnology. 2012; Available online 19 September 2012.

[39] Abo-Amer AE. Characterization of a strain of Pseudomonas putida isolated from agricultural soil that degrades cadusafos (an organophosphorus pesticide). World J Microbiol Biotechnol 2012; 28: 805-814.

[40] Vásquez M, Reyes W. *Degradación de Aroclor 1242 por Pseudomas sp.* Biblioteca Nacional del Perú, Perú. 2002.

[41] Bordjiba O, Steiman R, Kadri M, Semadi A, Guiraud P. Removal of herbicides from liquid media by fungi isolated from a contaminated soil. Journal of Environmental Quality, 2001; *30(2): 418-426.* DOI: 10.2134/jeq2001.302418x*Vol.*

[42] Senthilkumar S, Anthonisamy A, Arunkumar S, Sivakumari V. 2011. Biodegradation of methyl parathion and endosulfan using *Pseudomonas aeruginosa* and *Trichoderma viridae.* J Environ Sci Eng. 53(1):115-122.

[43] Fournier D, Halasz A, Spain J, Fiurasek P, Hawari J. Determination of key metabolites during biodegradation of hexahidro 1,3,5 Trinitro 1,3,5 Triazine with Rhodococcus sp Strain DN 22. Applied and Enviromental Microbiology 2002; 68: 166-172.

[44] Paul D, Pandey G, Pandey J, Jain RK. Accessing microbial diversity for bioremediation and environmental restoration. Trends Biotechnol 2005; 23:135–142.

[45] Megharaj M, Ramakrishnan B, Venkateswarlu K, Sethunathan N, Naidu R. Bioremediation approaches for organic pollutants: A critical perspective. *Environment International* 2011; 37: 1362–1375.

[46] Ramakrishnan B, Megharaj M, Venkateswarlu K, Naidu R, Sethunathan N. The impacts of environmental pollutants on microalgae and cyanobacteria. Crit Rev Environ Sci Tech. 2010; 40: 699–821.

[47] Ramakrishnan B, Megharaj M, Venkateswarlu K, Sethunathan N, Naidu R. Mixtures of Environmental Pollutants: Effects on Microorganisms and Their Activities in Soils Reviews of Environmental Contamination and Toxicology, 2011; 211: 63-120.

[48] Abraham WR, Nogales B, Golyshin PN, Pieper DH, Timmis KN. Polychlorinated biphenyl-degrading microbial communities and sediments. Curr Opin Microbiol 2002; 5:246–253.

[49] Briceño G, Palma G, Duran N. Influence of Organic Amendment on the Biodegradation and Movement of Pesticides. Critical Reviews in Environmental Science and Technology 2007; 37: 233-271, DOI: 10.1080/10643380600987406.

[50] Diez MC. Biological aspects involved in the degradation of organic pollutants *J. Soil. Sci. Plant. Nutr.* 2010,10(3): 244–267.

[51] Bass C, Field LM. Gene amplification and insecticide resistance. Pest Management Science. 2011; 67 (8): 886–890.

[52] Riya P, Jagatpati T. Biodegradation and bioremediation of pesticides in Soil: Its Objectives, Classification of Pesticides, Factors and Recent Developments. World Journal of Science and Technology 2012; 2(7): 36-41.

[53] Scott C, Pandey G, Hartley CJ, Jackson CJ, Cheesman MJ, Taylor MC., Pandey R, Khurana JL., Teese M, Coppin CW, Weir KM, Jain RK., Lal R, Russell RJ, Oakeshott JG. The enzymatic basis for pesticide bioremediation. Indian J. Microbiol 2008; 48: 65-79.

[54] Trigo A, Valencia A, Cases I Systemic approaches to biodegradation. FEMS Microbiol Rev, 2009; 33: 98-108.

[55] Li X, Schuler MA, Berenbaum MR. Molecular mechanisms of metabolic resistance to synthetic and natural xenobiotics. *Ann Rev Entomol* 2007; 52: 231-253.

[56] Van Eerd LL, Hoagland R E, Zablotowicz RM, Hall JC. Pesticide metabolism in plants and microorganisms. *Weed Science.* 2003; 51(4): 472–495.

[57] Yan Q-X, Hong Q, Han P, Dong X-J, Shen YJ. Li SP. Isolation and characterization of a carbofuran-degrading strain *Novosphingobium* sp. FND-3. FEMS Microbiol Lett. 2007; 271: 207-213.

[58] Chino-Flores C, Dantán-González E, Vázquez-Ramos A, Tinoco-Valencia R, Díaz-Méndez R, Sánchez-Salinas E, Castrejón-Godínez ML, Ramos-Quintana F, Ortiz-Hernández ML. Isolation of the *opdE* gene that encodes for a new hydrolase of *Enterobacter* sp. capable of degrading organophosphorus pesticides. Biodegradation 2012; 23: 387-397.

[59] Ghanem E, Raushel FM. Detoxification of organophosphate nerve agents by bacterial phosphotriesterase. Toxicol Appl Pharmacol 2005; 207: 459-470. DOI: 10.1016/j.taap. 2005.02.025.

[60] Porzio E, Merone L, Mandricha L, Rossia M, Manco G. A new phosphotriesterase from *Sulfolobus acidocaldarius* and its comparison with the homologue from *Sulfolobus solfataricus*. Biochemie, 2007; 89: 625-636. DOI:10.1016/j.biochi.2007.01.007.

[61] Theriot CM, Grunden AM. Hydrolysis of organophosphorus compounds by microbial enzymes. Appl Microbiol Biotechnol, 2010; 89: 35-43. DOI:10.1007/s00253-010- 589 2807-9.

[62] Shen YJ, Lu P, Mei H, Yu HJ, Hong Q, Li SP Isolation of a methyl-parathion degrading strain *Stenotrophomonas* sp. SMSP-1 and cloning of the *ophc2* gene. Biodegradation, 2010; 21: 785-792. DOI:10.1007/s10532-010-9343-2.

[63] Latifi AM, Khodi S, Mirzaei M, Miresmaeili M, Babavalian H. Isolation and characterization of five chlorpyrifos degrading bacteria. African Journal of Biotechnology 2012; 11(13): 3140-3146.

[64] Mulbry WW, Karns JS. Parathion hydrolase specified by the *Flavobacterium* opd gene: Relationship between the gene and protein. J. Bacteriol 1989;171: 6740-6746.

[65] Ortiz-Hernández ML, Quintero-Ramírez R, Nava-Ocampo AA, Bello-Ramírez AM. Study of the mechanism of *Flavobacterium* sp. for hydrolyzing organophosphate pesticides. Fundam Clin Pharmacol, 2003; 17(6): 717-23.

[66] Sethunathan N, Yoshida T. A *Flavobacterium* sp. that degrades diazinon and parathion. Can J Microbiol, 1973; 19: 873-875. DOI:10.1139/m73-138.

[67] Cui Z, Li S, Fu G. Isolation of methyl-parathion-degrading strain M6 and cloning of the methyl-parathion hydrolase gene. Appl Environ Microbiol, 2001; 67: 4922-4925. DOI:10.1128/AEM.67.10.4922-4925.2001.

[68] Horne J, Sutherland TD, Harcourt RL, Russell RJ, Oakesthott JG. Identification of an opd (organophosphate degradation) gene in an Agrobacterium isolate. Appl. Environ. Microbiol 2002; 68: 3371-3376.

[69] Kapoor M, Rajagopal R. Enzymatic bioremediation of organophosphorus insecticides by recombinant organophosphorous hydrolase. International Biodeterioration & Biodegradation 2011; 65: 896-901.

[70] Bansal OP. Degradation of pesticides Editorial CRC Press. 2012.

[71] López SN. Evaluación de mecanismos de resistencia a insecticidas en *Frankliniella occidentales* (Pergande): implicación de carboxilesterasas y acetilcolinesterasas. PhD Thesis, Universidad de Valencia, España; 2008.

[72] Rosario-Cruz R, García-Vázquez Z, George EJ. Deteccion inmunoquimica de esterasas en dos cepas de la garrapata *Boophilus microplus* (Acarii: Ixodidae) resistentes a Ixodicidas. Téc. Pecu Méx 2000; 38(2): 203-210.

[73] Galego LG, Ceron CR, Carareto MA. Characterization of esterases in a Brazilian population of *Zaprionus indianus* (Diptera: Drosophilidae). Genetica 2006; 126:89-99.

[74] Baffi MA, Rocha LG, Soares de Sousa C, Ceron CR, Bonetti AM. Esterase enzymes involved in pyrethroid and organophosphate resistance in a Brazilian population of *Riphicephallus (Boophilus) microplus* (Acari, Ixodidae). Molecular & Biochemical Parasitology 2008; 160: 70-73.

[75] Reiner E, Aldridge WN, Hoskin CG. Enzymes Hydrolyzing Organophosphorus Compounds John Wiley & Sons, Inc, New York. 1989.

[76] Dary O, Georghiou GP, Parsons E, Pasteur N. Microplate adaptation of Gomori's assay for quantitative determination of general esterase activity in single insects. Department of Entomology, University of California. Riverside, California 92521. Journal Economic Entomology 1990; 53(6): 2187-2192.

[77] Siddique T, Okeke BC, Arshad M, Frankenberger WT Jr. Enrichment and isolation of endosulfan-degrading microorganisms. J Environ Qual. 2003; 32(1): 47-54.

[78] Kwon GS, Sohn HY, Shin KS, Kim E. Seo BI. Biodegradation of the organochlorine insecticide, endosulfan, and the toxic metabolite, endosulfan sulfate, by *Klebsiella oxytoca* KE-8. Appl Envirnon Microbiol. 2005; 67(6): 845-50.

[79] Weir KM. Sutherland TD, Horne I, Russell RJ, Oakeshott JG. A single monooxygenase, *ese*, is involved in the metabolism of the organochlorides endosulfan and endosulfan in an *Arthrobacter* sp. Appl Envirnon Microbiol. 2006; 72(5): 3524-30.

[80] Kumar K, Devi SS, Krishnamurthi K, Kanade GS, Chakrabarti T. Enrichment and isolation of endosulfan degrading and detoxifying bacteria. Chemosfere. 2007; 68(2): 317-22. Epub 2007 Feb 7.

[81] Hussain S. Arshad M, Saleem M, Khalid A. Biodegradation of alpha- and beta-endosulfan by soil bacteria. Biodegradation. 2007; 18(6): 731-40.

[82] Khaled A, Miia T, Arja R, Jukka H, Olavi P. Metabolism of Pesticides by Human Cytochrome P450 Enzymes In Vitro-A Survey, Insecticides-Advances in Integrated Pest Management. InTech. 2012.

[83] Urlacher VB, Lutz-Wahl S, Schmid RD. Microbial P450 enzymes in biotechnology. Appl Microbiol Biotechnol 2004; 64: 317–325.

[84] Alzahrani AM. Insects Cytochrome P450 Enzymes: Evolution, Functions and Methods of Analysis. Global Journal of Molecular Sciences 2009; 4(2): 167-179.

[85] Shi H, Pei L, Gu S, Zhu S, Wang Y, Zhang Y, Li B. Glutathione S-transferase (GST) genes in the red flour beetle, *Tribolium castaneum*, and comparative analysis with five additional insects. Genomics 2012; 100(5): 327-335.

[86] Hussain S, Siddique T, Arshad M, Saleem M. Bioremediation and Phytoremediation of Pesticides: Recent Advances. *Critical Reviews in Environmental Science and Technology* 2009; 39(10): 843-907.

[87] Zhongli C, Shunpeng L, Guoping F. Isolation of methyl parathion-degrading strain m6 and cloning of the methyl parathion hydrolase gene. Apple. and Environ. Microbiol. 2001; 67(10): 4922-4925.

[88] Zhang R, Cui Z, Jiang J, Gu X, Li S. Diversity of organophosphorus pesticides degrading bacteria in a polluted soil and conservation of their organophosphorus hydrolase genes. Can. J. Microbiol. 2005; 5: 337-343.

[89] Shimizu H. Metabolic engineering–integrating methodologies of molecular breeding and bioprocess systems engineering. *J Biosci Bioeng* 2002; 94: 563–73.

[90] Pieper DH, Reineke W. Engineering bacteria for bioremediation. *Curr Opin Biotechnol* 2000; 11: 262–70.

[91] Cases I, de Lorenzo V. Promoters in the environment: transcriptional regulation in its natural context. *Nat Rev Microbiol* 2005; 3: 105–18.

[92] Wu NF, Deng MJ, Liang GY, Chu XY, Yao B, Fan YL. Cloning and expression of ophc2, a new organophosphorus hydrolase gene. *Chin Sci Bull* 2004; 49: 1245–1249.

[93] Fu GP, Cui Z, Huang T, Li SP. Expression, purification, and characterization of a novel methyl parathion hydrolase. *Protein Expr Purif* 2004; 36: 170–176.

[94] Yu H, Yan X, Shen W, Hong Q, Zhang J, Shen Y, Li S. Expression of methyl parathion hydrolase in Pichia pastoris. *Curr Microbiol* 2009; 59: 573–578.

[95] Wang XX, Chi Z, Ru SG, Chi ZM. Genetic surface-display of methyl parathion hydrolase on Yarrowia lipolytica for removal of methyl parathion in water. *Biodegradation* 2012; 23: 763–774 DOI: 10.1007/s10532-012-9551-z.

[96] Shen YJ, Lu P, Mei H, Yu HJ, Hong Q, Li SP. Isolation of a methyl parathion-degrading strain Stenotrophomonas sp. SMSP-1 and cloning of the ophc2 gene. *Biodegradation* 2009; DOI: 10.1007/s10532-010-9343-2.

[97] Eapen S, Singh S, D'Souza S F. Advances in development of transgenic plants for remediation of xenobiotic pollutants. *Biotechnol. Adv.* 2007; 25: 442-451.

[98] Singh S, Sherkhane PD, Kale SP, Eapen S. Expression of a human cytochrome P4502E1 in Nicotiana tabacum enhances tolerance and remediation of g-hexachlorocyclohexane. *New Biotechnology,* 2011; 28: 423-429.

[99]   Casey MT, Grunden AM. Hydrolysis of organophosphorus compounds by microbial enzymes. *Appl Microbiol Biotechnol* 2011; 89:35–43 DOI: 10.1007/s00253-010-2807-9.

[100] Van Dyk JS, Brett P. Review on the use of enzymes for the detection of organochlorine, organophosphate and carbamate pesticides in the environment. *Chemosphere* 2011; 82: 291-307.

[101] Cho MH, Mulchandani CA., Chen W. Bacterial cell surface display of organophosphorus hydrolase for selective screening of improved hydrolysis of organophosphate nerve agents. *Appl. Environ. Microbiol* 2002; 68: 2026–2030.

[102] Cho MH, Mulchandani CA, Chen W. Altering the Substrate Specificity of Organophosphorus Hydrolase for Enhanced Hydrolysis of Chlorpyrifos. *Applied and Environmental Microbiology* 2004; 70: 4681–4685 DOI: 10.1128/AEM.70.8.4681-4685.2004.

[103] Zhou J, He Z, Van Nostrand JD, Wu L, Deng Y. Apply-ing Geo Chip analysis to disparate microbial communities. *Microbe* 2010; 5: 60–65.

[104] Maphosa F, Lieten SH, Dinkla I, Stams AJ, Smidt H, Fennell DE. Ecogenomics of microbial communities in bioremediation of chlorinated contaminated sites. *Frontiers in Microbiology* 2012; 3: 1-14 DOI: 10.3389/fmicb.2012.00351.

[105] Rayu S, Karpouzas DG, Singh BK. Emerging technologies in bioremediation: constraints and opportunities. *Biodegradation* 2012; 23: 917–926 DOI: 10.1007/s10532-012-9576-3.

[106] Rajendhran J, Gunasekaran P. Strategies for accessing soil metagenome for desired applications. *Biotechnology Advances* 2008; 26: 576-590 DOI: 10.1016/j.biotechadv.2008.08.002.

[107] Zhao B, Poh LC. Insights into environmental bioremediation by microorganisms through functional genomics and proteomics. *Proteomics* 2008; 8: 874-881 DOI: 10.1002/pmic.200701005.

[108] Koffas M, Roberge C, Lee K, Stephanopoulos G. Metabolic engineering. *Annu Rev Biomed Eng* 1999; 1: 535–557.

[109] Richins R, Mulchandani A, Chen W. Expression, immobilization, and enzymatic characterization of cellulose-binding domain-organophosphorus hydrolase fusion enzymes, Biotechnol. Bioeng. 2000; 69: 591-596.

[110] Chen W, Georgiou G Cell-surface display of heterologousproteins: from high throughput screening to environmental applications. Biotechnol Bioeng. 2002; 5:496-503. DOI:10.1002/bit.10407.

[111] Martin M, Mengs G, Plaza E, Garbi C, Sánchez A, Gibello A, Gutierrez F, Ferrer E. Propachlor removal by Pseudomonas strain GCH1 in a immobilized-cell system. Appl Environ Microbiol. 2000; 66(3):1190-1194.

[112] Ha J, Engler CR, Lee SJ. Determination of diffusion coefficients and diffusion charac-
       teristics for chlorferon and diethylthiophosphate in Ca-alginate gel beads. Biotechnol
       Bioeng. 2008; 100(4): 698-706. DOI:10.1002/bit.21761.

[113] Ha J, Engler CR, Wild J. Biodegradation of coumaphos, chlorferon, and diethylthio-
       phosphate using bacteria immobilized in Ca-alginate gel beads. Bioresource Technol-
       ogy 2009; 100: 1138-1142.

[114] Kennedy JF, Cabral JMS. Solid Phase Biochemistry. Schouten, W.H. (ed.) Wiley Pub.,
       New York. 1983.

[115] Arroyo M. Inmovilización de enzimas. Fundamentos, métodos y aplicaciones. *Ars.*
       Pharmaceutica. 1998; 39(2): 23-39.

[116] Lusta KA, Starostina NG, Fikhte BA. Immobilization of microorganisms: cytophysio-
       logical aspects. In: De Bont, JAM, Visser J, Mattiasson B, Tramper J. (Ed). Proceedings
       of an International Symposium: Physiology of Immmobilized Cells. Amsterdam,
       Netherlands: Elsevier. 1990.

[117] Heitkamp M, Camel V, Reuter T, Adams W. Biodegradation of p-nitrophenol inan
       aqueous waste stream by immobilized bacteria. *Appl. Environ. Microbiol.* 1990; 56(10):
       2967-73.

[118] Wang P, Sergeeva M, Lim L, Dordick J. Biocatalytic Plastics as active and stable ma-
       terials for biotransformations. *Nature Biotechnol.* 1997; 15: 789-793.

[119] Karamanev G, Chavarie C, Samson R. Soil immobilization: new concept for biotreat-
       ment of soil contaminants. Biotechnol. Bioeng. 1998; 4 (57): 471-476.

[120] Pedersen S, Christensen M. Immobilized biocatalysts. En: Straathof, A. y Adler-
       creutz, P. (edts) Applied Biocatalysis. Harwood Academic Publishers. Netherlands.
       2000.

[121] Davey ME, O'Toole GA. Microbial Biofilms: from Ecology to Molecular Genetics. Mi-
       crobiology and Molecular Biology Reviews. 2000; 64(4): 847-867.

[122] Nicolella CM, van Loosdrecht, Heijnen JJ. Review article Wastewater treatment with
       particulate biofilm reactors. Journal of Biotechnology. 2000; 80: 1-33

[123] Yañez-Ocampo G., Penninckx M., Jiménez-Tobon G.A, Sánchez-Salinas E, Ortiz-Her-
       nández ML. Removal of two organophosphate pesticides employing a bacteria con-
       sortium immobilized in either alginate or tezontle. *Journal of Hazardous Materials*
       2009; 168: 1554-1561.

[124] Betancourth M, Botero JE, Rivera SP. Biopelículas: una comunidad microscópica en
       desarrollo. Colombia Médica. 2004; 35(3-1)

[125] Dorigo U, Leboulanger C, Bérard A, Bouchez A, Humbert J F, Montuelle B. Lotic bio-
       film community structure and pesticide tolerance along a contamination gradient in
       a vineyard área. Aquat Microb Ecol. 2007; 50: 91-102.

[126] Villena GK, Gutiérrez-Correa M. Biopelículas de *Aspergullus niger* para la producción de celulasas algunos aspectos estructurales y fisiológicos. Rev. Peru. Biol. 2003; 10(1): 78-87.

[127] Mansee AH, Chen W, Mulchandani A. Detoxification of the organophosphate nerve agent coumaphos using organophosphorus hydrolase immobilized on cellulose materials. J Ind Microbiol Biotechnol. 2005; 32: 554–560.

[128] Martín M, Mengs G, Plaza E, Garbi C, Sánchez M, Gibello A, Gutiérrez F, Ferrer E. Propachlor removal by Pseudomonas strain GCH1 in a immobilized-cell system. Appl. Environ. Microbiol. 2000; 66(3): 1190-1194.

[129] Ettayebi K, Errachidi F, Jamai L, Tahri-Jouti MA, Sendide K, Ettayebi M. Biodegradation of polyphenols with immobilized *Candida tropicalis* under metabolic induction. FEMS Microbiology Letters 2003; 223: 215-219.

[130] Chivita L, Dussán J. Evaluación de matrices para la inmovilización de *Pseudomonas spp.* En biorremediación de fenol. Rev. Colombiana de Biotecnología 2003. V(2); 5-10.

[131] Barragán HB, Costa C, Peralta J, Barrera J, Esparza F, Rodríguez R. Biodegradation of organochlorine pesticides by bacteria grown in microniches of the porous structure of green vean coffee. International Biodeterioration and Biodegradation. 2007; 59: 239-244.

[132] Garzón-Jiménez RC. Cinética de degradación de colorantes textiles de diferentes clases químicas por hongos y bacterias inmovilizados sobre fibra de *Agave tequilana* Webber var. Azul. Thesis. Pontificia Universidad Javeriana. 2009.

[133] Santacruz G, Bandala E, Torres LG. Chlorinated pesticidas (2,4 –D and DDT) biodegradation at high concentrations using immobilized *Pseudomonas fluorescens*. J. Environ. Sc. Heal. 2005; 40(4):571-583.

[134] Mazmanci M, Unyayara. Decolourisation of reactive black 5 by *Funalia trogii* immobilized on *Loofa cylindrical* sponge. Process Biochemistry. 2005; 40: 337-342.

[135] Akhtar N, Iqbal J, Iqbal M. Removal and recovery of nickel (II) from aqueous solution by Loofa sponge immobilized biomass of *Chlorella sorokiniana* characterization studies. J. Hazard. Mater. 2004; 108(1-2): 85-94.

[136] Iqbal M, Edyvean RGJ. Biosorption of lead, copper and zinc ions on loofa immobilized biomass of *Phanerochaete chrysosporium*. Mineral Engineering 2004; 17: 217-223.

[137] Saratale RG, Saratale GD, Chang JS. Govindwar SP. Decolorization and biodegradation of reactive dyes and dye wastewater by a developed bacterial consortium. Biodegradation. 2010; 21 (6): 999-1015. DOI: 10.1007/s10532-010-9360-1.

[138] Pattanasupong A, Naimogase H, Sugimoto E, Hori Y, Hirata K, Tani K, Nasu M, Miyamoto K. Degradation of carbendazim and 2,4-dichlorophenoxyacetic acid by

immobilized consortium on Loofa sponge. J. of Bioscience and Bioengineering. 2004; 98 (1): 28-33.

[139] Nagase H, Pattanasupong A, Sugimoto E, tani K, Nasu M, Hirata K, Miyamoto K. Effect of environmental factors on performance of immobilized consortium system for degradation of carbendazim and 2,4-dichlorophenoxyacetic acid in continuous culture. Biochemical Engineering Journal 2006; 29: 163-168.

[140] Nabizadeh R, Naddafi K, Mesdaghinia A, Nafez AH. Feasibility study of organic matter and Ammonium removal using loofa sponge as a supporting medium in an aerated submerged fixed-film reactor (ASFFR). Electronic Journal of Biotechnology 2008; 11(4). DOI: 10.2225/vol11-issue4-fulltext-8.

[141] Demir H; Top A; Balkose D, Ulku S. Dye adsorption behavior of Luffa cylindrica fibers. Journal of Hazardous Material, 2008; 153(1-2): 389-394.

[142] Annunciado TR, Sydenstricker THD, Amico SC. Experimental investigation of various vegetable fibers as sorbent materials for oil spills. Marine Pollution Bulletin. 2005; 50:1340-1346.

[143] Altinisik A, Gur E, Seki Y. A natural sorbent, *Luffa cylindrica* for removal of a model basic dye. Journal of hazardous materials, 2010; 179: 658-664.

Chapter 11

Biodegradation: Involved Microorganisms and Genetically Engineered Microorganisms

Nezha Tahri Joutey, Wifak Bahafid,
Hanane Sayel and Naïma El Ghachtouli

Additional information is available at the end of the chapter

1. Introduction

Biodegradation is defined as the biologically catalyzed reduction in complexity of chemical compounds [1]. Indeed, biodegradation is the process by which organic substances are broken down into smaller compounds by living microbial organisms [2]. When biodegradation is complete, the process is called "mineralization". However, in most cases the term biodegradation is generally used to describe almost any biologically mediated change in a substrate [3].

So, understanding the process of biodegradation requires an understanding of the microorganisms that make the process work. The microbial organisms transform the substance through metabolic or enzymatic processes. It is based on two processes: growth and cometabolism. In growth, an organic pollutant is used as sole source of carbon and energy. This process results in a complete degradation (mineralization) of organic pollutants. Cometabolism is defined as the metabolism of an organic compound in the presence of a growth substrate that is used as the primary carbon and energy source [4]. Several microorganisms, including fungi, bacteria and yeasts are involved in biodegradation process. Algae and protozoa reports are scanty regarding their involvement in biodegradation [5]. Biodegradation processes vary greatly, but frequently the final product of the degradation is carbon dioxide [6]. Organic material can be degraded aerobically, with oxygen, or anaerobically, without oxygen [4, 7].

Biodegradable matter is generally organic material such as plant and animal matter and other substances originating from living organisms, or artificial materials that are similar enough to plant and animal matter to be put to use by microorganisms. Some microorganisms have the astonishing, naturally occurring, microbial catabolic diversity to degrade, transform or accumulate a huge range of compounds including hydrocarbons (e.g. oil), polychlorinated biphenyls (PCBs), polyaromatic hydrocarbons (PAHs), radionuclides and metals [8].

The term biodegradation is often used in relation to ecology, waste management and mostly associated with environmental remediation (bioremediation) [2]. Bioremediation process can be divided into three phases or levels. First, through natural attenuation, contaminants are reduced by native microorganisms without any human augmentation. Second, biostimulation is employed where nutrients and oxygen are applied to the systems to improve their effectiveness and to accelerate biodegradation. Finally, during bioaugmentation, microorganisms are added to the systems. These supplemental organisms should be more efficient than native flora to degrade the target contaminant [9]. A feasible remedial technology requires microorganisms being capable of quick adaptation and efficient uses of pollutants of interest in a particular case in a reasonable period of time [10]. Many factors influence microorganisms to use pollutants as substrates or cometabolize them, like, the genetic potential and certain environmental factors such as temperature, pH, and available nitrogen and phosphorus sources, then, seem to determine the rate and the extent of degradation [4]. Therefore, applications of genetically engineered microorganisms (GEM) in bioremediation have received a great deal of attention. These GEM have higher degradative capacity and have been demonstrated successfully for the degradation of various pollutants under defined conditions. However, ecological and environmental concerns and regulatory constraints are major obstacles for testing GEM in the field [11].

In this chapter we will try to foster an in-depth understanding of biodegradation process by trying to cover all types of microorganisms implied in degradation of different pollutants. Moreover, although we are aware that the term biodegradation is often used in relation to ecology, waste management, biomedicine, and the natural environment (bioremediation) and is now commonly associated with environmentally friendly products, this chapter will mainly give attention to biodegradation in relation to bioremediation through describing processes (natural attenuation, biostimulation and bioaugmentation) utilizing degradation abilities of microorganisms in bioremediation and factors affecting this process. Microorganisms may be genetically engineered for many purposes. One such purpose is for the efficient degradation of pollutants. So, the second scope of this chapter is to demonstrate the importance of some GEM in this process and to describe obstacles for testing GEM in the field, which must be overcome before GEM can provide an effective clean-up process at lower cost. Figure 1 summarizes the contents of this chapter.

2. Role of microorganisms in biodegradation of pollutants

In this chapter, biodegradation is described associated with environmental bioremediation. Therefore, biodegradation is nature's way of recycling wastes, or breaking down organic matter into nutrients that can be used and reused by other organisms. In the microbiological sense, "biodegradation" means that the decaying of all organic materials is carried out by a

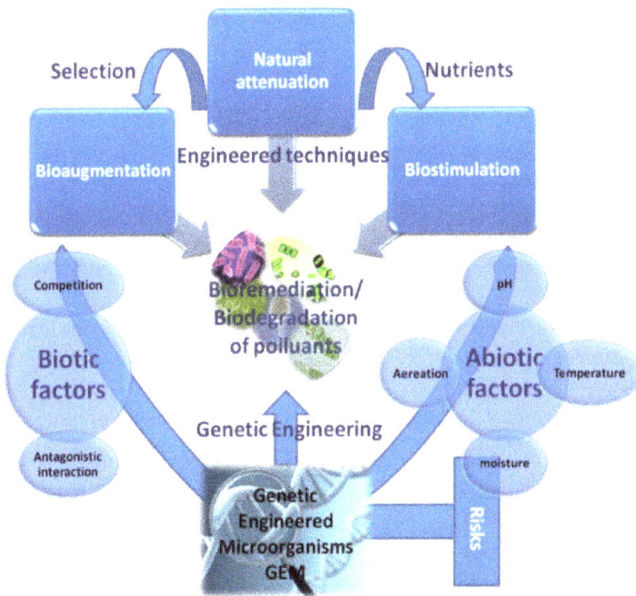

Figure 1. Bioremediation of polluants utilizing biodegradation abilities of microorganisms include the natural attenuation, although it may be enhanced by engineered techniques, either by addition of selected microorganisms (bioaugmentation) or by biostimulation, where nutrients are added. Genetic engineering is also used to improve the biodegradation capabilities of microorganisms by GEM. Nevertheless, there are many factors affecting the efficiency of this process and risks associated to the use of GEM in the field.

huge assortment of life forms comprising mainly bacteria, yeast and fungi, and possibly other organisms.

Bioremediation and biotransformation methods endeavour to harness the astonishing, naturally occurring, microbial catabolic diversity to degrade, transform or accumulate a huge range of compounds including hydrocarbons (e.g. oil), polychlorinated biphenyls (PCBs), polyaromatic hydrocarbons (PAHs), radionuclides and metals [12].

2.1. Some biodegradable pollutants

In the last few decades, highly toxic organic compounds have been synthesized and released into the environment for direct or indirect application over a long period of time. Fuels, polychlorinated biphenyls (PCBs), polycyclic aromatic hydrocarbons (PAHs), pesticides and dyes are some of these types of compounds [9]. Some other synthetic chemicals like radionuclides and metals are extremely resistant to biodegradation by native flora compared with the naturally occurring organic compounds that are readily degraded upon introduction into the environment.

Hydrocarbons: are organic compounds whose structures consist of hydrogen and carbon. Hydrocarbons can be seen as linear linked, branched or cyclic molecules. They are observed as aromatic or aliphatic hydrocarbons. The first one has benzene (C_6H_6) in its structure, while the aliphatic one is seen in three forms: alkanes, alkenes and alkynes [13].

Polycyclic aromatic hydrocarbons (PAHs): are important pollutants class of hydrophobic organic contaminants (HOCs) widely found in air, soil and sediments. The major source of PAH pollution is industrial production [7]. They have been studied with increasing interest for more than twenty years because of more findings about their toxicity, environmental persistence and prevalence [14]. PAHs can sorb to organic-rich soils and sediments, accumulate in fish and other aquatic organisms, and may be transferred to humans through seafood consumption [7]. The biodegradation of PAHs can be considered on one hand to be part of the normal processes of the carbon cycle, and on the other as the removal of man-made pollutants from the environment. The use of microorganisms for bioremediation of PAH-contaminated environments seems to be an attractive technology for restoration of polluted sites.

Polychlorinated biphenyls (PCBs): are mixtures of synthetic organic chemicals. Due to their non-flammability, chemical stability, high boiling point, and electrical insulating properties, PCBs were used in hundreds of industrial and commercial applications including electrical, heat transfer, and hydraulic equipment; as plasticizers in paints, plastics, and rubber products; in pigments, dyes, and carbonless copy paper; and many other industrial applications. Consequently, PCBs are toxic compounds that could act as endocrine disrupters and cause cancer. Therefore, environmental pollution with PCBs is of increasing concern [15].

Pesticides: are substances or mixture of substances intended for preventing, destroying, repelling or mitigating any pest. Pesticides which are rapidly degraded are called nonpersistent while those which resist degradation are termed persistent. The most common type of degradation is carried out in the soil by microorganisms, especially fungi and bacteria that use pesticides as food source [16].

Dyes: are widely used in the textile, rubber product, paper, printing, color photography, pharmaceuticals, cosetics and many other industries [17]. Azo dyes, which are aromatic compounds with one or more (–N=N–) groups, are the most important and largest class of synthetic dyes used in commercial applications [18]. These dyes are poorly biodegrabale because of their structures and treatment of wastewater containing dyes usually involves physical and / or chemical methods such as adsorption, coagulation-flocculation, oxidation, filtration and electrochemical methods [19].

The success of a biological process for color removal from a given effluent depends in part on the utilization of microorganisms that effectively decolorize synthetic dyes of different chemical structures.

Radionuclides: a radionuclide is an atom with an unstable nucleus, characterized by excess energy available to be imparted either to a newly created radiation particle within the nucleus or via internal conversion. During this process, the radionuclide is said to undergo radioactive decay, resulting in the emission of gamma ray(s) and/or subatomic particles such as alpha or beta particles [20].

Heavy metals: unlike organic contaminants, the metals cannot be destroyed, but must either be converted to a stable form or removed. Bioremediation of metals is achieved through biotransformation. Mechanisms by which microorganisms act on heavy metals include biosorption (metal sorption to cell surface by physicochemical mechanisms), bioleaching (heavy metal mobilization through the excretion of organic acids or methylation reactions), biomineralization (heavy metal immobilization through the formation of insoluble sulfides or polymeric complexes), intracellular accumulation, and enzyme-catalyzed transformation (redox reactions) [21]. The major microbial processes that influence the bioremediation of metals are summarized in Figure 2.

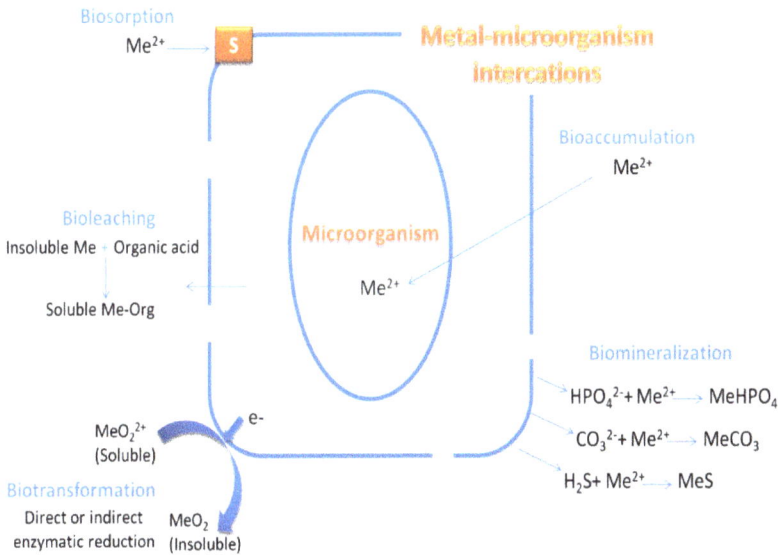

Biosorption
Me^{2+} → S

Metal-microorganism interactions

Bioaccumulation
Me^{2+}

Bioleaching
Insoluble Me — Organic acid

Microorganism
Me^{2+}

Soluble Me-Org

Biomineralization
$HPO_4^{2-} + Me^{2+}$ → $MeHPO_4$

MeO_2^{2+}
(Soluble)

e-

$CO_3^{2-} + Me^{2+}$ → $MeCO_3$

$H_2S + Me^{2+}$ → MeS

Biotransformation
Direct or indirect
enzymatic reduction

MeO_2
(Insoluble)

Figure 2. Microbial processes used in bioremediation technologies modified from Lloyd and Lovley [21].

2.2. Bacterial degradation

There are many reports on the degradation of environmental pollutants by different bacteria. Several bacteria are even known to feed exclusively on hydrocarbons [22]. Bacteria with the ability to degrade hydrocarbons are named hydrocarbon-degrading bacteria. Biodegradation of hydrocarbons can occur under aerobic and anaerobic conditions, it is the case for the nitrate reducing bacterial strains *Pseudomonas* sp. and *Brevibacillus* sp. isolated from petroleum contaminated soil [23]. However, data presented by Wiedemeier et al. [24] suggest that the anaerobic biodegradation may be much more important. 25 genera of hydrocarbon degrading bacteria were isolated from marine environment [25]. Furthermore, among 80 bacterial strains isolated by Kafilzadeh et al. [26] which belonged to 10 genus as follows: *Bacillus, Corynebacte-rium, Staphylococcus, Streptococcus, Shigella, Alcaligenes, Acinetobacter, Escherichia, Klebsiella* and *Enterobacter, Bacillus* was the best hydrocarbon degrading bacteria.

Bacterial strains that are able to degrade aromatic hydrocarbons have been repeatedly isolated, mainly from soil. These are usually gram negative bacteria, most of them belong to the genus *Pseudomonas*. The biodegradative pathways have also been reported in bacteria from the genera *Mycobacterium, Corynebacterium, Aeromonas, Rhodococcus* and *Bacillus* [7].

Although many bacteria are able to metabolize organic pollutants, a single bacterium does not possess the enzymatic capability to degrade all or even most of the organic compounds in a polluted soil. Mixed microbial communities have the most powerful biodegradative potential because the genetic information of more than one organism is necessary to degrade the complex mixtures of organic compounds present in contaminated areas [27].

Both, anaerobic and aerobic bacteria are capable of biotransforming PCBs. Higher chlorinat-ed PCBs are subjected to reductive dehalogenation by anaerobic microorganisms. Lower chlorinated biphenyls are oxidized by aerobic bacteria [28]. Research on aerobic bacteria isolated so far has mainly focused on Gram-negative strains belonging to the genera *Pseudomonas, Burkholderia, Ralstonia, Achromobacter, Sphingomonas* and *Comamonas*. Howev-er, several reports about PCB-degrading activity and characterization of the genes that are involved in PCB degradation indicated PCB-degrading potential of some Gram-positive strains as well (genera *Rhodococcus, Janibacter, Bacillus, Paenibacillus* and *Microbacterium*) [29]. Aerobic catabolic pathway for PCB degradation seems to be very similar for most of the bacteria and comprises four steps catalysed by the enzymes, biphenyl dioxygenase (BphA), dihydrodiol dehydrogenase (BphB), 2, 3-dihydroxybihenyl dioxygenase (DHBD) (BphC) and hydrolase (BphD) [30].

Successful removal of pesticides by the addition of bacteria had been reported earlier for many compounds, including atrazine [31]. Recent findings concerning pesticide degrading bacteria include the chlorpyrifos degrading bacterium *Providencia stuartii* isolated from agricultural soil [32] and isolates *Bacillus, Staphylococcus* and *Stenotrophomonas* from cultivated and uncultivated soil able to degrade dichlorodiphenyltrichloroethane (DDT) [33].

Researches on bacterial strains that are able to degrade azo dyes under aerobic and anaerobic conditions have been extensively reported [34]. Based on the available literature, it can be concluded that the microbial decolourization of azo dyes is more effective under anaerobic conditions. On the other hand, these conditions lead to aromatic amine formation, and these are mutagenic and toxic to humans requiring a subsequent oxidative (aerobic) stage for their degradation. In this context, the combined anaerobic/aerobic biological treatments of textile dye effluents using microbial consortia are common in the literature [35]. For exemple, Chaube et al. [36] have used the mix consortia of bacteria consisting of *Proteus* sp., *Pseudomonas* sp. and *Enterococcus* sp. in biodegradation and decolorisation of dye. However, several researchers have identified single bacterial strains that have very high efficacy for removal of azo dyes, it is the case of *Shewanella decolorations* [37]. In contrast to mixed cultures, the use of a pure culture has several advantages. These include predictable performance and detailed knowledge on the degradation pathways with improved assurance that catabolism of the dyes will lead to nontoxic end products under a given set of environmental conditions. Another advantage is that the bacterial strains and their activity can be monitored using culture-based or molecular methods to quantify population densities of the bacteria over time. Knowledge of the population density can be extrapolated to quantitative analysis of the kinetics of azo dye decoloration and mineralization [38].

Heavy metals cannot be destroyed biologically (no"degradation", change in the nuclear structure of the element, occurs) but are only transformed from one oxidation state or organic complex to another [39]. Besides, bacteria are also efficient in heavy metals bioremediation. Microorganisms have developed the capabilities to protect themselves from heavy metal toxicity by various mechanisms, such as adsorption, uptake, methylation, oxidation and reduction. Reduction of metals can occur through dissimilatory metal reduction [40], where bacteria utilize metals as terminal electron acceptors for anaerobic respiration. In addition, bacteria may possess reduction mechanisms that are not coupled to respiration, but instead are thought to impart metal resistance. For example, reduction of Cr(VI) to Cr(III) under aerobic [41] or anaerobic conditions [42], reduction of Se(VI) to elemental Se [43], reduction of U(VI) to U(IV) [44] and reduction of Hg(II) to Hg(0) [45]. Microbial methylation plays an important role in heavy metals bioremediation, because methylated compounds are frequently volatile. For example, Mercury, Hg(II) can be biomethylated by a number of different bacterial species *Alcaligenes faecalis, Bacillus pumilus, Bacillus* sp., *P. aeruginosa* and *Brevibacterium iodinium* to gaseous methyl mercury [46]. In addition to redox conversions and methylation reactions, acidophilic iron bacteria like *Acidithiobacillus ferrooxidans* [47] and sulfur oxidizing bacteria [48] are able to leach high concentrations of As, Cd, Cu, Co and Zn from contaminated soils. On the other hand metals can be precipitated as insoluble sulfides indirectly by the metabolic activity of sulphate reducing bacteria [48]. Sulphate reducing bacteria are anaerobic heterotrophs utilizing a range of organic substrates with SO_4^{2-} as the terminal electron acceptor. Heavy metal ions can be entrapped in the cellular structure and subsequently biosorbed onto the binding sites present in the cellular structure. This method of uptake is independent of the biological metabolic cycle and is known as biosorption or passive uptake. The heavy metal can also pass into the cell across the cell membrane through the cell metabolic cycle. This mode of metal uptake is referred as active uptake. *Pseudomonas* strain, characterized as part of a project to develop a biosorbent for removal of toxic radionuclides from nuclear waste streams, was a potent accumulator of uranium (VI) and thorium (IV) [49].

Most works on pollutants bioremediation uses pure microbial cultures. However, the use of mixed microbial cultures is undoubtedly advantageous. Some of the best examples of enrichment cultures comprising several specific consortia involve the bioremediation. In the case of heavy metals removal, Adarsh et al. [50] have used an environmental bacterial consortium to remove Cd, Cr, Cu, Ni and Pb from a synthetic wastewater effluent. For Cr(VI) removal we reported that the survival and stability of bacteria are better when they are present as a mixed culture, especially, in highly contaminated areas and in the presence of more than one type of metal [51]. Indeed, the indigenous bacteria enriched from chromium contaminated biotopes, were able to remove Cr(VI) successfully in multi-contaminated heavy metal solution [51]. A microbial consortium consisting of three bacterial Pseudomonas species originally obtained from dye contaminated sites was capable of decolorizing textile effluent and dye faster than the individual bacteria under static conditions [52].

2.3. PGPR and PGPB degradation

Plant associated bacteria, such as endophytic bacteria (non-pathogenic bacteria that occur naturally in plants) and rhizospheric bacteria (bacteria that live on and near the roots of plants), have been shown to contribute to biodegradation of toxic organic compounds in contaminated soil and could have potential for improving phytoremediation [53]. Plant growth promoting rhizobacteria (PGPR) are naturally occurring soil bacteria that aggressively colonize plant roots and benefit plants by providing growth promotion [54]. Some plants can release structural analogs of PAHs such as phenols, to promote the growth of hydrocarbon degrading microbes and their degradation on PAHs. For such plant/microbe systems, an important class of bacteria is *Pseudomonas* spp., have PGPR activity and hydrocarbon degrading capacity [55]. Furthermore, the rhizosphere of vegetation in contaminated field contains higher diversity of population of PAH-degrading bacteria, among which two *Lysini bacillus* strains were isolated [56]. Culturable PCB degraders were also associated with both the rhizosphere and root zone of mature trees growing naturally in a contaminated site, they were identified as members of the genus *Rhodococcus, Luteibacter* and *Williamsia,* which suggest that biostimulation through rhizoremediation is a promising strategy for enhancing PCB degradation in situ [57]. Also, the free living nitrogen fixer *Azospirillum lipoferum* generally found in the rhizoplane of the crop plants was used for Malathion degradation which is one of the largest organo phosphorus insecticides in the world [33]. Results from the literature suggest that heavy metals may be removed from contaminated soils using plant growth promoting rhizobacteria. The use of soil bacteria (often plant growth promoting bacteria (PGPB)) as adjuncts in metal phytoremediation can significantly facilitate the growth of plants in the presence of high (and otherwise inhibitory) levels of metals [58]. To increase the efficiency of contaminants extraction, it is interesting to apply plants combined to some microorganisms; such technique is called rhizoremediation [59].

2.4. Microfungi and mycorrhiza degradation

Microfungi are described as a group of organisms that constitute an extremely important and interesting group of eukaryotic, aerobic microbes ranging from the unicellular yeasts to the extensively mycelial molds [60]. Yeasts preferentially grow as single cells or form pseudomycelia, whereas molds typically grow as mycelia-forming real hyphae.

Fungi are an important part of degrading microbiota because, like bacteria, they metabolize dissolved organic matter; they are principal organisms responsible for the decomposition of carbon in the biosphere. But, fungi, unlike bacteria, can grow in low moisture areas and in low pH solutions, which aids them in the breakdown of organic matter [61]. Equipped with extracellular multienzyme complexes, fungi are most efficient, especially in breaking down the natural polymeric compounds. By means of their hyphal systems they are also able to colonize and penetrate substrates rapidly and to transport and redistribute nutrients within their mycelium [62]. Mycorrhiza is a symbiotic association between a fungus and the roots of a vascular plant. In a mycorrhizal association, the fungus colonizes the host plant's roots, either intracellularly as in arbuscular mycorrhizal fungi (AMF), or extracellularly as in ectomycorrhizal fungi. They are also an important component of soil life and soil chemistry. Bioremediation using mycorrhiza is named mycorrhizoremediation [63]. Fungi possess important degradative capabilities that have implications for the recycling of recalcitrant polymers (e.g., lignin) and for the elimination of hazardous wastes from the environment [27]. Below, some aspects of the microfungal degradation of some pollutants by unicellular and filamentous fungi are discussed.

2.4.1. Yeasts degradation

Several yeasts may utilize aromatic compounds as growth substrates, but more important is their ability to convert aromatic substances cometabolically. Some species such as the soil yeast *Trichosporon cutaneum* possess specific energy-dependent uptake systems for aromatic substrates (e.g., for phenol) [64].

Furthermore, biodegradation of aliphatic hydrocarbons occurring in crude oil and petroleum products has been investigated well, especially for yeasts. The n-alkanes are the most widely and readily utilized hydrocarbons, with those between C_{10} and C_{20} being most suitable as substrates for microfungi [65]. However, the biodegradation of n-alkanes having chain lengths up to n-C_{24} has also been demonstrated [27]. Typical representatives of alkane-utilizing yeasts include *Candida lipolytica, C. tropicalis, Rhodoturularubra,* and *Aureobasidion(Trichosporon) pullulans. Rhodotorula aurantiaca* and *C. ernobii* were found able to degrade diesel oil [66]. Yeasts are also reported for aniline biodegradation (a potential degradation product of the azo dye breakdown) it is the example of *C. methanosorbosa* BP-6 [67]. According to many authors, bacteria have been described as being more efficient hydrocarbon degraders than yeast, or at least that bacteria are more commonly used as a test microorganism. However, there is information that yeasts are better hydrocarbon degraders than bacteria [68].

In addition to aromatic and aliphatic hydrocarbons compounds, microfungi may transform numerous of other aromatic organopollutants cometabolically, including polycyclic aromatic hydrocarbons (PAHs) and biphenyls, dibenzofurans, nitro aromatics, various pesticides, and plasticizers [69]. There have also been studies of PCB metabolism by yeasts C. *boidinii* and C. lipolytica [70] and Saccharomyces cerevisiae [71]. Insecticides and fungicides can also be adsorbed by *S. cerevisiae* during aerobic fermentation [72].

Yeasts are known for playing an important role in the removal of toxic heavy metals. There are many reports on biosorption of heavy metals by yeasts. Several investigations demonstrated that yeasts are capable of accumulating heavy metals such as Cu(II), Ni(II), Co(II), Cd(II) and Mg(II) and are superior metal accumulators compared to certain bacteria [73]. In the case of hexavalent chromium (Cr(VI)) we found that *P. anomala* is able to remove Cr(VI) [74] and we studied the biosorption of Cr(VI) by live and dead cells of three yeasts species: *Cyberlindnera fabianii, Wickerhamomyces anomalus* and *C. tropicalis* [75]. Several yeast strains *S. cerevisiae, P. guilliermondii, Rhodotorula pilimanae, Yarrowiali polytica* and *Hansenula polymorpha* have been reported to reduce Cr(VI) to Cr(III) [76]. In addition, the tolerance of *P. guilliermondii* to chromate was found to depend on its capacity for extracellular reduction of Cr(VI) and Cr(III) chelation [77]. Most studies, have reported the efficiency of immobilized cells of yeasts in metals removal, one example is *Schizosaccharomyces pombe* for copper removal [78].

2.4.2. Filamentous fungi degradation

The attributes that distinguish filamentous fungi from other life forms determine why they are good biodegraders. First, the mycelial growth habit gives a competitive advantage over single cells such as bacteria and yeasts, especially with respect to the colonization of insoluble substrates. Fungi can rapidly ramify through substrates, literally digesting their way along by secreting a battery of extracellular degradative enzymes. Hyphal penetration provides a mechanical adjunct to the chemical breakdown affected by the secreted enzymes. The high surface to cell ratio characteristic of filaments maximizes both mechanical and enzymatic contact with the environment. Second, the extracellular nature of the degradative enzymes enables fungi to tolerate higher concentrations of toxic chemicals than would be possible if these compounds had to be brought into the cell. In addition, insoluble compounds that cannot cross a cell membrane are susceptible to attack [3].

Many workers divide bioremediation strategies into three general categories: 1) the target compound is used as a carbon source; 2) the target compound is enzymatically attacked but is not used as a carbon source (cometabolism) and 3) the target compound is not metabolized at all but is taken up and concentrated within the organism (bioaccumulation). Although fungi participate in all three strategies, they are often more proficient at cometabolism and bioaccumulation than at using xenobiotics as sole carbon sources [79]. The isolates identified as deuteromycetes belonging to the genera *Cladophialophora*, *Exophiala* and *Leptodontium* and the ascomycete *Pseudeurotium zonatum* are toluene-degrading fungi, they use toluene as sole carbon and energy source [80].

The majority of filamentous fungi are unable to totally mineralize aromatic hydrocarbons; they only transform them into indirect products of decreased toxicity and increased susceptibility to decomposition with the use of bacteria suggesting that the interaction among fungi and bacteria is profitable for the process of petroleum hydrocarbon mineralization. Among the filamentous fungi participating in aliphatic hydrocarbon biodegradation are *Cladosporium* and *Aspergillus*, whereas fungi belonging to *Cunninghamella*, *Penicillinum*, *Fusarium* and *Aspergillus* can take part in aromatic hydrocarbon decomposition [81]. Fungal genera, namely, *Amorphoteca*, *Neosartorya* and *Talaromyces* were isolated from petroleum contaminated soil and proved to be the potential organisms for hydrocarbon degradation [82]. A group of fungi, namely, *Aspergillus*, *Cephalosporium* and *Pencillium* was also found to be potential degrader of crude oil hydrocarbons [83]. Fungal potentiality in PCBs degradation has been rarely explored. Several studies revealed that filamentous fungi can degrade PCBs. Among the filamentous fungi, the ligninolytic ones have been specifically investigated because of their extracellular, aspecific oxido-reductive enzymes that have been already successfully exploited in the degradation of many aromatic pollutants [84]. There have also been studies of PCB metabolism by ectomycorrhizal fungi [85] and other fungi such as *Aspergillus niger* [86]. Fungi are also reported to degrade standing timber, finished wood products, fibers, and a wide range of non cellulosic products such as plastics, fuels, paints, glues, drugs, and other human artifacts. Fungi are known to tolerate and detoxify metals by several mechanisms including valence transformation, extra and intracellular precipitation and active uptake [87]. Many species can adsorb cadmium, copper, lead, mercury, and zinc into their mycelium and spores. Sometimes the walls of dead fungi bind better than living ones. Systems using *Rhiloprzs arrhizus* have been developed for treating uranium and thorium [88]. *Aspergillus niger* AB10 during cadmium and *R. arrhizus* M1 during lead biosorption indicated that the cell surface functional groups of the fungus might act as ligands for metal sequestration resulting in removal of the metals from the aqueous culture media [89]. Furthermore, the proteins in the cell walls of AMF appear to have similar ability to sorb potentially toxic elements by sequestering them. There is evidence that AMF can withstand potentially toxic elements and glomalin produced on hyphae of AMF can sequester them [90]. AMF plays a significant ecological role in the phytostabilization of potentially toxic trace element polluted soils by sequestration and, in turn, help mycorrhizal plants survive in polluted soils.

The most widely researched fungi in regard to dye degradation are the ligninolytic fungi [91]. Nine strains of filamentous fungi were isolated by Abruscia et al. [79] from cinematographic film consisted of three species of *Aspergillus* i.e. *A. ustus, A. nidulans* var. *nidulans, A. versicolor*, as well as, *Penicillium chrysogenum, Cladosporium cladosporioides, Alternaria alternata, Mucor racemosus, Phoma glomerata* and *Trichoderma longibrachiatum* were able to biodegrade the gelatin emulsion with different rates of metabolic CO_2 production. Filamentous fungi may degrade pesticides using two types of enzymatic systems: intracellular (cytochromes P450) and exocellular (lignin-degrading system mainly consisting in peroxidases and lactases). Each of these systems could also be induced or inhibited by pesticides, thus able to modulate their metabolism [92].

2.5. Degradative capacities of algae and protozoa

In spite of algae and protozoa are the important members of the microbial community in both aquatic and terrestrial ecosystems, reports are scanty regarding their involvement in hydrocarbon biodegradation [5]. Walker et al. [93] isolated an alga, *Prototheca zopfi* which was capable of utilizing crude oil and a mixed hydrocarbon substrate and exhibited extensive degradation of n-alkanes and isoalkanes as well as aromatic hydrocarbons. Cerniglia and Gibson [94] observed that nine cyanobacteria, five green algae, one red alga, one brown alga, and two diatoms could oxidize naphthalene. Protozoa, by contrast, had not been shown to utilize hydrocarbons, however, protozoa population have been shown to significantly reduce the number of bacteria available for hydrocarbon removal so their presence in a biodegradation system may not always be beneficial [95]. Rogerson and Berger [96] found no direct utilization of crude oil by protozoa cultured on hydrocarbon-utilizing yeasts and bacteria. Overall, the limited available evidence does not appear to suggest an ecologically significant role for algae and protozoa in the degradation of hydrocarbons in the environment [97]. Some research has demonstrated that certain fresh algae (e.g. *Chlorella vulgaris, Scenedesmus platydiscus, S. quadricauda* and *S. capricornutum*) are capable of uptaking and degrading PAHs [98].

Information on the interactions between pesticides andalgae were compiled by Kobayashi and Rittman [99], showing that not only algae were capable of bioaccumulating pesticides, but they were also capable of biotransforming some of these environmental pollutants.

Degradation of azo dyes by *C. vulgaris* and *C. pyrenoidosa* was made using dyes as carbon and nitrogen sources, but this was dependent on the chemical structure of the dyes [100]. The degradation was found to be an inducible catabolic process. *C. vulgaris, Lyngbyala gerlerimi, Nostoc lincki, Oscillatoria rubescens, Elkatothrix viridis* and *Volvox aureus* were able to decolorize and remove methyl red, orange II, G-Red (FN-3G), basic cationic, and basic fuchsin [101].

Species of *Chlorella, Anabaena inacqualis, Westiellopsis prolifica, Stigeoclonium lenue, Synechoccus* sp. tolerate heavy metals and several species of *Chlorella, Anabaena* and marine algae have been used for the removal of heavy metals, but the operational conditions limit the practical application of these organisms [102]. Metals are taken up by algae through adsorption. Metal chelation by unicellular algae has been reported. Biosorption of heavy metals by brown algae is also known since a long time, this includes sorption of heavy metals by a number of cell wall constituents such as alginate and fucoidan. Most of the research in this area has been carried out on marine and soil algae [103]. The microalga *S. incrassatulus* was reported to remove Cr(VI), Cd(II) and Cu(II) in continuous cultures [104]. Green algae were also reported in heavy metals bioremediation, *C. sorokiniana* for Cr(III) removal [105].

The protozoa are the main grazer on the degrading bacteria for organic contaminants, so the interaction between protozoa and degrading bacteria will affect the result of bacteria degradation directly. Mattison and Harayama [106] constructed a model for the food chain in order to study the influence of grazing bacteria of protozoa flagellate *Heteromita globosa* on the biodegradation of benzene and methylbenzene. They found that during the logarithmic growing period of flagellate population, the degrading rate of benzene and methylbenzene has improved 8.5 times by bacteria than before. The protozoa infusorians can obviously accelerate the biodegradation of heterogenous substances in the environment such as PAH. For example, the degradation rate of naphthalene can be improved by 4 times than before. There are several possible hypotheses about the mechanism of protozoa accelerating biodegradation of organic contaminants, which mainly include the following six parts: 1) the nutrient mineralization which improves the turnover of nutrients; 2) bacteria activation which controls the quantity, grazes the aged cells or excretes active substance; 3) selective grazing which reduces the competition to the resource and space and thus is good for the growth of degrading bacteria; 4) physical disturbance which can increase oxygen content and the surface of degraded matters; 5) direct degradation which can excrete special enzymes participating in the degradation; 6) sym-metabolism which offers energy and carbon resource for the bacteria during the degradation [107].

3. Bioremediation and biodegradation

The application of bioremediation as a biotechnological process involving microorganisms has become a crescent study field in microbiology, because of its increasing potential of solving the dangers of many pollutants through biodegradation. Microorganisms might be considered excellent pollutant removal tools in soil, water, and sediments, mostly due to their advantage over other bioremediation procedures [108]. Moreover, bioremediation using biodegradation represents a high impact strategy, but still a low cost way tool of removing pollutants, hence a very viable process to be applied. The principles of bioremediation are based on natural attenuation, bioaugmentation and biostimulation [109]. The simplest method of bioremediation is natural attenuation, in which soils are only monitored for variations in pollution concentrations to ensure that the pollutant transformation is active [110]. Bioaugmentation is usually applied in cases where natural active microbial communities are present in low quantities or even absent, wherein the addition of contaminant degrading organisms can accelerate the transformation rates [111]. In such cases, the adaptation of exogenous strains that exert highly efficient activities for pollutant transformation to new environments is a key challenge in implementation [112]. The capacity of a microbial population to degrade pollutants can be enhanced also by stimulation of the indigenous microorganisms by addition of nutrients or electron acceptors [109].

3.1. Natural attenuation

Natural attenuation or bioattenuation is the reduction of contaminant concentrations in the environment through biological processes (aerobic and anaerobic biodegradation, plant and animal uptake), physical phenomena (advection, dispersion, dilution, diffusion, volatilization, sorption/desorption), and chemical reactions (ion exchange, complexation, abiotic transformation). Terms such as intrinsic remediation or biotransformation are included within the more general natural attenuation definition [113]. Although, one of the most important components of natural attenuation is biodegradation, the change in form of compounds carried out by living creatures such as microorganisms. Under the right conditions, microorganisms can cause or assist chemical reactions that change the form of the contaminants so that little or no health risk remains. Natural attenuation occurs at most polluted sites. However, the right conditions must exist underground to clean sites properly. If not, cleanup will not be quick enough or complete enough. Scientists monitor these conditions to make sure natural attenuation is working. This is called monitored natural attenuation or (MNA). So, Monitored natural attenuation is a technique used to monitor or test the progress of natural attenuation processes that can degrade contaminants in soil and groundwater. It may be used with other remediation processes as a finishing option or as the only remediation process if the rate of contaminant degradation is fast enough to protect human health and the environment. Natural processes can then mitigate the remaining amount of pollution; regular monitoring of the soil and groundwater can verify those reductions [114].

When the environment is polluted with chemicals, nature can work in four ways to clean it up [115]: 1) Tiny bugs or microbes that live in soil and groundwater use some chemicals for food. When they completely digest the chemicals, they can change them into water and harmless gases. 2) Chemicals can stick or sorb to soil, which holds them in place. This does not clean up the chemicals, but it can keep them from polluting groundwater and leaving the site. 3) As pollution moves through soil and groundwater, it can mix with clean water. This reduces or dilutes the pollution. 4) Some chemicals, like oil and solvents, can evaporate, which means they change from liquids to gases within the soil. If these gases escape to the air at the ground surface, sunlight may destroy them.

If the natural attenuation is not quick enough or complete enough, bioremediation will be enhanced either by biostimulation or bioaugmentation.

3.2. Biostimulation

Biostimulation involving the addition of soil nutrients, trace minerals, electron acceptors, or electron donors enhances the biotransformation of a wide range of soil contaminants [115]. There are many examples of biostimulation of pollutants biodegradation by indigenous microorganisms. Trichloroethene and perchloroethene are reported to be completely converted to ethane by microorganisms in a short span of time with the addition of lactate during biostimulation [116]. Electron shuttles, such as humic substances (HS), may play a significant stimulation role in the anaerobic biotransformation of organic pollutants through enhancing the electron transfer speed. Anthraquinone-2,6-disulfonate (AQDS) from the category of HS can serve as an electron shuttle to promote the reduction of iron oxides and transformation of chlorinated organic contaminants [117]. Chen et al. [118] reported that the biostimulation of indigenous microbial communities by the addition of lactate and AQDS led to the enhanced rates of Pentachlorophenol PCP dechlorination by the dechlorinating and ironreducing bacteria in soils. Among various nutrient media, glycerol appeared to show the most favorable metabolic characteristics against phenol toxicity on the indigenous Rhizobium *Ralstonia taiwanensis*, leading to better degradation efficiency of the toxic pollutant [119]. Liliane et al. [120] observed that biostimulation was more efficient when compared to natural attenuation of biodiesel in contaminated soils. However, the comparative study of Bento et al. [121] revealed that bioaugmentation showed the greatest degradation potential and natural attenuation was more effective than biostimulation of soils contaminated with diesel oil. Results obtained by Yu et al. [122] indicate that autochthonous microbes may interact and even compete with the enriched consortium during polycyclic aromatic hydrocarbons biodegradation and the natural attenuation appeared to be the most appropriate way to remedy fluorene and phenanthrene contaminated mangrove sediments while biostimulation was more capable to degrade pyrene contaminated sediments.

3.3. Bioaugmentation

We can define bioaugmentation as the technique for improvement of the capacity of a contaminated matrix (soil or other biotope) to remove pollution by the introduction of specific competent strains or consortia of microorganisms [123]. The basic premise for this intervention is that the metabolic capacities of the indigenous microbial community already present in the biotope slated for cleanup will be increased by an exogenously enhanced genetic diversity, thus leading to a wider repertoire of productive biodegradation reactions [111]. Moreover, genetically engineered microorganisms (GEMs) exhibiting enhanced degradative capabilities encompassing a wide range of aromatic hydrocarbons have also potential for soil bioaugmentation [124]. It is thought that bioaugmentation approach should be applied when the biostimulation and bioattenuation have failed [111].

Many studies have shown that both abiotic and biotic factors influence the effectiveness of bioaugmentation, the most important abiotic factors are temperature, moisture, pH and organic matter content, however, aeration, nutrient content and soil type also determine the efficiency of bioaugmentation. Biotic factors, including competition between indigenous and exogenous microorganisms for limited carbon sources as well as antagonistic interactions and predation by protozoa and bacteriophages, also play essential roles in the final results of bioaugmentation [124].

The combination of bioaugmentation and biostimulation might be a promising strategy to speed up bioremediation. Both indigenous and exogenous microorganisms could benefit from biostimulation by the addition of energy sources or electron acceptors [111]. Bioaugmentation-assisted phytoextraction using PGPR or AMF is also a promising method for the cleaning-up of soils contaminated by metals [123].

4. Factors affecting microbial degradation

Microorganisms can degrade numerous of organic pollutants owing to their metabolic machinery and to their capacity to adapt to inhospitable environments. Thus, microorganisms are major players in site remediation. However, their efficiency depends on many factors, including the chemical nature and the concentration of pollutants, their availability to microorganisms, and the physicochemical characteristics of the environment [111]. So, factors that influence the rate of pollutants degradation by microorganisms are either related to the microorganisms and their nutritional requirements (biological factors) or associated to the environment (environmental factors).

4.1. Biological factors

A biotic factor is the metabolic ability of microorganisms. The biotic factors that affect the microbial degradation of organic compounds include direct inhibition of enzymatic activities and the proliferation processes of degrading microorganisms [125]. This inhibition can occur for example if there is a competition between microorganisms for limited carbon sources, antagonistic interactions between microorganisms or the predation of microorganisms by protozoa and bacteriophages [126]. The rate of contaminant degradation is often dependent on the concentration of the contaminant and the amount of "catalyst" present. In this context, the amount of "catalyst" represents the number of organisms able to metabolize the contaminant as well as the amount of enzymes(s) produced by each cell. Furthermore, the extent to which contaminants are metabolized is largely a function of the specific enzymes involved and their "affinity" for the contaminant and the availability of the contaminant. In addition, sufficient amounts of nutrients and oxygen must be available in a usable form and in proper proportions for unrestricted microbial growth to occur [126]. Other factors that influence the rate of biodegradation by controlling the rates of enzyme catalyzed reactions are temperature, pH and moisture. Biological enzymes involved in the degradation pathway have an optimum temperature and will not have the same metabolic turnover for every temperature [127]. Indeed, the rate of biodegradation is decreased by roughly one-half for each 10°C decrease in temperature [159]. Biodegradation can occur under a wide-range of pH; however, a pH of 6.5 to 8.5 is generally optimal for biodegradation in most aquatic and terrestrial systems. Moisture influences the rate of contaminant metabolism because it influences the kind and amount of soluble materials that are available as well as the osmotic pressure and pH of terrestrial and aquatic systems [128].

4.2. Environmental factors

Soil type and soil organic matter content affect the potential for adsorption of an organic compound to the surface of a solid. Absorption is an analogous process wherein a contaminant penetrates into the bulk mass of the soil matrix. Both adsorption and absorption reduce the availability of the contaminant to most microorganisms and the rate at which the chemical is metabolized is proportionately reduced [126]. Variations in porosity of the unsaturated and saturated zones of the aquifer matrix may influence the movement of fluids and contaminant migration in groundwater. The ability of the matrix to transmit gases, such as oxygen, methane and carbon dioxide, is reduced in fine grained sediments and also when soils become more saturated with water. This can affect the rate and type of biodegradation taking place [126].The oxidation-reduction potential of a soil provides a measurement of the electron density of the system. Biological energy is obtained from the oxidation of compounds in which electrons are transferred to various more oxidized compounds referred to as electron acceptors. A low electron density (Eh greater than 50 mV) indicates oxidizing, aerobic conditions, whereas high electron density (Eh less than 50 mV) indicates reducing, anaerobic conditions [126].

5. Degradation by genetically engineered microorganisms

As mentioned above, bioaugmentation and biostimulation are methods that can be applied to accelerate the recovery of polluted sites. In the late 1970s and early 1980s, bacterial genes encoding catabolic enzymes for recalcitrant compounds started to be cloned and characterized. Soon, many microbiologists and molecular biologists realized the potential of genetic engineering for addressing biodegradation [128]. A genetically engineered microorganism (GEM) or modified microorganism (GMM) is a microorganism whose genetic material has been altered using genetic engineering techniques inspired by natural genetic exchange between microorganisms. These techniques are generally known as recombinant DNA technology. Genetically engineered microorganisms (GEMs) have shown potential for bioremediation of soil, groundwater and activated sludge, exhibiting the enhanced degrading capabilities of a wide range of chemical contaminants [129]. As soon as the prospect of releasing genetically modified microorganisms for bioremediation became a reality, much of the research effort in the field was aimed at biosafety and risk assessment [128].

There are at least four principal approaches to GEM development for bioremediation appli-cation [11]. These include: 1) Modification of enzyme specificity and affinity; 2) Pathway construction and regulation; 3) Bioprocess development, monitoring and control; 4) Bioaffinity bioreporter sensor applications for chemical sensing, toxicity reduction and end point analysis.

5.1. Genetically engineered microorganisms

Molecular biology offers the tools to optimize the biodegradative capacities of microorgan-isms, accelerate the evolution of "new" activities, and construct totally "new" pathways through the assemblage of catabolic segments from different microbes [130].

Genes responsible for degradation of environmental pollutants, for example, toluene, chloro-benzene acids, and other halogenated pesticides and toxic wastes have been identified. For every compound, one separate plasmid is required. It is not like that one plasmid can degrade all the toxic compounds of different groups. The plasmids are grouped into four categories: 1) OCT plasmid which degrades, octane, hexane and decane; 2) XYL plasmid which degrades xylene and toluenes, 3) CAM plasmid that decompose camphor and 4) NAH plasmid which degrades naphthalene [130].

The potential for creating, through genetic manipulation, microbial strains able to degrade a variety of different types of hydrocarbons has been demonstrated by Friello et al. [131]. They successfully produced a multiplasmid-containing *Pseudomonas* strain capable of oxidizing aliphatic, aromatic, terpenic and polyaromatic hydrocarbons.

Pseudomonas putida that contained the XYL and NAH plasmid as well as a hybrid plasmid derived by recombining parts of CAM and OCT developed by conjugation could degrade camphor, octane, salicylate, and naphthalene [129] and could grew rapidly on crude oil because it was capable of metabolizing hydrocarbons more efficiently than any other single plasmid [131].

This product of genetic engineering was called as superbug (oil eating bug). The plasmids of *P. putida* degrading various chemical compounds are TOL (for toluene and xylene), RA500 (for 3, 5-xylene) pAC 25 (for 3-cne chlorobenxoate), pKF439 (for salicylate toluene). Plasmid WWO of *P. putida* is one member of a set of plasmids now termed as TOL plasmid. It was the first living being to be the subject of an intellectual property case. At that point, it seemed that molecular techniques, either through plasmid breeding or sheer genetic engineering, could rapidly produce microbes with higher catalytic abilities, able to basically degrade any environmental pollutant [129].

Reports on the degradation of environmental pollutants by genetically engineered microor-ganisms are focused on genetically engineered bacteria using different genetic engineering technologies: Pathway modification, modification of substrate specificity by *Comamonas testosteroni* VP44 [132].

The application of genetic engineering for heavy metals removal has aroused great interest. For example, *Alcaligenes eutrophus* AE104 (pEBZ141) was used for chromium removal from industrial wastewater [133] and the recombinant photosynthetic bacterium, *Rhodopseudomonas palustris*, was constructed to simultaneously express mercury transport system and metallo-thionein for Hg^{2+} removal from heavy metal wastewater [134].

For polychlorinated biphenyls degradation, chromosomally located PCB catabolic genes of *R. eutropha* A5, *Achromobacter* sp. LBS1C1, and *A. denitrificans* JB1 were transferred into a heavy metal resistant strain *R. eutropha* CH34 through natural conjugation [11].

Genetic engineering of endophytic and rhizospheric bacteria for use in plant associated degradation of toxic compounds in soil is considered one of the most promising new technol-ogies for remediation of contaminated environmental sites [53]. To select a suitable strain for gene recombination and inoculation into the rhizosphere, there are three criteria that has been recommended: first, the strain should be stable after cloning and the target gene should have a high expression, second, the strain should be tolerant or insensitive to the contaminant; and third, some strains can survive only in several specific plant rhizosphere [135]. Many bacteria in the rhizosphere show only limited ability in degrading organic pollutants. With the development of molecular biology, the genetically engineered rhizobacteria with the contam-inant-degrading gene are constructed to conduct the rhizoremediation [58]. Examples about the molecular mechanisms involved in the degradation of some pollutants such as trichloro-ethylene (TCE) and PCBs has been studied [136].

For heavy metals, Sriprang et al. [136] introduced *Arabidopsis thaliana* gene for phytochelatin synthase (PCS; PCSAt) into *Mesorhizobium huakuii* subsp. *rengei* strain B3 and then established the symbiosis between *M. huakuii* subsp. *rengei* strain B3 and *Astragalus sinicus*. The gene was expressed to produce phytochelatins and accumulate Cd^{2+}, under the control of bacteroid-specific promoter, the nifH gene [137].

Finally, the use of GEM strains as an inoculum during seeding would preclude the problems associated with competition between strains in a mixed culture. However, there is considerable controversy surrounding the release of such genetically engineered microorganisms into the environment, and field testing of these organisms must therefore be delayed until the issues of safety and the potential for ecological damage are resolved [138].

5.2. Obstacles associated with the use of GEM in bioremediation applications

While genetic engineering has produced numerous strains able to degrade otherwise intract-able pollutants in a Petri dish or in a bioreactor, the practical translation of this research into actual in situ bioremediation practices has been quite scanty [129]. One major issue in this respect is the growing realization that the strains and bacterial species that most frequently appear in traditional enrichment procedures are not the ones performing the bulk of biode-gradation in natural niches and may not even be any good as bioremediation mediators. The use of stable isotope probing (SIP) and equivalent methods in microbial ecology have revealed that *Pseudomonas, Rhodococcus,* and the typical aerobic fast growers that are widely favored as hosts of biodegradation related recombinant genes are far less significant under natural conditions [138]. Furthermore, using fast-growers as agents for biodegradation is the inevitable buildup of unwelcome biomass. As an alternative, the optimal clean-up agent would be the one that displays a maximum catalytic ability with a minimum of cell mass. The expression of biodegradation genes can be artificially uncoupled from growth with the use of stationary-phase promoters or starvation promoters [139]. In addition, the recent advances in the area of recombinant DNA technologies have paved the way for conceptualizing "suicidal genetically engineered microorganisms" (S-GEMS) to minimize such anticipated hazards and to achieve efficient and safer decontamination of polluted sites [140]. In some cases, whether the intro-duced bacterium is recombinant or not makes little difference, because the problem is that of implantation of foreign microbes in an unfamiliar territory. The introduction of bacterial biomass in an existing niche may create a palatable niche for protozoa that prevents the bacterial population to grow beyond a certain level [141]. Ingenious approaches have been developed to circumvent this problem, including encapsulation of the inoculum in a polymeric matrix or protection in plastic tubing [142].

The efficacy of a desired in-situ catalytic activity (biodegradation or otherwise) depends first on its presence in the target site. One key enzyme may not be there, or it may preexist in the site but not be manifested. Alternatively, it can be hosted by just a very minor part of the whole microbial population, so that its factual expression in the site might not be significant [129].

A field release of *P. fluorescens* HK44 for bioremediation application has been successfully conducted on moderately large-scale and controlled field condition [143]. However, the future application of genetically engineered bacteria for pollution remediation will not be free from the risks associated with their release in the environment. The future risk regarding use of other engineered bacteria is still unclear. Therefore, the future perspectives of engineered bacterial strains under the field conditions will be the focus of review, which may help us to assess the obstacles related with application of genetically engineered bacteria in environ-mental bioremediation.

The major problem encountered in successful bioremediation technology pertains to hostile field conditions for the engineered microbes. Besides, the molecular applications are mainly confined to only few well characterized bacteria such as *E. coli, P. putida, B. subtilis*, etc. Other bacterial strains need to be tried for developing the engineered microbes. The specific charac-teristic of open biotechnological applications has clearly necessitated the development of engineered bacterial strains to meet the new challenges. The main concern is to construct GE bacteria for field release in bioremediation with an adequate degree of environmental certain-ty. Efforts should be made to examine the performance of engineered bacteria in terms of their survival, potential of horizontal gene transfer, which may affect the indigenous microflora within a complex environmental situation. Often the novel scientific researches always give rise to still more fascinating questions pertaining to public concern. In the majority cases, the bacteria designed for bioremediation processes have been designed for specific purpose under the laboratory conditions, ignoring the field requirement and other complex situations. However, there is no evidence that the deliberate release of GE bacteria for bioremediation has caused a measurable adverse impact on the natural microbial community. At least the overstated idea of risk appraisal has fuelled so much debate and triggered so many research efforts, which have immensely contributed a lot in the field of environmental microbiology. However, survival of the GE bacteria in complex environmental situations is still a big question, needs to be addressed in the light of latest findings [144].

6. Conclusion

Microbial activities are very important for the renewal of our environment and maintenance of the global carbon cycle. These activities are included in the term biodegradation. Amid the substances that can be degraded or transformed by microorganisms are a huge number of synthetic compounds and other chemicals having ecotoxicological effects like hydrocarbons and heavy metals. However, in most cases this statement concerns potential degradabilities which were estimated in the laboratory by using selected cultures and under ideal growth conditions. Due to a whole range of factors: competition with microorganisms, insufficient supply with essential substrates, unfavorable external conditions (aeration, moisture, pH, temperature), and low bioavailability of the pollutant, biodegradation in natural conditions is lesser. So, environmental biotechnology has the objective of tackling and solving these problems so as to permit the use of microorganisms in bioremediation technologies. For this reason, it is necessary to support the activities of the indigenous microorganisms in polluted biotopes and to enhance their degradative abilities by bioaugmentation or biostimulation. Genetic engineering is also used to improve the biodegradation capabilities of microorganisms. Nevertheless, there are many risks associated to the use of GEM in the field. Whether or not such approaches are ultimately successful in bioremediation of pollutants may make a difference in our ability to reduce wastes, eliminate industrial pollution, and enjoy a more sustainable future.

Acknowledgements

The authors are grateful for the financial and scientific support rendered by Microbial Biotechnology Laboratory of Faculty of Sciences and Innovation City, SMBA University, Fez, Morocco.

Author details

Nezha Tahri Joutey, Wifak Bahafid, Hanane Sayel and Naïma El Ghachtouli*

*Address all correspondence to: naimaelghachtouli@yahoo.com

Microbial Biotechnology Laboratory, Sidi Mohammed Ben Abdellah University, Faculty of Science and Technics, Fez, Morocco

References

[1] Alexander M. Biodegradation and Bioremediation, San Diego CA 1994. Academic Press.

[2] Marinescu M, Dumitru M and Lacatusu A. Biodegradation of Petroleum Hydrocarbons in an Artificial Polluted Soil. Research Journal of Agricultural Science 2009; 41(2).

[3] Bennet JW, Wunch KG and Faison B D. Use of fungi biodegradation. Manual of environmental microbiology 2002, 2nd ed., ASM Press: Washington, D.C., 960-971.

[4] Fritsche W and Hofrichter M. Aerobic Degradation by Microorganisms 2008 in Biotechnology Set, Second Edition (eds H.-J. Rehm and G. Reed), Wiley-VCH Verlag GmbH, Weinheim, Germany. doi: 10.1002/9783527620999.ch6m

[5] Das N and Chandran P. Microbial Degradation of Petroleum Hydrocarbon Contaminants: An Overview SAGE-Hindawi Access to Research Biotechnology. Research International. Volume 2011, Article ID 941810, 13 pages. Doi:10.4061/2011/941810.

[6] Pramila R., Padmavathy K, Ramesh KV and Mahalakshmi K. *Brevibacillus parabrevis*, *Acinetobacter baumannii* and *Pseudomonas citronellolis* - Potential candidates for biodegradation of low density polyethylene (LDPE). Journal of Bacteriology Research 2012; 4(1) 9-14.

[7] Mrozik A, Piotrowska-Seget Z and Labuzek S. Bacterial degradation and bioremediation of Polycyclic Aromatic Hydrocarbons. Polish Journal of Environmental Studies 2003; 12 (1) 15-25.

[8] Leitão AL. Potential of Penicillium Species in the Bioremediation Field. International Journal of Environmental Resarch. Public Health 2009; 6 1393-1417; doi:10.3390/ijerph6041393.

[9] Diez MC. Biological Aspects involved in the degradation of organic pollutants. Journal of Soil Science and Plant Nutrition. [online] 2010; 10 (3) 244-267. ISSN 0718-9516.

[10] Seo JS, Keum YS and Li QX. Bacterial Degradation of Aromatic Compounds. International Journal of Environmental Resarch. Public Health 2009, 6, 278- 309; doi:10.3390/ijerph6010278.

[11] Menn FM, Easter JP and Sayler GS. Genetically Engineered Microorganisms and Bioremediation, in Biotechnology: Environmental Processes II 2008; 11b, Second Edition (eds H.-J. Rehm and G. Reed), Wiley-VCH Verlag GmbH, Weinheim, Germany. doi: 10.1002/9783527620951.ch21.

[12] Lesley A. Ogilvie and Penny R. Hirsch. 2012. Microbial ecological theory: Current perspectives.

[13] McMurry J (2000). In Aromatic hydrocarbons. Organic Chmistry.5th ed brooks/cole: Thomas learning. New York, pp. 120-180.

[14] Okere UV and Semple KT. Biodegradation of PAHs in 'Pristine' Soils from Different Climatic Regions. Okere and Semple. J BioremedBiodegrad 2012, S:1. http://dx.doi.org/10.4172/2155-6199.S1-006.

[15] Seeger M, Hernández M, Méndez V, Ponce B, Córdoval M and González M. Bacteri-
      al degradation and bioremediation of chlorinated herbicides and biphenyls. Journal
      of Soil Science and Plant Nutrition 2010; 10 (3) 320-332.

[16] Vargas JM. Pesticide degradation. Journal of Arboriculture 1975; 1(12) 232- 233.

[17] Raffi F, Hall JD and Cernigila, CE. Mutagenicity of azo dyes used in foods, drugs and
      cosmetics before and after reduction by Clostridium species from the human intesti-
      nal tract. Food and chemical Toxicology 1997; 35 897-901.

[18] Vandevivere PC, Bianchi R, Verstraete W. Treatment and reuse of wastewater from
      the textile wet-processing industry: review of emerging technologies. Journal of
      Chemical Technology and Biotechnology 1998; 72 289-302.

[19] Verma P, Madamwar, D. Decolorization of synthetic dyes by a newly isolated strain
      of Serratia marcescers. World Journal of Microbiology and Biotechnology 2003; 19 615-
      618.

[20] Petrucci RH, Harwood WS and Herring FG, General Chemistry (8th ed., Prentice-
      Hall 2002), p.1025-26.

[21] Lloyd JR and Lovley DR. Microbial detoxification of metals and radionuclides. Cur-
      rent Opinion in Biotechnology 2001; 12 248–253.

[22] Yakimov MM, Timmis KN and Golyshin PN. Obligate oil-degrading marine bacteria.
      Current Opinion in Biotechnology, 2007; 18(3) 257-266.

[23] Grishchenkov VG, Townsend RT, McDonald TJ, Autenrieth RL, Bonner JS, Boronin
      AM Degradation of petroleum hydrocarbons by facultative anaerobic bacteria under
      aerobic and anaerobic conditions. Process Biochemistry 2000; 35(9) 889-896.

[24] Wiedemeier TH, Miller RN and Wilson JT. Significance of anaerobic processes for the
      Intrinsic bioremediation of fuel hydrocarbons: In, Proceedings of the Petroleum Hy-
      drocarbons and Organic Chemicals in Groundwater - Prevention, Detection, and Re-
      mediation Conference, November 29 - December 1, 1995, Houston Texas.

[25] Floodgate G. The fate of petroleum in marine ecosystems. In Petroleum Microbiolo-
      gy, R. M. Atlas, Ed., pp. 355–398, Macmillion, New York, NY, USA, 1984.

[26] Kafilzadeh F, Sahragard P, Jamali H and Tahery Y. Isolation and identification of hy-
      drocarbons degrading bacteria in soil around Shiraz Refinery. African Journal of Mi-
      crobiology Research 2011; 4(19) 3084-3089.

[27] Fritsche W and Hofrichter M. Aerobic degradation of recalcitrant organic com-
      pounds by microorganisms, in environmental biotechnology: Concepts and applica-
      tions (eds H.-J. Jördening and J. Winter), Wiley-VCH Verlag GmbH & Co. KGaA,
      Weinheim, FRG, 2005 doi: 10.1002/3527604 206.ch7.

[28] Seeger M, Cámara B, Hofer B. Dehalogenation, denitration, dehydroxylation, and angular attack on substituted biphenyls and related compounds by a biphenyl dioxygenase. Journal of Bacteriology 2001; 183 3548-3555.

[29] Petric I, Hrak D, Fingler S, Vonina E, Cetkovic H, Begonja A Kolar and UdikoviKoli N. Enrichment and characterization of PCB-Degrading bacteria as potential seed cultures for bioremediation of contaminated soil. Food Technology and Biotechnology 2007; 45(1)11-20.

[30] Taguchi K, Motoyama M and Kudo T. PCB/biphenyl degradation gene cluster in *Rhodococcus rhodochrous*K37, is different from the well-known *bph* gene clusters in *Rhodococcus* sp. P6, RHA1, and TA42. RIKEN Review 2001; 42. Focused on Ecomolecular Science Research.

[31] Struthers JK, Jayachandran K, Moorman TB. Biodegradation of atrazine by *Agrobacterium radiobacter* J14a and use of this strain in bioremediation of contaminated soil. Applied of Environonmental Microbiology 1998; 64 3368-3375.

[32] Surekha Rani M, Lakshmi V, Suvarnalatha Devi KP, Jaya Madhuri R, Aruna S, Jyothi K, Narasimha G and Venkateswarlu K. Isolation and characterization of a chlorpyrifos degrading bacterium from agricultural soil and its growth response. African Journal of Microbiology Research 2008; (2) 026-031.

[33] Kanade SN, Ade1 AB and Khilare VC. Malathion Degradation by *Azospirillum lipoferum* Beijerinck. Science Research Reporter 2012; 2(1) 94-103.

[34] Dos Santos AB, Cervantes JF, Van Lier BJ. Review paper on current technologies for decolourisation of textile wastewaters: perspectives for anaerobic biotechnology. Bioresource Technology 2007; 98 2369-2385.

[35] Lodato A, Alfieri F, Olivieri G, Di Donato, Marzocchella A, Salatino A, P. Azo-dye conversion by means of *Pseudomonas* sp. OX1. Enzyme and Microbial Technology 2007; 41 646-652.

[36] Chaube P, Indurkar H, Moghe S. Biodegradation and decolorisation of dye by mix consortia of bacteria and study of toxicity on *Phaseolus mungo* and *Triticum aestivum*. Asiatic Journal of Biotechnology Resources 2010; 01 45-56.

[37] Hong Y, Xu M, Guo J et al. Respiration and growth of *Shewanella decolorations* S12 with an azo compound as the sole electron acceptor. Applied and Environmental Microbiology 2007; 73 64-72.

[38] Khalid A, Arshad M, and Crowley D. Bioaugmentation of Azo Dyes. The Handbook of Environmental Chemistry 2010; 9.

[39] Garbisu C, Alkorta I. Phytoextraction: A cost-effective plant based technology for the removal of metals from the environment. Bioresource Technology 2001; 77(3) 229-236.

[40] Fern´andez PM, Martorell MM, Fari˜na JI, and Figueroa LIC., Removal Efficiency of Cr⁶⁺ by Indigenous *Pichia* sp. Isolated from Textile Factory Effluent. The Scientific World Journal. 2012, Article ID 708213, 6 pages doi:10.1100/2012/708213.

[41] Sayel H, Bahafid W, Tahri-Joutey N, Derraz K, Fikri-Benbrahim K, Ibnsouda-Koraichi S and El Ghachtouli N. Cr(VI) reduction by *Enterococcus gallinarum* isolated from tannery waste-contaminated soil. Annal of Microbiology. DOI 10.1007/s13213-011-0372-9.

[42] Zhu W., Chai L., Ma Z., Wang Y., Xiao H. and Zhao K. Anaerobic reduction of hexavalent chromium by bacterial cells of *Achromobacter* sp. StrainCh1. Microbiological Research 2008; 163 616-623.

[43] Yee N, Ma J, Dalia A, Boonfueng T, and Kobayashi DY. Se(VI) Reduction and the Precipitation of Se(0) by the Facultative Bacterium *Enterobacter cloacae* SLD1a-1 Are Regulated by FNR. Applied and environmental microbiology 2007; 73(6) 1914-1920.

[44] Gao W and Francis AJ. Reduction of Uranium(VI) to Uranium(IV) by Clostridia. Applied and environmental microbiology 2008; 74(14) 4580-4584.

[45] Brim H, McFarlan SC, Fredrickson JK, Minton KW, Zhai M, Wackett LP, Daly MJ. Engineering Deinococcusradiodurans for metal remediation in radioactive mixed waste environments. Nature Biotechnology 2000; 18(1) 85-90. doi:10.1038/71986. PMID 10625398.

[46] De Jaysankar, Ramaiah N and Vardanyan L. Detoxification of toxic heavy metals by marine bacteria highly resistant to mercury. Marine Biotechnology 2008; 10(4) 471-477.

[47] Takeuchi F and Sugio T. Volatilization and Recovery of Mercury from Mercury-Polluted Soils and Wastewaters Using Mercury-Resistant *Acidithiobacillus ferrooxidans* strains SUG 2-2 and MON-1.Environmental Sciences 2006; 13(6) 305-316.

[48] White C, Sharman AK, Gadd GM: An integrated microbial process for the bioremediation of soil contaminated with toxic metals. Nature Biotechnology1998; 16 572-575.

[49] PinakiSar, Sufia K. Kazy, S.F. D'Souza. Radionuclide remediation using a bacterial biosorbent. International Biodeterioration& Biodegradation 2004; 54(2-3) 193-202.

[50] Adarsh VK, Madhusmita Mishra, SanhitaChowdhury, 2M. Sudarshan, A.R. Thakur and 1S. Ray Chaudhuri. Studies on Metal Microbe Interaction of Three Bacterial Isolates From East Calcutta Wetland OnLine Journal of Biological Sciences 2007; 7(2) 80-88.

[51] Tahri-Joutey N, Bahafid W, Sayel H, El Abed S and El Ghachtouli N. Remediation of hexavalent chromium by consortia of indigenous bacteria from tannery waste-contaminated biotopes in Fez, Morocco. International Journal of Environmental Studies 2011; (686) 901-912.

[52] Jadhav JP, Kalyani DC, Telke AA, Phugare SS, Govindwar SP. Evaluation of the efficacy of a bacterial consortium for the removal of color, reduction of heavy metals, and toxicity from textile dye effluent. Bioresource Technology 2010; 101(1) 165-173.

[53] Divya B and Deepak Kumar M. Plant–Microbe Interaction with Enhanced Bioremediation. Research Journal of BioTechnology. 2011; 6 (4).

[54] Saharan BS, Nehra V. Plant Growth Promoting Rhizobacteria: A Critical Review Life Sciences and Medicine Research, Volume 2011: LSMR-21.

[55] Hontzeas N, Zoidakis J and Glick BR. Expression and characterization of 1-aminocyclopropane-1-carboxylate deaminase from the rhizobacterium *Pseudomonas putida* UW4: A key enzyme in bacterial plant growth promotion. Biochimica et Biophysica Acta 2004; 1703 11-19.

[56] Ma B, Chen HH, He YandXu JM. Isolations and consortia of PAH-degrading bacteria from the rhizosphere of four crops in PAH-contaminated field. 2010 19th World Congress of Soil Science, Soil Solutions for a Changing World.

[57] Leigh BM, Prouzova P, Mackova M, Macek T, Nagle DP and Fletcher Kinuthia Mwangi JS. Polychlorinated Biphenyl (PCB)-Degrading Bacteria Associated with Trees in a PCB-Contaminated Site. Applied and Environmental Microbiology 2006; 4(72) 2331-2342.

[58] Glick BR. Using soil bacteria to facilitate phytoremediation. Biotechnology Advances. 2010; 28 367-374.

[59] Jing Y, He Z, Yang X. Role of soil rhizobacteria in phytoremediation of heavy metal contaminated soils. Journal of Zhejiang University Sciences 2007; 8 192-207.

[60] Rossman AY. Microfungi: Molds, mildews, rusts and smuts in our living resources. Washington, DC: U.S. Department of the Interior National Biological Service 1995.

[61] Spellman F. R. Ecology for non ecologists. 2008. Page 176.

[62] Matavuly MN and Molitoris HP. Marine fungi degraders of poly-3-hydroxyalkanoate based plastic materials. Proceedings for Natural Sciencespublished by MaticaSrpska 2009; 116 253-265.

[63] Khan AG. Mycorrhizoremediation an enhanced form of phytoremediation. Journal of Zhejiang University Science 2006 7(7) 503-514.

[64] Mörtberg M, Neujahr HY, Uptake of phenol in *Trichosporon cutaneum*, Journal of Bacteriology1985; 161 615-619.

[65] Bartha R. Biotechnology of petroleum pollutant degradation, Microbial Ecology1986. 12 155-172.

[66] De Cássia Miranda R, de Souza CS, de Barros Gomes E, Lovaglio RB, Lopes CE and de Fátima Vieira de Queiroz Sousa M. Biodegradation of Diesel Oil by Yeasts Isolat-

ed from the Vicinity of Suape Port in the State of Pernambuco –Brazil. Brazilian archives of biology and technology 2007; 50(1)147-152.

[67] Mucha K, Kwapisz E, Kucharska U and Okruszeki A. Mechanism of aniline degradation by yeast strain *Candida methanosorbosa* BP-6. Polish journal of microbiology 2010; 59(4) 311-315.

[68] Ijah UJJ. Studies on relative capabilities of bacterial and yeast isolates from tropical soil in degradating crude oil. Waste Manage. 1998; 18 293.

[69] Fritsche W, Hofrichter M, Aerobic degradation by microorganisms, in: Biotechnology. Vol. 11b, Environmental Processes, (-Rehm, H.-J., Reed, G., Eds.), Wiley-VCH, Weinheim 2000, pp. 145–167.

[70] Sasek V, Volfova O, Erbanova P, Vyas B R M, Matucha M. Degradation of PCBs by white rot fungi, methylotrophic and hydrocarbon utilizing yeasts and bacteria. Biotechnology Letters 1993; 15 521-526.

[71] Eaton D C. Mineralization of polychlorinated biphenyls by Phanerochaete chrysosporium: a ligninolytic fungus. Enzymeand Microbial Technology 1985; 7 194-196.

[72] P. Cabras, M. Meloni, F. M. Pirisi, G. A. Farris and F. Fatichenti. Yeast and pesticide interaction during aerobic fermentation. Applied Microbiology and Biotechnology 1988; 29(2-3) 298-301, DOI:10.1007/BF01982920.

[73] Wang J, Chen C. Biosorption of heavy metals by *Saccharomyces cerevisiae*: A Review. Biotechnology Advances 2006; 24: 427-451.

[74] Bahafid W, Sayel H, Tahri-Joutey N and EL Ghachtouli N. Removal Mechanism of Hexavalent Chromium by a Novel Strain of *Pichia anomala* Isolated from Industrial Effluents of Fez (Morocco). Journal of Environmental Science and Engineering 2011; 5 980-991.

[75] Bahafid W, Tahri-Joutey N, Sayel H, Iraqui-Houssaini M and El Ghachtouli N. 2012.Chromium adsorption by three yeast strains isolated from sediments in Morocco. DOI: 10.1080/01490451.2012.705228.

[76] KsheminskaH P., Taras M. Honchar, Galyna Z. Gayda, Mykhailo V. Gonchar. Extracellular chromate-reducing activity of the yeast cultures. Central European Science Journals 2006; 1(1) 137-149.

[77] Ksheminska H, Fedorovych D, Honchar T, Ivash M and Gonchar M. Yeast Tolerance to Chromium Depends on Extracellular Chromate Reduction and Cr(III) Chelation. Food Technology and Biotechnology 2008; 46 (4) 419-426.

[78] SaiSubhashini S, Kaliappan S, Volan M. Removal of heavy metal from aqueous solution using *Schizosaccharomyces pombe* in free and alginate immobilized cells. 2nd International Conference on Environmental Science and Technology 2011; 6107-111.

[79] Abruscia C., Marquinaa D., Del Amob A., Catalina F. Biodegradation of cinemato-
      graphic gelatin emulsion by bacteria and filamentous fungi using indirect impedance
      technique. International Biodeterioration and Biodegradation 2007; 60 137-14.

[80] Francesc X. Prenafeta-Boldu, Andrea Kuhn, Dion M. A. M. Luykx, HeidrunAnke, Jo-
      han W. van Groenestijn and Jan A. M. de Bont. Isolation and characterisation of fungi
      growing on volatile aromatic hydrocarbons as their sole carbon and energy source.
      Mycological Research 2001; 105(4) 477-484.

[81] Steliga T. Role of fungi in biodegradation of Petroleum Hydrocarbons in Drill Waste.
      Polish Journal of Environmental Studies 2012; 21(2) 471-479.

[82] Chaillan F, Le Flèche A, Bury E et al. Identification and biodegradation potential of
      tropical aerobic hydrocarbon degrading microorganisms. Research in Microbiology
      2004; 155(7) 587-595.

[83] Singh H, Mycoremediation: Fungal bioremediation, Wiley-Interscience, New York,
      NY, USA, 2006.

[84] Tigini V, Di Toro S, Belardo A, Prigione V, Fava F and Varese GC. Identification and
      characterization of the active mycoflora of a consortium in the bioremediation of a
      soil historically contaminated by polychlorinated biphenyls (PCBs). 4th European bio-
      remediation conference.

[85] Donnelly PK, Fletcher JS. PCB metabolism by ectomycorrhizal fungi. Bulletin of En-
      vironmental Contamination and Toxicology1995; 54 507-513.

[86] Dmochewitz S, Ballschmiter K. Microbial transformation of technical mixtures of pol-
      ychlorinated biphenyls (PCB) by the fungus *Aspergillus niger*. Chemosphere 1988; 17
      111-121.

[87] Gadd GM. Interaction of fungi with toxic metals. New Phytologist 1993; 124 25-60.

[88] Treen-Sears ME, Martin SM and Volesky B. Propagation of *Rhiloprzs juvanicus* bio-
      sorbent. Applied and Environmental Microbiology 198; 448 137-141.

[89] Pal TK, Bhattacharyya S, Basumajumdar A. Cellular distribution of bioaccumulated
      toxic heavy metals in *Aspergillus niger* and *Rhizopus arrhizus*. International Journal of
      pharma and bio sciences 2010; 1(2) 1-6.

[90] Gonzalez-Chavez MC, Carrillo-Gonzalez R, Wright SF, Nichols K. The role of gloma-
      lin, a protein produced by arbuscularmycorrhizal fungi, in sequestering potentially
      toxic elements. Environmental Pollution 2004; 130(3) 317-323. [doi:10.1016/j.envpol.
      2004.01.004].

[91] Bumpus JA. Biodegradation of azo dyes by fungi. In: Arora, D.K. (Ed.), Fungal Bio-
      technology in Agricultural, Food and Environmental Applications. Marcel Dekker,
      New York, pp. 2004 457-480.

[92] Chaplain V, Mamy L, Vieublé-Gonod L, Mougin C, Benoit P, Barriuso E and Nélieu
      S. Fate of Pesticides in Soils: Toward an Integrated Approach of Influential Factors,

Pesticides in the Modern World - Risks and Benefits, Margarita Stoytcheva (Ed.) 2011 ISBN: 978-953-307-458-0.

[93] Walker JD, Colwell RR, Vaituzis Z and Meyer SA. Petroleum-degrading achlorophyllous alga *Protothecazopfi*. Nature (London) 1975; 254 423-424.

[94] Cerniglia CE and Gibson DT. Metabolism of naphthalene by *Cunninghamella elegans*. Applied and Environmental Microbiology 1977; 34 363-370.

[95] Stapleton RD and Jr.,V.P. Singh. Biotransformations: Bioremediation Technology for Health and Environmental protection. 2002.

[96] Rogerson A, and Berger J. Effect of crude oil and petroleum-degrading microorganisms on the growth of freshwater and soil protozoa. Journal of General Microbiology 1981; 124 53-59.

[97] Bossert I, and Bartha R. 1984. The fate of petroleum in soil ecosystems, p. 434-476. In R. M. Atlas (ed.), Petroleum microbiology. Macmillan Publishing Co., New York.

[98] X.-C. Wang and H.-M. Zhao. 2007. Uptake and Biodegradation of Polycyclic Aromatic Hydrocarbons by Marine Seaweed. Journal of Coastal Research, SI 50 (Proceedings of the 9th International Coastal Symposium), 1056 – 1061. Gold Coast, Australia, ISSN 0749.0208.

[99] Kobayashi H and Rittman BE. Microbial removal of hazardous organic compounds. Environmental Science and Technology 1982; 16 170A-183A.

[100] Jinqi I and Houtian O. Degradation of azo dyes by algae. Environmental Pollution 1992; 75, 273-278.

[101] El-Sheekh MM, Gharieb MM and Abou-El-Souod GW. Biodegradation of dyes by some green algae and cyanobacteria. International biodeterioration and biodegrdation 2009; 63 699-704.

[102] Dwivedi S. Bioremediation of heavy metal by algae: Current and future perspective. Journal of Advanced Laboratory Research in Biology 2012; 3(3).

[103] Davis TA, Volesky B, Mucci A. A review of the biochemistry of heavy metal biosorption brown algae. Water Research 2003;37:4311-30.

[104] Peña-Castro JM, Martínez-Jerónimo F, Esparza-García F, Cañizares-Villanueva RO. Heavy metals removal by the microalga *Scenedesmus incrassatulus* in continuous cultures. Bioresource Technology 2004; 94 219-222.

[105] Akhtar N, Iqbal M, Iqbal ZS and Iqbal J. Biosorption characteristics of unicellular green alga *Chlorella sorokiniana* immobilized in loofa sponge for removal of Cr(III). Journal of Environmental Sciences 2008; 20 231-239.

[106] Mattison RG, Taki H, Harayama S. The soil flagellate *Heteromita globosa* accelerates bacterial degradation of alkylbenzenesthrough grazing and acetate excretion in batch culture.Microbial Ecology 2005; 49: 142-150.

[107] Chen X., Liu M., Hu F., Mao X. and Li H. Contributions of soil micro-fauna (protozoa and nematodes) to rhizosphere ecological functions. ActaEcologicaSinica 2007; 27(8) 3132-3143.

[108] Demnerova K, Mackova M, Spevakova, V, Beranova K, Kochankova L, Lovecka P, Ryslava, E, Macek T. Two approaches to biological decontamination of groundwater and soil polluted by aromatics characterization of microbial populations. International Microbiology 2005; 8 205-211.

[109] Olaniran AO, Pillay D, Pillay B. Biostimulation and bioaugmentation enhances aerobic biodegradation of dichloroethenes. Chemosphere 2006; 63 600-608.

[110] Kaplan CW, Kitts CL. Bacterial succession in a petroleum land treatment unit. Applied and Environmental Microbiology 2004; 70 1777-1786.

[111] El Fantroussi S and Agathos SN. Is bioaugmentation a feasible strategy for pollutant removal and site remediation? Current Opinion in Microbiology 2005; 8 268-275.

[112] El Fantroussi S, Belkacemi M, Top EM, Mahillon J, Naveau H, Agathos SN. Bioaugmentation of a soil bioreactor designed for pilot-scale anaerobic bioremediation studies. Environmental Science & Technology 1999; 33 2992-3001.

[113] Army US. Interim army policy on natural attenuation for environmental restoration: Washington, DC, Department of the Army, Assistant Chief of Staff for Installation Management, DAIM-ED-R (200-1c), September 12, 1995.

[114] EPA (U.S. Environmental Protection Agency). 2001. A Citizen's Guide to Monitored Natural Attenuation. EPA 542-F-01-004, U.S. Environmental Protection Agency, Washington, D.C.

[115] Li CH, Wong YS, Tam NF. Anaerobic biodegradation of polycyclic aromatic hydrocarbons with amendment of iron(III) in mangrove sediment slurry. Bioresource Technology 2010; 101 8083–8092.

[116] Shan HF, Kurtz HD, Freedman D L. Evaluation of strategies for anaerobic bioremediatidon of high concentrations of halomethanes. Water Research 2010; 44 1317-1328.

[117] Bond DR, Lovley DR. Reduction of Fe(III) oxide by methanogens in the presence and absence of extracellular quinines. Environmental Microbiology 2002; 4 115-124.

[118] Chen M., Shih K., Hu M., Li F., Liu C., Wu W. and Tong H. Biostimulation of Indigenous Microbial Communities for Anaerobic transformation of pentachlorophenol in paddy soils of southern China. Journal of Agricultural and Food Chemistry 2012; 60 2967-2975.

[119] Chen BY., Chen WM. nd Chang JS. Optimal biostimulation strategy for phenol degradation withindigenous rhizobium *Ralstonia taiwanensis*. Journal of Hazardous Materials; 2007 B139 232-237.

[120] Liliane RR, Meneghetti, AntônioThomé, Fernando Schnaid, Pedro D.M. Prietto and Gabriel Cavelhão. Natural Attenuation and Biostimulation of Biodiesel Contaminated Soils from Southern Brazil with Different Particle Sizes. Journal of Environmental Science and Engineering 2012; B1 155-162.

[121] Fatima M. Bento, Fla'vio A.O. Camargo, Benedict C. Okeke, William T. Frankenberger. Comparative bioremediation of soils contaminated with diesel oil by natural attenuation, biostimulation and bioaugmentation. Bioresource Technology 2005; 96 1049-1055.

[122] Yu KSH, Wong AHY, Yau KWY, Wong YS, Tam NFY. Natural attenuation, biostimulation and bioaugmentation on biodegradation of polycyclic aromatic hydrocarbons (PAHs) in mangrove sediments. Marine Pollution Bulletin 2005; 51(8-12)1071-1077.

[123] Thierry L., Armelle B. and Karine J. Performance of bioaugmentation-assisted phytoextraction applied to metal contaminated soils: A review . Environmental Pollution 2008; 153 497-522.

[124] Mrozik A, Piotrowska-Seget Z. Bioaugmentation as a strategy for cleaning up of soils contaminated with aromatic compounds. Microbial Research 2006; (2009), doi: 10.1016/j.micres.2009.08.001.

[125] Riser-Roberts E. Remediation of petroleum contaminated soils. Biological, physical, and chemical process. Lewis Publishers Inc (ed). CRC Press LLC. USA. 1998 1-542.

[126] ERD: Environmental Response Division. 1998. Fundamental principles of bioremediation (An aid to the development of bioremediation proposals).

[127] van der Heul RM. Environmental Degradation of petroleum hydrocarbons. 2009

[128] Cases I, de Lorenzo V. Genetically modified organisms for the environment: stories of success and failure and what we have learned from them. International microbiology 2005; 8 213-222.

[129] Sayler GS, Ripp S. Field applications of genetically engineered microorganisms for bioremediation processes. Current Opinion in Biotechnology 2000; 11 286-289.

[130] Ramos JL, Díaz E, Dowling D, de Lorenzo V, Molin S, O'Gara F, Ramos C, Timmis KN. The behavior of bacteria designed for biodegradation. Biotechnology (N Y); 1994 12(13) 1349-56.

[131] Markandey DK and Rajvaidya N. Environmental Biotechnology, 1st edition, APH Publishing corporation 2004; 79

[132] Hrywna Y, Tsoi TV, Maltseva OV, Quensen JF, and Tiedje JM. Construction and characterization of two recombinant bacteria that grow on ortho- and para-substituted chlorobiphenyls. Applied and Environmental Microbiology 1999; 65(5) 2163-2169.

[133] Srivastava NK, Jha MK, Mall ID, Singh D. Application of Genetic Engineering for Chromium Removal from Industrial Wastewater. International Journal of Chemical and Biological Engineering 2010; 3 3.

[134] Xu D and Pei J. Construction and characterization of a photosynthetic bacterium genetically engineered for Hg^{2+} uptake. Bioresource Technology 2011; 102 3083-3088.

[135] Huang XD, El-Alawi Y, Penrose DM, Glick BR, Greenberg BM. Responses of three grass species to creosote during phytoremediation. Environmental Pollution 2004; 130 453-63.

[136] Sriprang R, Hayashi M, Ono H, Takagi M, Hirata K, Murooka Y. Enhanced accumulation of Cd^{2+} by a Mesorhizobium sp. transformed with a gene from Arabidopsis thaliana coding for phytochelatin synthase. Applied and Environmental Microbiology 2003; 69: 1791-1796.

[137] Sussman MC, Collins H, Skinner FA, and Stewart-Tull DE (ed.). 1988. Release of genetically-engineered micro-organisms. Academic Press, Inc. (London), Ltd., London.

[138] Wackett LP. Stable isotope probing in biodegradation research. Trends Biotechnol 2004; 22:153-154.

[139] Matin A. Starvation promoters of Escherichia coli. Their function, regulation, and use in bioprocessing and bioremediation. Annals of the New York Academy of Sciences 1994; 721: 277-291.

[140] Pandey G, Paul D, Jain RK. Conceptualizing "suicidal genetically engineered microorganisms" for bioremediation applications. Biochemical and Biophysical Research Communications 2005; 327 637-639.

[141] Iwasaki K, Uchiyama H, Yagi O. Survival and impact of genetically engineered Pseudomonas putida harboring mercury resistance gene in aquatic microcosms. Bioscience Biotechnology & Biochemistry1993; 57 1264-1269.

[142] Foster L, John R, Kwan Boon N, De Gelder L, Lievens H, Siciliano SD, Top EM, Verstraete W. Bioaugmenting bioreactors for the continuous removal of 3-chloroaniline by a slow release approach. Environmental Science & Technology 2002; 36 4698-46704.

[143] Ripp S, Nivens DE, Ahn Y, Werner C, Jarrel J, Easter JP, Cox CD, Burlage RS, Sayler GS: Controlled field release of a bioluminescent genetically engineered microorganism for bioremediation process monitoring and control. Environmental Science & Technology 2000; 34 846-853.

[144] Singh JS, Abhilash PC, Singh HB, Singh RP, Singh DP. Genetically engineered bacteria: An emerging tool for environmental remediation and future research perspectives. Gene 2011; 480 1-9.

Chapter 12

Phenolic Extractives and Natural Resistance of Wood

M. S. Nascimento, A. L. B. D. Santana,
C. A. Maranhão, L. S. Oliveira and L. Bieber

Additional information is available at the end of the chapter

1. Introduction

Wood is a natural organic material that consists mainly of two groups of organic compounds: carbohydrates (hemicelluloses and cellulose) and phenols (lignin), that correspond to (65-75%) and (20-30%), respectively (Pettersen 1984). The wood is also constituted of minor amounts of extraneous materials, mostly in the form of organic extractives (usually 4–10%) and inorganic minerals (ash), mainly calcium, potassium, and magnesium, besides manganese and silica.

Generally, wood has an elemental composition of about 50% carbon, 6% hydrogen, 43% oxygen, trace amounts of nitrogen and several metal ions.

Cellulose is a long-chain linear polymer exclusively constructed of β-1,4-linked D-glucose units which can appear as a highly crystalline material (Fan et al, 1982). Often 5000 to 15000 glucose rings are polymerized into a single cellulose molecule.

Hemicelluloses consist of relatively short heteropolymers consisting of the pentoses D-xylose and L-arabinose and the hexoses, D-glucose, D-mannose, D-galactose, D-rhamnose and their corresponding uronic acids. It is composed of only 500-3000 sugar units, and thus has a shorter chain than cellulose (Saka 1991)

Lignin, the third cell wall component, is an aromatic polymer synthesized from phenylpropanoid precursors (Adler 1977). It is a three-dimensional polymer formed of coniferyl, syringyl, and coumaryl alcohol units with many different types of linkages between the building blocks and by far the most complex of all natural polymers.

Extractives are chemical constituents residing in the lignocellulosic tissue that contains an higher diversity of organic compounds, for example triglycerides, steryl esters, fatty acids, sterols, neutral compounds, such as fatty alcohols, sterols, phenolic compounds such as tannins (Fava et al, 2006), quinones (Carter et al, 1978; Ganapaty et al, 2004), flavonoids (Reyes-Chilpa et al, 1995; Ohmura et al, 2000; Chen et al, 2004; Morimoto et al, 2006; Sirmah et al, 2009), besides terpenoids (Kawaguchi et al, 1989; Chang et al, 2000; Watanabe et al, 2005) and alkaloids (Kawaguchi et al, 1989).

2. Extractives and natural resistance of wood

Cellulose is the major structural component of wood and also the major food of insects and decay fungi. Termites, like fungi, are important biological agents in the biodegradation of wood (Syofuna et al, 2012).

Extractives are low molecular weight compounds present in wood (Chang et al, 2001), also called secondary metabolites, and are indeed crucial for many important functional aspects of plant life. The relationship between extractives and natural durability of wood was first reported by Hawley et al (1924). The natural durability of wood is often related with its toxic extractive components (Scheffer and Cowling 1966; Carter et al, 1978; Hillis 1987; McDaniel 1992; Taylor 2006; Santana et al, 2010).

Heartwood extractives retard wood decay can protect the wood against decay organisms (Walker 1993, Hinterstoisser et al, 2000; Schultz and Nicholas 2002), but the natural durability is extremely complex and additional factors such as density of wood and lignin content, besides this dual fungicidal and antioxidant action, may be involved (Schultz and Nicholas 2002).

Several studies have shown that after removal of extractives, durable wood loses its natural resistance and makes them more susceptible to decay (Ohmura, 2000; Taylor et al, 2002; Oliveira et al, 2010). Several authors investigated the relationships between the wood properties and extractives (Carter et al, 1978; Schultz et al, 1990; Reyes-Chilpa et al, 1998; Chang et al, 1999; Morimoto et al, 2006).

One of the most limiting factors for the commercial utilization of wood is its low resistance to fungi and termites, especially in the semi-arid and sub-humid tropics. The biodegradation is supposed to be one of the major challenges to incur the heavy economic loss. Wood decay fungi and some species of termites are important and potent wood-destroying organisms attacking various components of the wood (Istek et al, 2005; Gonçalves and Oliveira 2006).

The largest group of fungi that degrades wood is the basidiomycetes and is divided into: white-rot, brown- rot and soft-rot fungi (Anke et al, 2006). Brown-rot fungi occurs most often in buildings, can degrade only structural carbohydrates (cellulose and hemicellulose), leaving lignin essentially undigested, whereas white-rot fungi utilize all wood constituents including both the carbohydrates and the lignin. Soft-rot fungi utilize preferably carbohydrates, but also degrade lignin (Belie et al, 2000). They hydrolyze and assimilate as food the lignocellulose components by injecting enzymes into the wood cells (Erickson et al, 1990).

Termites cause significant losses to annual and perennial crops and damage to wooden components in buildings (Verma 2010). Damage caused by subterranean termites, *Nasuti-termes*, *Coptotermes* and *Reticulitermes* historically has been a concern of researchers worldwide. Korb (2007) estimated annual damage caused by termites at about U.S. $50 billion worldwide. In the city of Sao Paulo, Brazil, alone, a 20-year loss of $3.5 billion was incurred (Lelis, 1994).

The concentration of extractives varies among species, between individual trees of the same species and within a single tree. Some of these extractives render the heartwood unpalatable to wood destroying organisms. Factors affecting wood consumption by termites and fungi are numerous and complexly related. The amount however can vary from season to season even in the same tissue or are restricted in certain wood species (Taylor et al, 2006).

Several woods contain extractives which are toxic or deterrent for termites, bacteria and fungi resistance (Maranhão 2013; Taylor et al, 2006). Termite resistance of wood is a function of heartwood extractive variability while individual extractives inhibit fungal growth (Neya at al, 2004; Arango et al, 2006).

Biological deterioration of wood is of concern to the timber industry due to the economic losses caused to wood in service or in storage. Fungi, insects, termites, marine borers and bacteria are the principal wood biodegraders. They attack different components of wood at different rates giving rise to a particular pattern of damage (Sirmah 2009). Degradation is influenced by environmental conditions of the wood; whether in storage or in use. The degraded wood material is returned into the soil to enhance its fertility (Silva et al, 2007).

The proposal of this study is to demonstrate the importance of phenolic compounds in natural resistence of wood biodegradation. We collected information of the most representative phenolic compounds (flavonoids, stilbenes, quinones and tannins) found in wood, responsible for resistance of some wood species to bio-degraders (Toshiaki 2001; Windeisen et al, 2002).

3. Flavonoids

Flavonoids are secondary metabolites that occur naturally in all plant families (Harbone 1973). Widely distributed in all parts of plants, these compounds afford protection against ultraviolet radiation, pathogens, and herbivores (Harbone and Willians 2000). The general structure includes a C15 (C_6-C_3-C_6) skeleton joined to a chroman ring (benzopyran moiety), classified into flavanones, flavones, chalcones, dihydroflavonols, flavonols, aurones, flavan-3-ols, flavan-3,4-diols, anthocyanidins, isoflavonoids, and neoflavonoids. Some examples of each class of flavonoids are described in figure 1.

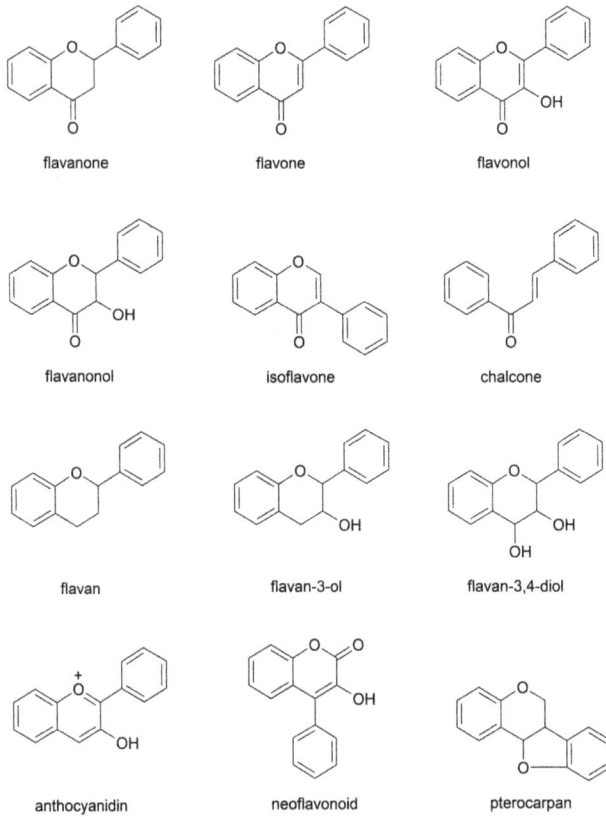

Figure 1. Classification of flavonoids

Flavonoids have an important effect on the durability of wood (Chang et al, 2001; Wang et al, 2004). Accord to Schultz and Nicholas (2000) flavonoids protect heartwood against fungal colonization by a dual function: fungicidal activity and being excellent free radical scavengers (antioxidants). Flavonoids are natural antioxidants and have received attention due to their role in the neutralization or scavenging of free radicals (Gupta and Prakash 2009). Pietarinen (2006) showed that the radical scavenging activity is particularly important because both white-rot and brown-rot fungi are believed to use radicals to disrupt cell walls.

The heartwood of *Lonchocarpus castilloi* Standley (Leguminosae) is highly resistant to attack by the dry wood termites *Cryptotermes brevis* (Walker) (Isoptera: Kalotermitidae). Two flavonoid isolated from the heartwood of this plant, castillen D and castillen E (Figure 2), that presented feeding deterrent activity to *C. brevis* (Reves-Chilpa et al, 1995).

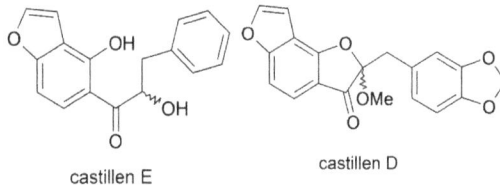

Figure 2. Structure of castillen E and castillen D

Ohmura et al (2000) reported that flavonoids present in *Larix leptolepis* (Pinaceae) wood, principally taxifolin and aromadedrin, showed strong feeding deterrent activities against the subterranean termite, *Coptotermes formosanus* Shiraki (*Isoptera*: Rhinotermitidae) and suggested that some flavonoids such as quercetin and taxifolin (Figure 3) might be useful for termite control agents considering their abundance in plants.

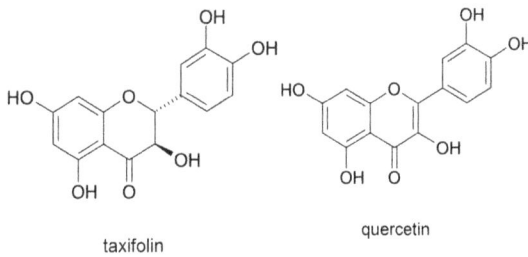

Figure 3. Structure of quercetin and taxifolin

The heartwood of *Acacia auriculiformis* (Leguminosae) has been shown to contain a number of different flavonoids and proanthocyanidins content (Sarai et al, 1980; Barry et al, 2005). According to Schultz et al (1995) the durability of *Acacia* species was attributed the presence of dihydromorin and aromadedrin (Figure 4).

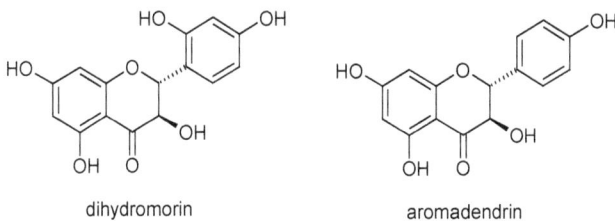

Figure 4. Structure of dihydromorin and aromadendrin

From heartwood of *Morus mesozygia* (Moraceae), besides dihydromorin, were isolated morin and pinobanksin (Figure 5), but the resistance against *wood* destroying *basidiomycetes, Coriolus versicolor, Lentinus squarrosulus* and *Poria* spp. was related to the presence of dihydromorin (Toirambe Bamoninga and Ouattara, 2008).

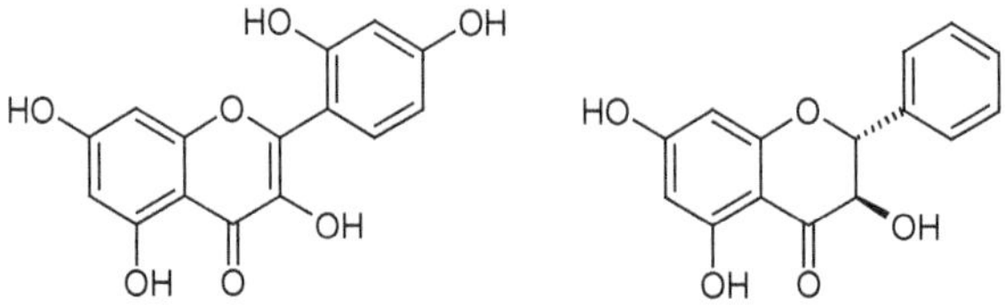

morin   pinobanksin

Figure 5. Structure of morin and pinobanksin

According to Sirmah et al (2009) the durability of *Prosopis juliflora* wood (Leguminosae) was assigned to (–)-mesquitol (Figure 6), but Pizzo et al (2011) related that (-)-mesquitol alone cannot be considered the single most important factor in determining the durability of the *Prosopis* species. Laboratory tests indicated that the heartwood of *P. juliflora* was resistance against to both white- and brown-rot fungi (Sirmah 2009).

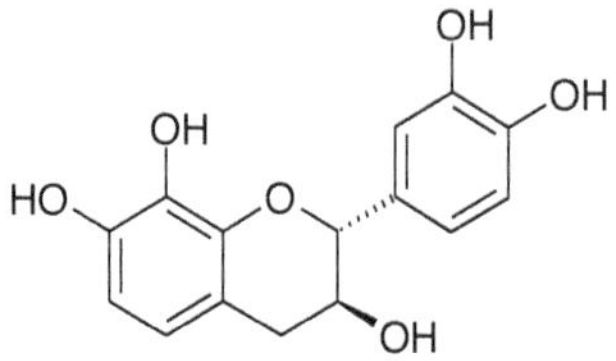

(-)-mesquitol

Figure 6. Structure of mesquitol

The antifeedant activity of some flavonoids against the subterranean termite *Coptotermes formosanus* Shiraki was examined with no-choice tests and two-choice tests (Ohmura et al, 2000). The structure-activity relationships of these flavonoids (Figure 7) were evaluated and it was found that flavonoids containing hydroxyl groups at C-5 and C-7 in A-rings showed high antifeedant activity. Furthermore, the presence of a carbonyl group at C-4 in the pyran rings of the compounds was necessary for the occurrence of high activity. 3-hydroxyflavones and 3-hydroxyflavanones with 3', 4'- dihydroxylated B-rings exhibited higher activity than those with 4'-hydroxylated B-rings.

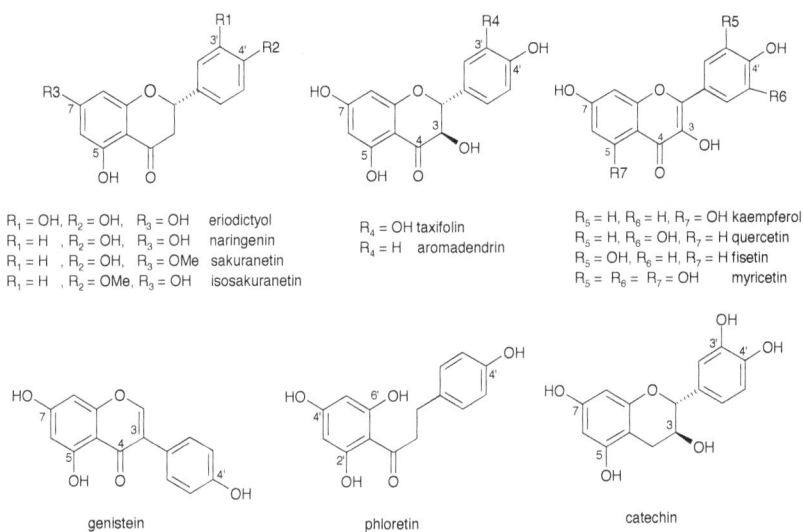

Figure 7. Flavonoids and antifeedant activity against the subterranean termite *C. formosanus*

The antifeedant activities of pterocarpans isolated from the heartwood of *Pterocarpus macro-carpus* Kruz. (Leguminosae) were evaluated against the subterranean termite, *Reticulitermes speratus* Kolbe (Isoptera: Rhinotermitidae). Three isolated pterocarpans, (-)-homopterocarpin, (-)-pterocarpin, and (-)-hydroxyhomopterocarpin were tested (Figure 8). The most active antifeedant against *R. speratus* was (-)-homopterocarpin. However, all pterocarpans showed antifeedant activity against *R. speratus* (Morimoto et al, 2006).

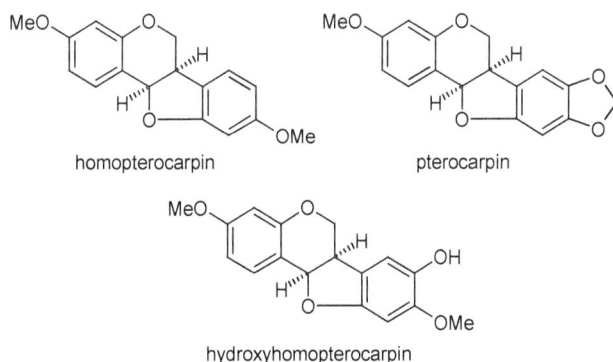

Figure 8. Structure of homopterocarpin, pterocarpin and hydroxyhomopterocarpin

From the heartwood of *Dalbergia latifolia* (Leguminosae) were isolated and identified as active against termites and fungi, the neoflavonoids, latifolin, dalbergiphenol, and 4-methoxydalbergione (Figure 9).

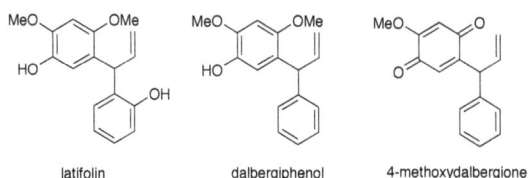

latifolin   dalbergiphenol   4-methoxydalbergione

Figure 9. Structure of latifolin, dalbergiphenol, and 4-methoxydalbergione

With respect to activity against *Trametes versicolor*, a white-rot basidiomycete, latifolin and 4-methoxydalbergione showed activity. Dalbergiphenol exhibited relatively high antifungal activity against the brown-rot basidiomycete, *Fomitopusis palustris* (Sekine et al, 2009).

Latifolin showed high termiticidal activity and termite-antifeedant against *Reticulitermes speratus* (Kolbe). Dalbergiphenol and 4-methoxydalbergione exhibited moderate termite-antifeedant activity (Sekine et al, 2009).

The structure-activity relationships of latifolin (Figure 10) and its derivatives were analyzed to check if there was a correlation between antitermitic and antifungal activity. It was found that the termite mortality in response to the derivatives 2'-O-methyllatifolin, latifolin dimethyl ether, and latifolin diacetate increased 2-fold compared to latifolin. No difference was presented in mortality of termites in the presence of 5-O-methyllatifolin and latifolin. The results indicate that the phenolic hydroxyl group at C-5 of the A ring provides antitermitic activities.

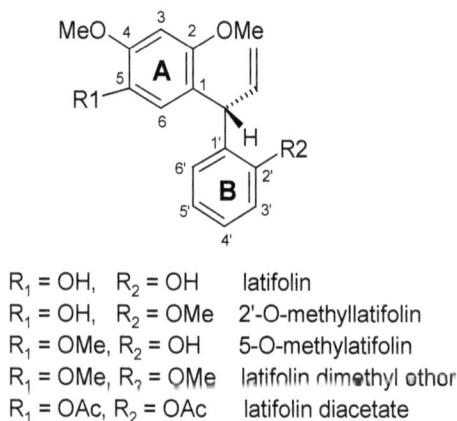

R_1 = OH, R_2 = OH   latifolin
R_1 = OH, R_2 = OMe   2'-O-methyllatifolin
R_1 = OMe, R_2 = OH   5-O-methylatifolin
R_1 = OMe, R_2 = OMe   latifolin dimethyl ether
R_1 = OAc, R_2 = OAc   latifolin diacetate

Figure 10. Structure of latifolin and its derivatives

With respect to antifungal activity of these compounds, it was found that all compounds presented less activity against white- and brown-rot fungi than latifolin. In addition, both C-5 and C-2' phenolic hydroxyl groups in the A and B rings have antifungal activity against white- and brown-rot fungi. In conclusion, the bioactivity of latifolin depends upon the position of phenolic hydroxyl groups (Sekine et al, 2009).

The heartwood of *Dalbergia congestiflora* Pittier (Leguminosae) tree presented natural resistance to fungal attack. The antifungal effect of various extracts from the *D. congestiflora* heartwood was evaluated against *Trametes versicolor* fungus (Martínez-Sotres et al, 2012). The major component of hexane extract that caused 100% growth inhibition from tested fungi was characterized as (-)-Medicarpin (Figure 11). Medicarpin also isolated from heartwood of *Platymiscium yucatanum* (Leguminosae) was identified active against *T. versicolor* (Reyes-Chilpa et al, 1998).

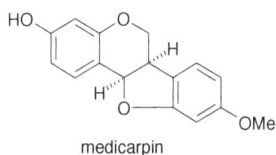

medicarpin

Figure 11. Structure of medicarpin

4. Quinones

Various types of quinones (benzoquinones, naphthoquinones, or anthraquinones) occur in many plant families (Toshiaki 2001). The above mentioned classification of quinones is described in Figure 12. Termite resistant woods are said to contain allelochemicals such as quinones that possess natural repellent and toxic properties (Carter et al, 1978; Scheffrahn 1991; Ganapy et al, 2004; Dungani et al, 2012).

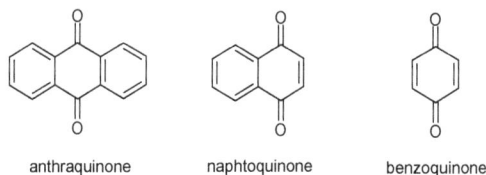

anthraquinone   naphtoquinone   benzoquinone

Figure 12. Classification of quinones

The heartwood of *Tectona grandis* L. f. (Lamiaceae) contains a large amount of quinones that possess considerable influence on the natural durability of teak wood. The naphthoquinone, 4', 5'–dihydroxyepiisocatalponol (Figure 13) plays a key role in the resistance of teak against fungi attack. In-vitro bioassays indicated that this compound acted as a fungicide against the White-rot fungi *Trametes versicolor* (Niamké et al, 2012). Tectoquinone (Figure 13), a anthraquinone, presented strong antitermitic activity and is assumed to be at the origin of the resistance of teak wood to termites (Haupt et al, 2003; Kokutse et al, 2006). According to Wolcott (1955) this substance is highly repellent to the dry-wood termite *Cryptotermes brevis* (Walker) and Sandermann and Dietrichs (1957) demonstrated its toxicity to subterranean termite *Reticulitermes flavipes*.

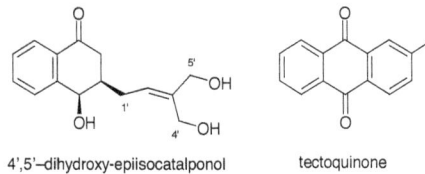

4',5'-dihydroxy-epiisocatalponol   tectoquinone

Figure 13. Structure of 4', 5'–dihydroxyepiisocatalponol and tectoquinone

Castillo and Rossini (2010) isolated naphthoquinones from heartwood of *Catalpa bignonioides* (Bignoniaceae) that showed activity against the termite *Reticulitermes flavipes*. The most abundant and active termiticidal compounds were catalponol and catalponone (Figure 14).

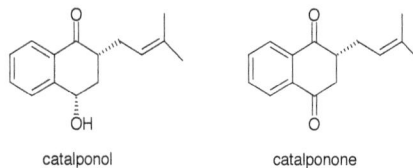

catalponol   catalponone

Figure 14. Structure of catalponol and catalponone

From heartwood of *Tabebuia impetiginosa* (Bignoniaceae) were isolated naphthoquinones, mainly lapachol (Figure 15), that showed no repellent activity to *Reticulitermes* termites but it was repellent to two other termites, *Microcerotermes crassus* (Isoptera: Termitidae) and *Kalotermes flavicollis* (Isoptera: Kalotermitidae) (Becker et al, 1972).

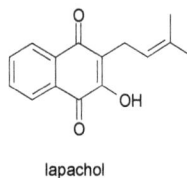

lapachol

Figure 15. Structure of lapachol

The naphthoquinone, 7-methyljuglone (Figure 16) was isolated and identified as termicidal constituent of heartwood of *Diospyros virginiana* L. (Ebenaceae). Its dimer, isodiospyrin possess also termicidal activity against *Reticulirmes flavipes*, but to a lesser extent (Carter et al, 1978).

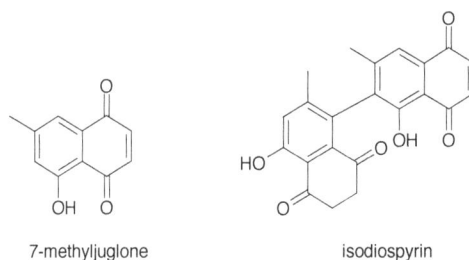

7-methyljuglone   isodiospyrin

Figure 16. Structure of 7-methyljuglone and isodiospyrin

5. Stilbenes

Stilbenes are compounds possessing the 1,2-diphenylethene structure, as well as bibenzyls and phenanthrenes, which are composed of C_6-C_2-C_6 skeleton. Stilbenes derivatives of 1,2-diphenylethlene, process a conjugated double bond system. There are two isomeric forms of 1,2-diphenylethylene: *trans*-stilbene and *cis*-stilbene, and the chemical structure of these two stilbenes are shown in Figure 17.

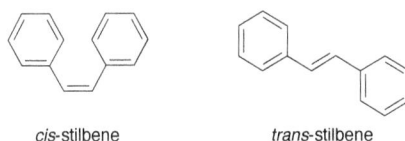

cis-stilbene   *trans*-stilbene

Figure 17. The chemical structure of stilbenes

Hydroxylated trans-stilbene has an important role in heartwood durability, especially for a resistance to fungal decay. The durability and resistance to decay by *Pinus sylvestris* (Pinaceae) is due to pinosylvins (Figure 18). Pinosylvin present in the heartwood of *Pinus* species is formed as a response to external stress such as fungal infections or UV light. The 2, 4, 3′, 5′-tetra and 3, 4, 5, 3′, 5′-pentahydroxystilbenes are responsible for wood resistance against Brown-rot and whit-rot fungi (Schultz et al, 1995).

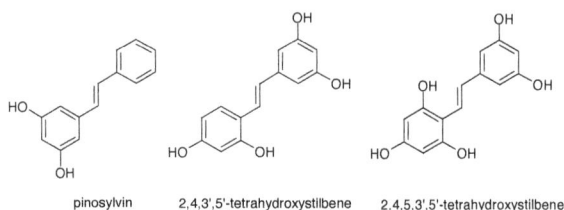

pinosylvin   2,4,3',5'-tetrahydroxystilbene   2,4,5,3',5'-tetrahydroxystilbene

Figure 18. The chemical structure of pinosylvin and derivates

From the heartwood of *Bagassa guianensis* (Moraceae) was isolated moracins including others polyphenols such as flavonoids and stilbenoids (Figure 19), that presented activity against *Pycnoporus sanguineus*, a white-rot fungus. Possible synergism between compounds have been hypothesized (Royer et al, 2012).

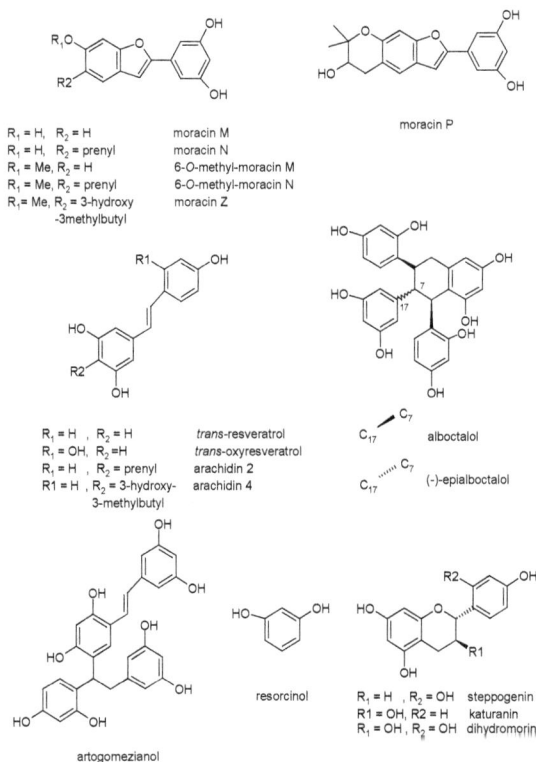

Figure 19. The chemical structure polyphenols from *B. guianensis*

6. Tannins

Tannins constitute a distinctive and unique group of higher plant metabolites. They presented polyphenolic character and relatively large molecular size (from 500 to >20,000). They are thought by some to constitute one of the most important groups of higher plant defensive secondary metabolites (Haslam 1989).

The designation of tannin includes compounds of two distinct chemical groups: hydrolysable tannins (Figure 20) and condensed tannins (Figure 21).

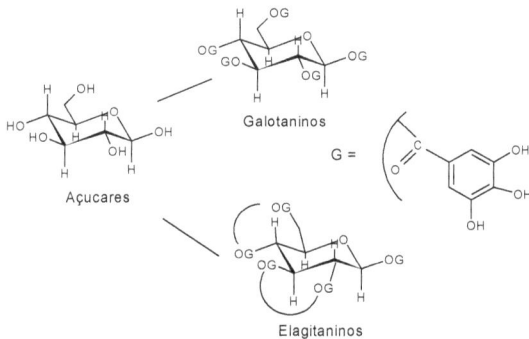

Figure 20. Structure of hydrolysable tannins

Hydrolysable tannins are molecules with a polyol (D-glucose) as a central core. The hydroxyl groups of these carbohydrates are partially or totally esterified with phenolic groups like gallic acid (gallotannins) or ellagic acid (ellagitannins). Hydrolysable tannins are usually present in low amounts in plants.

Condensed tannins are probably the most ubiquitous of all plant phenolics, and presented exceptional concentrations in the barks and heartwoods of a variety of tree species. They are oligomers or polymers of flavonoid units (flavan-3-ol) linked by carbon-carbon bonds not susceptible to cleavage by hydrolysis (Sirmah 2009).

Condensed tannins are natural preservatives and antifungal agents, found in high concentrations in the bark and wood of some tree species (Zucker 1983). Most plant-pathogenic fungi excrete extracellular enzymes such as cellulases and lignases, involved in the invasion and spread of the pathogen. Condensed tannins most likely act as inhibitors of these enzymes by complexing, blocking their action (Peter et al, 2008). For this reason, extract from various woods and barks rich in tannin have been used as adhesives and wood preservatives for a long time (Brandt 1952; Plomely 1966; Mitchell and Sleeter 1980; Pizzi and Merlin 1981; Laks et al, 1988; Lotz and Hollaway 1988; Toussaint 1997; Thevenon 1999).

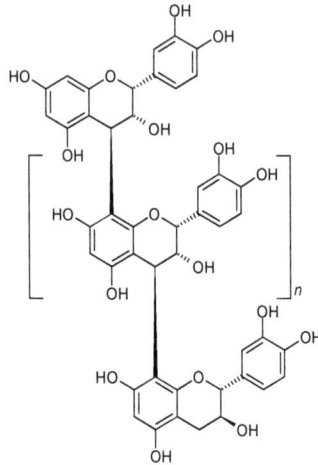

Figure 21. Structure of condensed tannins

7. Conclusions

The protection of wood against biodeterioration is related to its chemical composition, mainly due to the accumulation of extractives in the heartwood. Wood extractives are nonstructural wood components that play a major role in the susceptibility of wood against wood decay organisms. The attack of these organisms in general can be prevented with synthetic organic and inorganic preservatives; however, such products are very harmful to human health and the environment. Several studies have considered that, it is possible the application of wood extractives as natural preservatives. The main components of wood extractives that confers natural resistance against biodeterioration agents are, tannins, flavonoids, quinones and stilbenes.

- Frequently, condensed tannin can be obtained inexpensively by extracting the bark materials with hot water solvent and has been used as preservatives for a long time.

- Flavonoids exhibit antifungical activity as well as feeding deterrent activities against subterranean termites.

- Quinones – possess natural repellent and toxic properties, mainly against termites.

- Stilbenes has an important role in heartwood durability, especially for a resistance to fungal decay.

The characteristics of all wood species are described in Table 1.

Scientific name	Familie name	Common name	Resistance	Origin
Acacia auriculiformis	Leguminosae	Australian wattle	Durable wood (Ashaduzzaman et al, 2011)	Australia, Indonesia, Papua New Guinea
Bagassa guianensis	Moraceae	Tatajuba	Very resistant (Rover et al. 2012)	Guianas and Brazil
Catalpa bignonioides	Bignoniaceae	Common Catalpa Indian Bean	Highly decay resistant heartwood (Muñoz-Mingarro et al, 2006)	North America
Dalbergia congestifolia Pittier	Leguminosae	Rosewood	Resistant wood (Martínez-sotres et al, 2012)	Central America
Dalbergia latifolia	Leguminosae	Indian rosewood	Resistant wood (Lemmens, 2008)	Asia
Diospyros virginiana	Ebenaceae	Common persimmon	-	Africa, Asia
Larix leptolepis	Pinaceae	Japanece larch	resistant (Schaffer and Morrell 1998)	Japan
Lonchocarpus castilloi	Leguminosae	Black cabbage bark	very resistant (Schaffer and Morrell 1998)	Latin America
Morus mesozygia	Moraceae	Mulberry	Non-resistant (Schaffer and Morrell 1998)	Africa
Pinus sylvestris	Pinaceae	Redwood, Scots pine	Non-resistant (Schaffer and Morrell 1998)	Europe, Asia
Platymiscium yucatanum	Leguminosae	Granadillo	very resistant (Schaffer and Morrell 1998)	Latin America
Prosopis juliflora	Leguminosae	Mesquite, algarroba	Resistant (Ramos et al, 2006)	South and Central America
Pterocarpus macrocarpus	Leguminosae	Burma padauk	very resistant (Schaffer and Morrell 1998)	Native to Thailand and Myanmar
Tabebuia impetiginosa		Brazil wood	Very resistant (Paes et al, 2005)	Latin America
Tectona grandis L. f.	Lamiaceae	teak	Very resistant (Kokutse et al, 2006)	Native to southern Asia

Table 1. List of wood species with their family, common names, resistance and distribution.

Author details

M. S. Nascimento[1*], A. L. B. D. Santana[2], C. A. Maranhão[3], L. S. Oliveira[4] and L. Bieber[4]

*Address all correspondence to: msn@ufpe.br.

1 Departamento de Antibióticos, Universidade Federal de Pernambuco, UFPE, Recife, PE, Brazil, Brazil

2 Departamento de Química, Universidade Federal Rural de Pernambuco, UAST, Serra Talhada, PE, Brazil

3 Instituto Federal Educação, Ciência e Tecnologia de Pernambuco, IFPE, Campus Recife, Recife, PE, Brazil

4 Departamento de Química Fundamental, Universidade Federal de Pernambuco, UFPE, Recife, PE, Brazil, Brazil

References

[1] Adfa, M, Yoshimura, T, Komura, K, & Koketsu, M. (2010). Antitermite activities of coumarin derivatives and scopoletin from *Protium javanicum* Burm. f. Journal of Chemical Ecology , 36, 720-726.

[2] Adler, E. (1977). Lignin chemistry-past, present and future. Wood Sci. Technol. , 11, 169-218.

[3] Amusant, N, Moretti, C, Richard, B, Prost, E, Nuzillard, J. M, & Thevenon, M. F. (2007). Chemical compounds from Eperua falcata and *Eperua grandiflora* heartwood and their biological activities against wood destroying fungus (*Coriolus versicolor*). Holz als Roh- und Werkstof , 65, 23-28.

[4] Anke, H, Roland, W, & Weber, S. (2006). White-rots, chlorine and the environment- a tale of many twists. Mycologist, 20(3), 83-89.

[5] Arango, R. A, Green III, F, Hintz, K, Lebow, P. K, & Miller, R. B. (2006). Natural durability of tropical and native wood against termite damage by *Reticulitermes flavis* (Kollar). International Biodeterioration & Biodegradation, 57, 146-150.

[6] Ashaduzzaman, M, Das, A. K, & Shams, M. I. (2011). Natural Decay Resistance of Acacia auriculiformis Cunn. ex. Benth and *Dalbergia sissoo* Roxb. Bangladesh J. Sci. Ind. Res., , 46, 225-230.

[7] Barry, K. M, Mihara, R, Davies, N. W, Mitsunaga, T, & Mohammed, C. L. (2005). Pol-
     yphenols in *Acacia mangium* and *A. auriculiformis* heartwood with reference to heart
     rot. J. Wood Sci. , 51, 615-621.

[8] Becker, G, Lenz, M, & Dietz, S. (1972). Unterschiede im Verhalten und der Giftemp-
     findlichkeit verschiedener Termiten-Arten gegenuber einigen Kernholzstoffen. Z.
     Angew. Entomol. , 71, 201-214.

[9] Beckwith, J. R. (1998). Durability of Wood. University of Georgia School of Forest Re-
     sources Extension Publication for , 98-026.

[10] Belie, N. D, Richardson, M, Braam, C. R, Svennerstedt, B, Lenehan, J. J, & Sonck, B.
     (2000). Durability of Building Materials and Components in the Agricultural Envi-
     ronment: Part I, The agricultural environment and timber structures. J. agric. Engng
     Res. , 75, 225-241.

[11] Brandt, T. G. (1952). Mangrove tannin-formaldehyde resins as hot-pressed plywood
     adhesives. Tectona, 42, 137.

[12] Carter, F. L, Garlo, A. M, & Stanely, J. B. (1978). Termiticidal components of wood
     extracts: 7-methyl-juglone from *Diospyros virginiana*. Journal of Agricultural and
     Food Chemistry , 26, 869-873.

[13] Castillo, L, & Rossini, C. (2010). Bignoniaceae Metabolites as Semiochemicals. Mole-
     cules , 2010(15), 7090-7105.

[14] Chang, S. T, Wang, J-H, Wu, C. L, Chen, P. F, & Kuo, Y. H. (2000). Comparison of the
     antifungal activity of cadinane skeletal sesquiterpenoid from Taiwania (*Tawania Cry-
     promerioides* Hayara) heartwood. Holzforschung , 54(3), 241-245.

[15] Chang, S, Wang, T, Wu, S. -Y, Su, C. -L, & Kuo, Y. -C. Y.-H. (1999). Antifungal com-
     pounds in the ethyl acetate soluble fraction of the extractives of Taiwania (*Taiwania
     cryptomerioides* Hayata) heartwood. Holzforschung , 53(5), 487-490.

[16] Chen, K, Ohmura, W, Doi, S, & Aoyama, M. (2004). Termite feeding deterrent from
     Japanese larch wood. Resource Technology , 95, 129-134.

[17] Dungani, R, Bhat, I. U. H, Abdul Khalil, H. P. S, Naif, A, & Hermawan, D. (2012).
     Evaluation of Antitermitic Activity of Different Extract Obtained from Indonesian
     Teakwood (*Tectona grandis* L.f). Journal of Bioresources , 7(2), 1452-1461.

[18] Ericksson, K. E. L, Blanchette, R. A, & Ander, P. (1990). Microbial and enzymatic deg-
     radation of wood and wood components. Springer-Verlag, Berlin, Germany, 407p.

[19] Fan, L. T, Lee, Y. H, & Gharpuray, M. M. (1982). The nature of lignocellulosics and
     their pretreatments for enzymatic hydrolysis. Adv. Biochem. Eng. , 23, 158-187.

[20] Fava, F, Monteiro de Barros, M, Stumpp, E, & Ramão Marceli Jr., F. (2006). Aqueous
     extract to repel or exterminate termites. Patent Application WO BR173 20050824.,
     2005.

[21] Ganapaty, S, Thomas, P. S, Fotso, S, & Laatsch, H. (2004). Antitermitic quinones from *Disopyros sylvatica*. Phytochemistry, 65, 1265-1271.

[22] Gonçalves, F. G, & Oliveira, J. T. S. (2006). Resistência ao ataque de cupim-de-madeira seca (*Cryptotermes brevis*) em seis espécies florestais. Cerne, , 12(1), 80-83.

[23] Gupta, S, & Prakash, J. (2009). Studies on Indian green leafy vegetables for their antioxidant activity. Plant Foods and Human Nutrition , 64, 39-45.

[24] Harborne, J. B. (1973). Phytochemical Methods. Chapman and Hall Ltd., London , 49-188.

[25] Harborne, B. J, & Williams, A. C. Advances in flavonoids research since (1992). Phytochemistry. , 55, 481-504.

[26] Haupt, M, Leithoff, H, Meier, D, Puls, J, Richter, H. D, & Faix, O. (2003). Heartwood extractives and natural durability of plantation-grown teakwood (*Tectona grandis* L.)- a case study. Holz als Roh- und Werkst. , 61(6), 473-474.

[27] Hawley, L. F, Fleck, L. C, & Richards, C. A. (1924). The relation between durability and chemical composition in wood. Industrial & Engineering Chemistry. , 16(7), 699-700.

[28] Hinterstoisser, B, Stefke, B, & Schwanninger, M. (2000). Wood: Raw material- material-Source of Energy for the future. Lignovisionen , 2, 29-36.

[29] Istek, A, Sivrikaya, H, Eroglu, H, & Gulsoy, S. K. (2005). Biodegradation of *Abies bornmülleriana* (Mattf.) and *Fagus orientalis* (L.) by the white rot fungus *Phanerochaete chrysosporium*. International Biodeterioration & Biodegradation , 55, 63-67.

[30] Kawaguchi, H, Kim, M, Ishida, M, Ahn, Y. J, Yamamoto, T, Yamaoka, R, Kozuka, M, Goto, K, & Takhashi, S. (1989). Several antifeedants from *Phellodendron amurense* against Reticulitermes speratus. Agricultural Biology and Chemistry , 53, 2635-2640.

[31] Kokutse, A. D, Stokes, A, Bailleres, H, Kokou, K, & Baudasse, C. (2006). Decay resistance of Togolese teak (*Tectona grandis* L.) heartwood and relationship with colour. Trees 20, 219 223.

[32] Korb, J. (2007). Termites. Current Biology , 17, 995-999.

[33] Laks, P. E, Mckaig, P. A, & Hemingway, R. W. (1988). Flavanoid biocides: wood preservatives based on condensed tannins. Holzforschung , 42, 299-306.

[34] Lelis, A. T. (1994). Termite problem in São Paulo City-Brazil. In: Lenoir, A., Arnold, G., Lepage, M. (Eds.), Proceedings of the 12th Congress of the International Union for the Study of Social Insects-IUSSI, Paris, 42-46.

[35] Lemmens, R. H. M. J. (2008). *Dalbergia latifolia* Roxb. In: Louppe, D.; Oteng-Amoako, A. A.; Brink, M. (Editors). Prota 7(1): Timbers/Bois d'œuvre 1. [CD-Rom]. PROTA, Wageningen, Netherlands.

[36] Lotz, R. W, & Hollaway, D. F. (1988). Wood preservation. US patent (4732817)

[37] Maranhão, C. A, Pinheiro, I. O, Santana, L. B. D. A, Oliveira, L. S, Nascimento, M. S, & Bieber, L. W. (2013). Antitermitic and antioxidant activities of heartwood extracts and main flavonoids of *Hymenaea stigonocarpa* Mart. International Biodeterioration & Biodegradation, 79, 9-13.

[38] Martínez-Sotres, C, López-Albarrán, P, Cruz-de-León, J, García-Moreno, T, Rutiaga-quiñones, J. G, Vázquez-Marrufo, G, Tamariz-Mascarúa, J, & Herrera-Bucio, R. (2012). Medicarpin, an antifungal compound identified in hexane extract of *Dalbergia congestiflora* Pittier heartwood. International Biodeterioration & Biodegradation, 69, 38-40.

[39] Mcdaniel, C. A. (1992). Major antitermitic components of the heartwood of *Southern Catalpa*. Journal of Chemical Ecology, 18(3), 359-369.

[40] Mitchell, R, & Sleeter, T. D. (1980). Protecting wood from wood degrading organisms. US patent (4220688)

[41] Morimoto, M, Fukumoto, H, Hiratani, M, Chavasir, W, & Komai, K. (2006). Insect Antifeedants, Pterocarpans and Pterocarpol in Heartwood of *Pterocarpus macrocarpus* Kruz. Biosci. Biotechnol. Biochem. , 70, 1864-1868.

[42] Muñoz-mingarro, D, Acero, N, Llinares, F, Pozuelo, J. M, Galán de Mera, A, Vicenten, J. A, Morales, L, Alguacil, L. F, & Pérez, C. (2003). Biological activity of extracts from *Catalpa bignonioides* Walt. (Bignoniaceae) Journal of Ethnopharmacology, 87, 163-167.

[43] Scheffer, T. C, & Morell, J. J. (1998). Natural Durability of Wood: A Worldwide Checklist of Species. Forest Research Laboratory, Oregon State University; College of Forestry, Research Contribution 22, 45.

[44] Neya, B, Hakkou, M, Pétrissans, M, & Gérardin, P. (2004). On the durability of *Burkea Africana* heartwood: evidence of biocidal and hydrophobic properties responsible for durability," Annals of Forest Science, , 61(3), 277-282.

[45] Niamké, F. B, Amusant, N, Stien, D, Chaix, G, Lozano, Y, Kadio, A. A, Lemenager, N, Goh, D, Adima, A. A, Kati-coulibaly, S, & Jay-allemand, C. (2012). Dihydroxy-epiisocatalponol, a new naphthoquinone from *Tectona grandis* L.f. heartwood, and fungicidal activity. International Biodeterioration & Biodegradation , 74, 93-98.

[46] Ohmura, W, Doi, S, Aoyama, M, & Ohara, S. (2000). Antifeedant activity of flavonoids and related compounds against the subterranean termite *Coptotermes formosanus* Shiraki. Journal of Wood Science , 46, 149-153.

[47] Oliveira, L. S, Santana, A. L. B. D, Maranhão, C. A, Miranda, R. C. M, Galvão de Lima, V. L. A, Silva, S. I.; Nascimento, M. S, & Bieber, L. (2010). Natural resistance of five woods to *Phanerochaete chrysosporium* degradation. International Biodeterioration & Biodegradation , 64, 711-715.

[48] Paes, J. B, Morais, V. M, & Lima, C. R. (2005). Resistência natural de nove madeiras do semi-árido brasileiro a fungos causadores da podridão-mole. R. Árvore, 29(3), 365-371.

[49] Pettersen, R. C. (1984). The chemical composition of wood. In: Rowel, R.M. (Ed.), The Chemistry of Wood. Advances in Chemistry Series 207, American Chemical Society, Washington, DC, USA, 57-126.

[50] Pietarinen, S. P, Willfor, S. M, Virkstrom, F. A, & Holmbom, B. R. (2006). Aspen knots, a rich source of flavonoids. Journal of Wood Chemistry and Technology , 26, 245-258.

[51] Pizzi, A, & Merlim, M. (1981). A new class of tannin adhesives for exterior particle-board. Int. J. Adhes. Adhes. 1, 261.

[52] Pizzo, B, Pometti, C. L, Charpentier, J, Boizot, P, & Saidman, N. B. O. (2011). Relation-ships involving several types of extractives of five native argentine wood species of genera *Prosopis* and *Acacia*. Industrial Crops and Products, 34(1), 851-859.

[53] Plomely, K. F. (1966). Tannin-formaldehyde adhesives for wood. II Wattle tannin ad-hesives. CSIRO Division of Forest Products Technological Paper 39, 1-16.

[54] Ramos, I. E. C, Paes, J. B, Farias Sobrinho, D. W, & Santos, G. J. C. (2006). Efficiency of CCB on resistance of *Prosopis juliflora* (Sw.) D.C. wood in accelerated laboratory test decay. R. Árvore, 30(5), 811-820.

[55] Reyes-chilpa, R, Viveros-rodriguez, N, Gomez-garibay, F, & Alavez-solano, D. (1995). Antitermitic activity of *Lonchocarpus castilloi* flavonoids and heartwood extracts. Jour-nal of Chemical Ecology, 21(4), 455-463.

[56] Reyes-chilpa, R, Gomez-Garibay, F, Moreno-Torres, G, Jimenez-Estrada, M, & Quiroz Vaásquez, R. I. (1998). Flavonoids and isoflavonoids with antifungal properties from *Platymiscium yucatanum* heartwood. Holzforschung, 52(5), 459-462.

[57] Royer, M, Rodrigues, A. M. S, Herbette, G, Beauchêne, J, Chevalier, M, Hérault, B, Thibaut, B, & Stiena, D. (2012). Efficacy of *Bagassa guianensis* Aubl. extract against wood decay and human pathogenic fungi. International Biodeterioration & Biode-gradation, 70, 55-59.

[58] Sahai, R, Agarwal, S. K, & Rastogi, R. P. (1980). Auriculoside, a new flavan glycoside from *acacia auriculiformis*. Phytochemistry, 19, 1560-1562.

[59] Saka, S. (1991). Chemical composition and Distribution. Dekker, New York, 3-58.

[60] Sandermann, W, & Dietrichs, H. H. (1957). Investigations on termite-resistant wood. Holz als Roh-und Werkstoff, 15, 281.

[61] Santana, A. L. B. D, Maranhão, C. A, Santos, J. C, Cunha, F. M, Conceição, G. M, Bieber, L. W, & Nascimento, M. S. (2010). Antitermitic activity of extractives from

three Brazilian hardwoods against *Nasutitermes corniger*. International Biodeterioration & Biodegradation, 64, 7-12.

[62] Scheffrahn, R. H. (1991). Allelochemical resistance of wood to termites. Sociobiology , 19, 257-281.

[63] Scheffer, T. C, & Morrell, J. J. (1998). Natural Durability of Wood: a Worldwide Checklist of Species. Oregon State University College of Forestry, Forest Research Laboratory Research Contribution 22.

[64] Schultz, T. P, Harms, W. B, Fisher, T. H, Mcmurtrey, K. D, Minn, J, & Nicholas, D. D. (1995). Durability of angiosperm heartwood: The importance of extractives. Holzforschung, 49(1), 29-34.

[65] Schultz, T. P, & Nicholas, D. D. (2002). Naturally durable heartwood: evidence for a proposed dual defensive function of the extractives. Phytochemistry , 54, 47-52.

[66] Schultz, T. P, Hubbard Jr., T. F, Jin, L, Fisher, T. H, & Nicholas, D. D. (1990). Role of stilbenes in the natural durability of wood: fungicidal structure-activity relationships. Phytochemistry , 29, 1501-1507.

[67] Sekine, N, Ashitani, T, Murayama, T, Shibutani, S, Hattori, S, & Takahashi, K. (2009). Bioactivity of latifolin and its derivatives against termites and fungi. Journal of Agricultural and Food Chemistry , 57, 5707-5712.

[68] Silva, C. A, Monteiro, M. B. B, Brazolin, S, Lopez, G. A. C, Richter, A, & Braga, M. R. (2007). Biodeterioration of brazilwood *Caesalpinia echinata* Lam. (Leguminosae-Caesalpinioideae) by rot fungi and termites. International Biodeterioration & Biodegradation , 60, 285-292.

[69] Sirmah, P, Dumarçay, S, & Gérardin, P. (2009). Effect Unusual amount of (-)-mesquitol of from the heartwood of *Prosopis juliflora* Natural Product Research , 23, 183-189.

[70] Sirmah, P. K. (2009). Valorisation du *"Prosopis juliflora"* comme alternative à la diminution des ressources forestières au Kenya. Thesis-Université Henri Poincaré, Nancy I.

[71] Syofuna, A, Banana, A. Y, & Nakabonge, G. (2012). Efficiency of natural wood extractives as wood preservatives against termite attack. Maderas, Ciencia y Tecnología, 14(2), 155-163.

[72] Taylor, A. M, Gartner, B. L, & Morrell, J. J. (2006). Efects of Heartwood Extractive fractions of *Thuja plicata* and *Chamaecyparis nootkatensison* wood degradation by termites or Fungi. Journal of Wood Science, 52, 147-153.

[73] Thevenon, M. F. (1999). Formulation of long-term, heavy-duty and lowtoxicwood preservatives. Application to the associations boric acid-condensed tannins and boric acid-proteins. Ph.D. Thesis, University of Nancy I, France.

[74] Toirambe Bamoninga, B, & Ouattara, B. (2008). *Morus mesozygia* Stapf. In: Louppe, D, Oteng-Amoako, A. A, & Brink, M. (Editors). Prota 7(1): Timbers/Bois d'œuvre 1. [CD-Rom]. PROTA, Wageningen, Netherlands.

[75] Toussaint, L. (1997). Utiliser les tannins pour la protection du bois. Telex bois, 4, 12

[76] Toshiaki, U. (2001). Chemistry of extractives. In: "Wood and cellulosic chemistry". Ed Marcel Dekker, Inc. New York, , 213-241.

[77] Wang, Q. A, Zhou, B, & Shan, Y. (2004). Progress on antioxidant activation and extracting technology of flavonoids. Chem. Product. Technol. , 11, 29-33.

[78] Watanabe, Y, Mihara, R, Mitsunaga, T, & Yoshimura, T. (2005). Termite repellent sesquiterpenoids from *Callitris glaucophylla* heartwood. Journal of Wood Science , 51, 514-519.

[79] Verma, M, Sharma, S, & Prasad, R. (2010). Biological alternatives for termite control: A review International Biodeterioration & Biodegradation , 63(8), 959-972.

[80] Zucker, W. V. (1983). Tannins: Does structure determine function? An ecological perspective. Am. Nat. , 121, 335-365.

[81] Walker, J. C. F. (1993). Primary Wood Processing. Principles and Practice.1st Edition. Chapman and Hall. 285 pp.

[82] Windeisen, E, Wegener, G, Lesnino, G, & Schumacher, P. (2002). Investigation of the correlation between extractives content and natural durability in 20 cultivated larch trees. Holz als Roh- und Werkstoff , 60, 373-374.

[83] Wolcott, G. N. (1955). Organic termite repellents tested against *Cryptotermes brevis*. J. Agric. Univ. Puerto Rico, 39, 115.

Chapter 13

Microbial Reduction of Hexavalent Chromium as a Mechanism of Detoxification and Possible Bioremediation Applications

Silvia Focardi, Milva Pepi and Silvano E. Focardi

Additional information is available at the end of the chapter

1. Introduction

1.1. Characteristics of chromium

Chromium (Cr) is a naturally occurring element with atomic number 24 and atomic mass of 51.996 amu. The element belongs to the group of transition metals and in the oxidation state elementary presents an electronic configuration (Ar) 4d5s1. Chromium is naturally present in the environment, it is widespread in rocks, animal, plants and soil, and is the seventh most abundant element on Earth's crust, at concentrations ranging from 100 to 300 $\mu g\ g^{-1}$. In nature, Cr is found in the form of its compounds, and the most important chromium ore is chromite, (Fe, Mn)Cr_2O_4 [1,2].

Chromium exists in different oxidation states, the most stable and common forms are the trivalent [Cr(III)] and the hexavalent [Cr(VI)] species, which display quite different chemical properties [1]. Cr(III) in the form of oxides, hydroxides or sulfates, exists mostly bound to organic matter in soil and aquatic environments. Cr(VI) is usually associated with oxygen as chromate (CrO_4^{2-}) or dichromate ($Cr_2O_7^{2-}$) ions [1]. Cr(VI) is a strong oxidizing agent and in the presence of organic matter is reduced to Cr(III); this transformation is faster in acid environments such as acidic soils [1]. However, high levels of Cr(VI) may overcome the reducing capacity of the environment and thus persist in this form [3].

Chromium represents an essential micronutrient for living organisms, considering that Cr(III) is an essential trace element known for its particular role in the maintenance of normal carbohydrate metabolism in mammals and yeasts [4]. Moreover, it has also been suggested that Cr(III) is involved in the tertiary structure of proteins and in the conformation of cell RNA and DNA [5,6].

1.2. Toxicity of chromium

Cr(VI) exposure in humans can induce allergies, irritations, eczema, ulceration, nasal and skin irritations, perforation of eardrum, respiratory track disorders and lung carcinoma [7,8,9]. Moreover, Cr(VI) evidences the capability to accumulate in the placenta, damaging fetal development [10]. Cr(VI) pollution in the environment alters the structure of soil microbial communities [11], reducing microbial growth and related enzymatic activities, with a consequent persistence of organic matter in soils and accumulation of Cr(VI) [12].

The toxic action of Cr(VI) is due to its capability to easily penetrate cellular membranes, and cell membrane damages caused by oxidative stress induced by Cr(VI) have been extensively reported, both in eukaryotic and prokaryotic cells, with effects such as loss of membrane integrity or inhibition of the electron transport chain [13,14]. Moreover, Cr(VI) enters cells using the sulfate transport system of the membrane in cells of organisms that are able to use sulfate [15,16,17,18,19,20].

Once Cr(VI) entered into cells, spontaneous reactions occur with the intracellular reductants as ascorbate and glutathione, generating the short-lived intermediates Cr(V) and/or Cr(IV), free radicals and the end-product Cr(III) [21,22,23]. In the cytoplasm, Cr(V) is oxidized to Cr(VI) and the process produces a reactive oxygen species, referred as ROS, that easily combines with DNA–protein complexes. On the other hand, Cr(IV) is able to bind to cellular materials, altering their normal physiological functions [24,25]. It is known that Cr(VI) species and hydroxyl radicals cause DNA lesions in vivo [26]. The intermediates that originated from the action of Cr(VI) are dangerous to cell organelles, proteins and nucleic acids [27,28,29]. Cr(VI) is a very dangerous chemical form on biological systems as it can induce mutagenic, carcinogenic and teratogenic effects. Moreover, Cr(VI) is able to induce oxidative stress in cells, damaging its DNA [30]. Inside of cells, the Cr(III)-DNA adducts and related hydroxyl radical oxidative DNA damages have a central role in originating the genotoxic and mutagenic effects [31]. Moreover, the formation of Cr(III)-DNA binary adducts and L-cysteine-Cr(III)-DNA and ascorbate-Cr(III)-DNA ternary adducts likely increase both genotoxicity and mutagenicity in human cells [32,33]. Again the formation of DNA protein cross-linking, a process favoured by Cr(VI), induces a significant promutagenic effect [33].

Considering the dangerous effects Cr(VI) can cause to human health, Cr(VI) has been comprised among priority pollutants and listed as a class A human carcinogen by the US Environmental Protection Agency (USEPA) [34].

The cell membrane is nearly impermeable to Cr(III), Cr(III) has thus only about one thousandth of the toxicity of Cr(VI) [35,36]. Taking into account these considerations, it is possible to conclude that, depending on its oxidation state, chromium can have different biological effects, with Cr(VI) that is highly toxic to most organisms, and Cr(III) that is relatively innocuous [37,38].

2. Use of chromium and environmental contamination

Chromium enters in the anthropogenic activities, it is used in stainless steel plant, preparation of alloys, chrome plating, leather tanning, production of refractories, dye industry, industrial water cooling, paper pulp production, petroleum refining, wood preservation and nuclear power [1,39].

As consequence of its broad use, chromium is present in effluents originated from the different activities and represents a serious pollutant of sediments, soil, water and air [40]. Wastewaters have resulted in significant quantities of Cr(VI) in the environment, which may constitute toxicological risk to humans, animals, and plants [41]. Cr(VI) is introduced in the environment mainly as a consequence of its industrial use, while chromium in its trivalent form, Cr(III), naturally predominates in the environment [42].

Cr(VI) is highly dispersed in sediments and surface waters, and it is characterized by a much greater solubility, mobility and bioavailability than Cr(III) and all the forms of chromium [43,44]. As consequence of this high water solubility and elevated mobility, Cr(VI) diffuses easily away from the native site of contamination. Moreover, the increase in soil pH increases the leachability of Cr(VI). Cr(III) shows a low mobility and is relatively inert, and easily absorbable on mineral surfaces and solid-phase organic ligands, thus resulting less bioavailable in the environment. Additionally, Cr(III) is quite insoluble at environmentally significant pH values, since in these conditions there are formation of insoluble hydroxide and oxide compounds. Mobility of Cr(III) decreases with absorption of clays and oxide minerals below pH 5. Binding of Cr(III) by iron oxides can be considered an example of these mechanisms, as this feature can decrease the solubility of this form of chromium [43,45]. Again, the characteristic of insolubility of Cr(III) diminish its bioavailability and mobility of Cr(III) toxicity in saltwater exposures [46].

In the presence of oxidizing conditions Cr(VI), in forms of the anions chromate (CrO_4^{2-}) and bichromate ($HCrO^{4-}$), is extremely soluble and mobile (Barnhart 1997). In anaerobic environments, under reducing conditions, in the presence of reducing agent as sulfides, ferrous iron, and organic matter, that are several of the organic and inorganic constituents, Cr(VI) may rapidly convert to Cr(III) [47]. Again, bacterially mediated reduction of Cr(VI) has also been considered in the chromium biogeochemical cycle [48].

Cr(III) is stable in aquatic environments and its oxidation to Cr(VI) is improbable, even in the presence of dissolved oxygen [49,50,51]. Different factors affect Cr(III) oxidation to Cr(VI), depending on the presence and mineralogy of Mn(III, IV) hydroxides, pH, and the form and solubility of Cr(III) [52]. Oxidation of Cr(III) is improbable to occur in aquatic environments because aged waste materials containing Cr(III) are typically less soluble and more inert to oxidation, and $Cr(OH)_3$ precipitates may form on surfaces of Mn(III, IV) hydroxide [53]. Besides, possible Cr(III) oxidants are scarcer and less abundant than potential Cr(VI) reductants in natural sediments, and Cr(III) oxidation is slower than Cr(VI) reduction [54].

3. Microbial resistance to Cr(VI) and microbial Cr(VI)-reduction

Despite the toxicity of Cr(VI), some microorganisms evidence resistance to this heavy metal, showing the capability to reduce Cr(VI) to Cr(III), as was first reported for *Pseudomonas* spp., and a characterization of bacteria capable of reducing the Cr(VI) was reported successively in 1979 [55,56]. Since then it has been evidenced the presence of numerous bacteria capable of transforming the Cr(VI) to Cr(III) under different conditions [57,58,59]. Recent isolation and purification of Cr(VI) reductases from aerobic bacteria and the fact that the process involved in Cr(VI) reduction occurring under anaerobic conditions is starting to be understood, allowed knowledge of biological processes of chromium resistance [60]. Numerous bacteria have then been reported evidencing their capability to reduce Cr(VI) to Cr(III) as a mechanism of resistance to Cr(VI) [61,62]. Further studies evidenced different bacteria able to reduce Cr(VI), including *Escherichia coli* [63], *Pseudomonas putida* [64], *Desulfovibrio* sp. [65], *Bacillus* sp. [66], *Shewanella* sp. [67], *Arthrobacter* sp. [68], *Streptomyces* sp. MC1 [36] and *Microbacterium* sp. CR-07 [69].

The *chrBCAF* operon from transposable elements confers resistance to Cr(VI), synthesizing for ChrA and ChrB, a protein acting as chromate-sensitive regulator [70]. Enzymatic reduction of Cr(VI) was evidenced in an *Halomonas* sp. strain TA-04, isolated from polluted marine sediments, in the presence of 8.0% NaCl suggesting new insights for metal reduction at halophilic conditions [71]. Investigations on mechanisms of resistance to chromate, in particular at level of the bacterial cells have been evidenced. The best characterized mechanisms that have been reported include the efflux of chromate ion from the cell cytoplasm and the reduction of Cr(VI) to Cr(III). The efflux by the transport protein CHRA has been identified in *Pseudomonas aeruginosa* and *Cupriavidus metallidurans* (formerly *Alcaligenes eutrophus*) and consists of an energy-dependent process driven by a membrane potential. Moreover, the reduction of chromate is completed by chromate-reductase from different bacterial species generating Cr(III) that may be the object of detoxification due to other mechanisms. The most specific enzymes belong to the large family of flavoprotein reductase NAD(P)H-dependent. Other mechanisms of bacterial resistance to the chrome were evidenced, and these mechanisms were related to the expression of components of the systems of mechanism for the DNA repair and are related to the mechanisms of homeostasis of iron and sulfur [72].

Both aerobic and anaerobic microorganisms are able to reduce Cr(VI) to Cr(III). In aerobic conditions it is possible to observe the bio-reduction of Cr(VI) that can be obtained directly as a result of microbial metabolism [73]. In the presence of oxygen, microbial reduction of Cr(VI) is commonly catalyzed by soluble enzymes, except in *Pseudomonas maltophilia* O-2 and *Bacillus megaterium* TKW3, which utilize membrane-associated reductases. Soluble Cr(VI) reductase ChrR was purified from *Pseudomonas putida* MK1 and reductase YieF purified from *Escherichia coli*. Enzyme ChrR catalyzes an one-electron shuttle followed by a two-electron transfer to Cr(VI), with the formation of intermediate(s) Cr(V) and/or Cr(IV) before further reduction to Cr(III). Reductase YieF displays a four-electron transfer that reduces Cr(VI) directly to Cr(III). The membrane-associated Cr(VI) reductase was isolated from *B. megaterium* TKW3, without any characterization of related reduction kinetics. In the absence of oxygen, Cr(VI) reduction was evidenced both by soluble and membrane-associated enzymes, and Cr(VI) functions as the terminal electron acceptor of an electron transfer chain that frequently involves cytochromes. Researches on Cr(VI) reductases focusses on enzymes with higher reductive activity. In anaerobic conditions, the reduction processes uses a broad range of compounds, involving carbohydrates, proteins, fats, hydrogen, NAD(P)H and endogenous electron reserves, that can function as electron donors in the reduction processes [74]. According to the advancement in technology for enzyme immobilization, the direct application of Cr(VI) reductases could be an important approach for bioremediation of Cr(VI) in different environments, in particular where whole cells are difficult to apply [75].

4. Bioremediation of Cr(VI) by microorganisms

Conventional methods for removing metals from contaminated sites include chemical precipitation, oxidation/reduction, ion exchange, filtration, use of membranes, evaporation and adsorption on activated coal, alum, kaolinite, and ash [15,41]. However, most of these methods require high energy or large quantities of chemical reagents, with possible production of secondary pollution [76,77]. Concerning removal of Cr(VI), conventional approaches include chemical reduction followed by precipitation, ion exchange and adsorption on activated carbon, alum, kaolinite and of ashes, and most of these methods require a high energy and large amounts of chemical reagents [76]. Moreover, costly safe disposal of toxic sludge, incomplete reduction of Cr(VI) and high cost for Cr(VI) reduction, especially for the removal of relatively low concentrations of Cr(VI) are non-convenient from the economical point of view [78,79].

An innovative technology is represented by bioremediation, which uses the metabolic potential of microorganisms to remove toxic metals, in order to decontaminate the polluted areas. Bioremediation techniques can be classified as *in situ* or *ex situ* depending, respectively, on whether the intervention is carried out with suitable bacteria directly on the polluted site, or on portions of environmental matrices, such as water, sediment or soil, after being removed and transported in appropriate facilities for treatment [76,79].

Cr(VI)-resistant microorganisms represent an important opportunity to have safe, economical and environmentally friendly methods for reducing Cr(VI) to Cr(III), for possible bioremediation applications [27]. The reduction of Cr(VI) to Cr(III) is then a potential useful process for the recovery of sites contaminated by Cr(VI) [36]. Cr(VI)-removal based on microorganisms is now considered to be an effective alternative method to the conventional processes, and is receiving great attention for potential application in bioremediation [76,80]. Taking into account that the insolubility of Cr(III) facilitates its precipitation and removal, the biotransformation of Cr(VI) to Cr(III) has been considered as an alternative process for treating Cr(VI)-contaminated wastes [81,82]. Among biotechnological approaches, microbial reduction of Cr(VI) is cost-effective and eco-friendly and can offer a viable alternative [80,83,84].

Chromium resistant microorganisms are responsible of the biological reduction of Cr(VI) into the less mobile Cr(III), and its consequent precipitation, could represent an effective method for detoxification of Cr(VI) contaminated sites and have a potential use in bioremediation [85].

Included in the bioremediation technologies, phycoremediation is the use of photosynthetic microorganisms as microalgae, macroalgae and cyanobacteria for the removal of pollutants as metals. Furthermore, it is essential to understand the distribution of the metal adsorbed onto the surface in relation to the metal accumulated inside the cell, in order to understand the predominant removal mechanisms and to make decisions of the viability of the recovery of the adsorbed metals [86].

Biosorption and bioaccumulation of chromium for bioremediation purposes have been demonstrated. Yeasts and especially molds have been most widely investigated from this aspect, and the mechanisms of chromium tolerance or resistance of selected microbes are of particular importance in bioremediation technologies. The mechanisms of chromium toxicity and detoxification have been studied extensively in yeasts and fungi, and some promising results have emerged in this area [87].

The ability existing in a number of environmental microorganisms, known for their capability to bind metals, can be evidenced in human gastrointestinal bacteria. Bacterial species belonging to the genus *Lactobacillus*, resident in different districts as the human body and in fermented foods, have the ability to bind metals, including Cr(VI), and to detoxify them from different districts [88].

A method for bioremediation of sites contaminated by metals, including chromium, is represented by bioaugmentation-assisted phytoextraction, in which bacteria and fungi, associated with plants able to accumulate metals were analyzed on the basis of a proposed as bioprocess for a bioremediation approach. The implementation of bioaugmentation to favour the microbial survival, was suggested in order to enhance the microbial-plant association and the efficiency of the process [89].

The process of biomineralization is a process by which microorganisms transform aqueous metal ions, including chromium, into amorphous or crystalline precipitates. Biomineralization is regarded as a promising and cost-effective strategy for remediating chromium contamination. An example of arsenic precipitation was considered as a possible mechanism for arsenic bioremediation of sediments contaminated by arsenic [90]. Biologically mediated transformation, immobilization, and mineralization of toxic metals may represent an important perspective for bioremediation [91].

5. Case study: Cr(VI)-reduction by Actinobacteria isolated from polluted sediments near a stainless steel plant

Wastes from stainless steel plants produce soluble Cr(VI) contaminating sediments, soils and water bodies. Chromium at high concentrations are widespread in sediments of industrialized areas because of industrial discharges [92]. In a previous study, carried out from polluted marine sediments near a stainless steel plant in Southern Italy, near the industrialized area of Taranto, an halophilic Cr(VI)-resistant bacterial strain *Halomonas* sp. TA-04 was isolated. The isolated strain showed a MIC at 200 µg ml^{-1} Cr(VI), and the reduction of Cr(VI) in the presence of 80 g l^{-1} NaCl. Cr(VI) was removed from sediment leachate by immobilized cells and the cell free extract reduced Cr(VI) with a maximum of activity at pH 6.5, at 28°C. These results suggest the possible use of the isolated strain in bioremediation processes, in particular concerning detoxification of saline polluted environments [71].

The aim of the present investigation was the isolation of bacterial strains from chromium-polluted sediments and their characterization in terms of phylogenetic and physiological features. The description of two Cr(VI)-reducing microorganisms isolated from polluted sediments and included into Actinobacteria was carried out, for their possible use in biore-mediation applications.

6. Materials and methods

6.1. Study area

The microbiological study with the isolation of the bacterial strains investigated in this study was conducted in sediment samples collected from a polluted site near a stainless steel plant in the Bagnoli area, Naples (Southern Italy). The site was characterized by a total chromium content corresponding to 34 ± 0.23 mg kg^{-1}.

6.2. Sediment sampling

Sediment samples for microbiological analyses were collected manually using Plexiglas tubes (i.d. 10 cm), in June 2008. Collected samples were maintained at 4 °C and transported to the laboratory. Sterile sediment subsamples (0-10 cm) were collected and processed within twelve hours for microbiological analyses.

6.3. Enrichment cultures and isolation of the bacterial strain

Enrichment cultures were grown in flasks containing the complex YPEG medium, containing 5.0 g of tryptone, 2.5 g of yeast extract, and 1.0 g of D-glucose per litre of distilled water, in the presence of 5.0 mM of Cr(VI), inoculated with 0.5 g of sediment samples, and incubated at 28 °C in the dark. From flasks showing turbidity, a 100 µl aliquot was spread on Petri dishes containing the complex solid medium in the presence of the same initial concentration of Cr(VI) and incubated at 28°C for 48 hours. Colonies showing different morphologies were selected and subcultured at least three times. Isolated strains were stored in the presence of 30% sterile glycerol (v/v) in liquid nitrogen.

6.4. Isolates characterization and identification by 16S rRNA gene sequencing

The bacterial isolates were observed under a stereomicroscope (Optika, mod 620). Gram reactions were determined following the standardized method of bacterial cells staining (Gram stain kit, Carlo Erba). Catalase and oxidase activities were determined following Smibert and Krieg [93]. For 16S rDNA sequencing of the isolated bacterial strain, a single colony was suspended in 50 µl double-distilled water and treated for 5 min at 100°C. Amplification and sequencing of 16S rRNA gene was performed as previously reported [94]. Partial 16S rDNA sequences were determined for the bacterial isolates chr 2 and chr 3, and the sequences were deposited in the GenBank database with the accession numbers: HQ609600 and HQ609601, respectively. The consensus sequences of the isolates were compared with those deposited in GenBank using the BLAST program [95].

6.5. Analysis of sequence data

The 16S partial sequences were compared at the prokaryotic small subunit rDNA on the Ribosomal Database Project II website [96]. The 16S rDNA sequences retrieved from the databases were aligned using ClustalW included in the MEGA software, version 4.1 [97]. The phylogenetic trees were inferred by MEGA 4.1 (neighbour-joining method) [98]. Sequence divergences between strains were quantified using the Kimura-2-parameter distance model [99]. The "Complete Deletion" option was chosen to deal with gaps. Bootstrap analysis (1000 replicates) was used to test the topology of the neighbour-joining method data. The trees were unrooted.

6.6. Minimum Inhibitory Concentrations (MICs)

One ml aliquots of overnight cultures were incubated in 99.0 ml of YEPG-NaCl broth, and 10 ml were distributed in 18 ml test tubes sealed with radial caps. MIC tests were carried out at different concentrations of Cr(VI). Tubes were incubated in a rotary drum at 30°C for 24 hours. The optical density of the cultures, used as a measure of microbial growth, was detected at a wavelength of 600 nm by an UV-visible spectrophotometer (Jenway, mod. AC30); a blank with the culture medium alone (without bacteria) was also analysed. Experiments were carried out in duplicate.

6.7. Chromium (VI) assay

Hexavalent chromium was determined colorimetrically using the 1,5-diphenylcarbazide (DPC) (Sigma-Aldrich, Milan, Italy) method [100].

6.8. Effect of chromium concentration on bacterial growth and Cr(VI)-reduction

The Cr(VI)-resistant isolates were grown over-night in YEPG-NaCl medium, in the presence of 0.2 mM Cr(VI). The pre-culture were used for inocula in different cultures at the same conditions, and incubated at 28°C in the presence of Cr(VI) concentrations: 0, 10, 25, 50, 75, 10 and 150 µg ml^{-1}. At different times (0, 0.5, 3, 6, 12, 18 and 24 hours), aliquots were harvested in order to measure the absorbance at 600 nm spectrophotometrically, and to evaluate Cr(VI) reduction according to the DPC method. For each series, experiments were conducted in triplicate.

6.9. Effect of temperature on bacterial growth and Cr(VI)-reduction

Cultures of the isolates were incubated in a temperature range from 4 to 42°C, with an inoculum prepared by an overnight pre-culture in YEPG medium containing 25 μg ml^{-1} of Cr(VI). After different times of incubation of (0, 6, 12 and 24 hours), the effect of different temperatures was detected by harvesting two aliquots of 1 ml for each series, one to evaluate the biomass, revealing absorbance at 600 nm spectrophotometrically, the other to estimate Cr(VI) reduction according to the method of DPC. Experiments were conducted in triplicate.

6.10. Extraction of plasmids from the cells of the bacterial strains Cr (VI)-resistant isolates

The two isolated bacterial strains Cr(VI)-resistant were grown in liquid medium YEPG-NaCl in the presence of 25 μg ml^{-1} of Cr(VI). Aliquots of 2 ml were centrifuged at 15,000 × g and the pellet washed twice in saline (0.8% NaCl). The pellets were resuspended in 200 μl of solution I (25 mM Tris, 50 mM Glucose, 10 mM EDTA, pH 8.0) in the presence of 4 μg ml^{-1} of lysozyme (Sigma, Milan). The pellets were kept at room temperature (RT) for 5 min. Then were added 200 μl of solution II [0.2 M NaOH, 1% (w/v) sodium dodecyl sulphate (SDS)], and the pellets were homogenized gently and kept on ice for 5 min. Then were added 300 μl of solution III (5M potassium acetate, glacial acetic acid 11.5%, deionized water to make up to 100 ml) [101]. The suspension was then centrifuged at 15,000 × g for 5 min. and the supernatant was transferred to a new tube and were added 0.6 volumes of isopropanol, mixed and left at RT for 10 min. The suspension was then centrifuged at 15,000 × g for 5 min., and the pellet was washed using 400 μl of 70% ethanol, centrifuged at 15,000 × g for 5 min. The pellet was dried, then resuspended in 30 μl of deionized water, filtered and sterilized. Four microliters were run on agarose gel at 1,2% (w/v) (Flash Gel® System, Lonza) for testing the purity and quality of the plasmid DNA extracted. One milliter of standard DNA Marker 100-4000 bp (Flash Gel ®, Lonza) was also added to the gel. The gel image was acquired using the system Flash ® Gel Room (Lonza).

7. Results and discussion

From enrichment cultures arranged from samples of polluted sediments collected near the industrial area, including metallurgical plants, of Bagnoli (Naples, Italy), two Cr(VI)-resistant bacterial strains were isolated and named chr2 and chr3. A microbiological characterization of the isolated strains is reported in Table 1.

Bacterial strain	Gram staining	Oxidase test	Catalase test	Colony morphology
chr2	positive	+	+	Ø 1.0 mm; beige color; regular margins; flat; moist
chr3	positive	-	+	Ø 1.5 mm; yellow color; regular margins; convex; mat

Table 1. Characterization of the Cr(VI)-resistant isolated bacterial strains.

BLAST analysis evidenced a similarity of 100% for strain chr2 with strains *Cellulomonas* sp. DS04-T (GQ274926), able to produce a lactic acid depolimerase; *Cellulosimicrobium cellulans* strains DQ-4 (EU816697) and *C. cellulans* AS4.1333 (AY114178), isolated from soil, and *Cellulosimicrobium* sp. 87N50-1 (EU196469) originating from marine sediments. Strain Cellulomonas sp. chr2 did not show a proximity with the Cr(VI)-reducing *Cellulomonas* spp. WS01 (AY617101), ES6 (AY617099) and ES5 (AY617098) [102]. Moreover, BLAST analysis evidenced a similarity of 100% of the strain chr3 with strains *Microbacterium oxydans* spp. XH0903 (GQ279110), isolated from soil; WT141 (GQ152132) and with clones of non-cultivated bacteria nbw 291fa2c1 (GQ086586), nbw 289a11c1 (GQ086396) and nbt 38e04 (FJ894305), obtained from samples of animal origin.

Phylogenetic analysis of the strain *Cellulomonas* sp. chr2 highlighted its position close to the strain *Cellulomonas* sp. DS04-T (GQ274926), separated from neighboring strains *Cellulosimicrobium cellulans* DSM 43879 (X83809) and *C. funkei* DSM 16025 (AY501364) (Fig. 1).

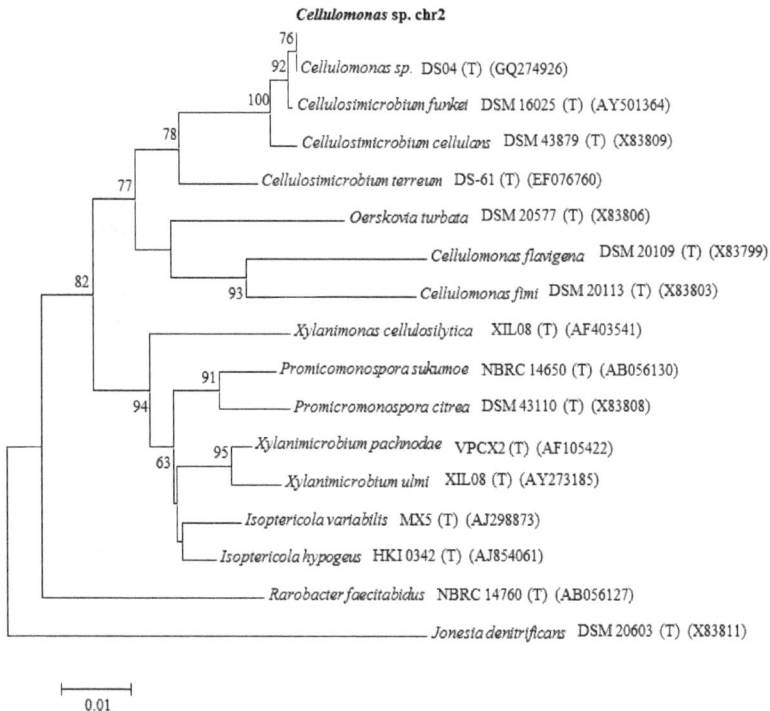

Figure 1. Unrooted phylogenetic tree based on 16S rDNA sequence comparisons showing the position of the isolate *Cellulomonas* sp. chr2. The sequence of *Jonesia denitrificans* DSM 20603(T) has been used as *outgroup*. The branching pattern was generated by neighbour-joining methods. Bootstrap values, shown at the nodes, were calculated from 1000 replicates. Bootstrap values lower than 60% are not shown. The scale bar indicates substitutions per nucleotide. The GenBank accession numbers for the 16S rDNA sequences are given in parentheses after the strain.

Strain *Microbacterium* sp. chr3 evidenced a proximity with strain *M. oxydans* DSM 20578, formerly known as *Brevibacterium oxydans*, included in a cluster comprehending strain *M. liquefaciens* DSM 20638 also (Fig. 2). Moreover, phylogenetic analyses showed that the position of the bacterial strains *Microbacterium* sp. chr3 was located quite far from the bacterial strains known for their capability to grow in the presence of Cr(VI) as the Cr(VI)-resistant strain *M. foliorum* DSM 12966 (AJ249780) [103] (Fig. 2)

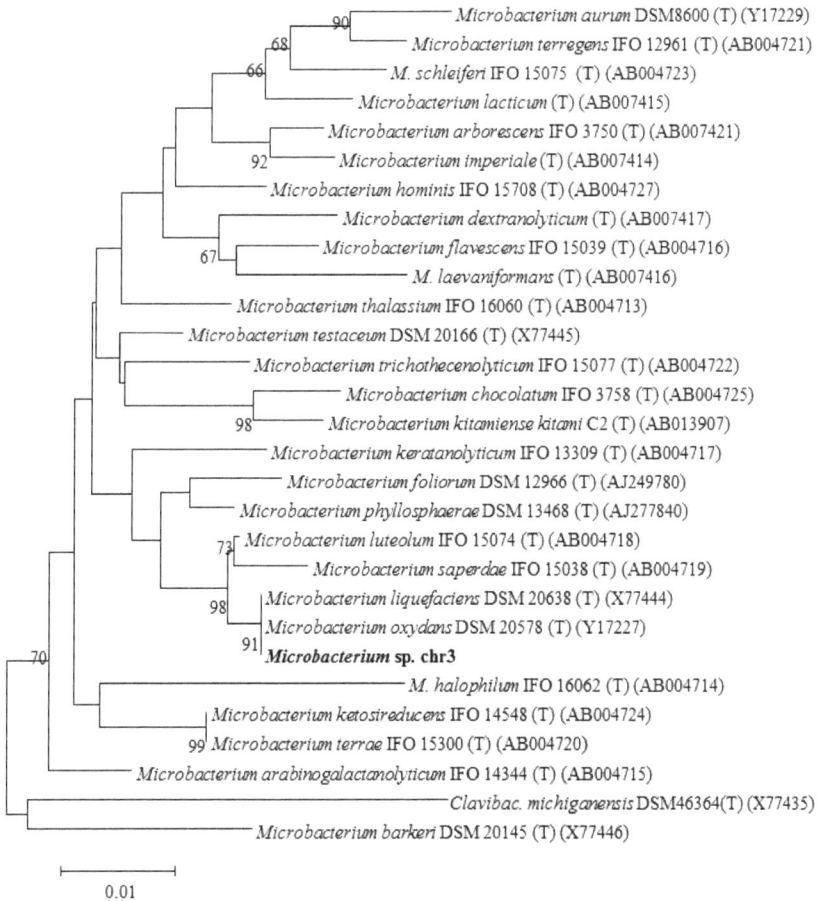

Figure 2. Unrooted phylogenetic tree based on 16S rDNA sequence comparisons showing the position of the *Micro-bacterium* sp. chr3 isolate and representative species of the genus *Microbacterium*. The sequence of *Clavibacter michiganiensis* DSM 46364 has been used as *outgroup*. The branching pattern was generated by neighbour-joining methods. Bootstrap values, shown at the nodes, were calculated from 1000 replicates. Bootstrap values lower than 60% are not shown. The scale bar indicates substitutions per nucleotide. The GenBank accession numbers for the 16S rDNA sequences are given in parentheses after the strain.

The Cr(VI)-resistant bacterial strains isolated in this study were assigned to the genera *Cellulomonas* and *Microbacterium*, known for including Cr(VI)-resistant bacterial strains. Eight Cr(VI)-reducing bacterial strains belonging to the genus *Cellulomonas* were isolated from polluted sediments [102]. A dissimilar reduction of Cr(VI) was evidenced in three bacterial strains *Cellulomonas* spp. isolated from environment [104]. A bacterial strain of the genus *Microbacterium* was able to reduce Cr(VI), included in a mixed culture [105], moreover, immobilized cells of a strain *Microbacterium* sp. showed the capability to reduce Cr(VI) [106] [56]. The bacterial strain *Microbacterium* sp. CR-07, isolated from a mud sample of iron ore, evidenced its characteristic of resistance to Cr(VI) its capability to reduce chromate [69].

The two genera *Cellulomonas* and *Microbacterium* are included into the Actinobacteria, that represents a significant component of microbial communities present mostly in soils, nevertheless they were isolated also from sediments of marine areas [107]. It is known that bacterial strains belonging to these two genera evidenced their capability to resist to heavy metals, in conjunction with particular growth characteristics, as a rather rapid colonization of selective substrates, suggesting them as right microorganisms to be used for bioremediation processes [36,85].

The bacterial strains showed MIC values in the presence of Cr(VI) of 150 µg ml^{-1} for the bacterial strain *Cellulomonas* sp. chr 2, and of 250 µg ml^{-1} for *Microbacterium* sp. chr 3 (Tab. 2). The isolated bacterial strains evidenced levels of resistance to Cr(VI) similar to those reported in literature [108,109]. Levels of resistance to Cr(VI) similar to those evidenced in this study were highlighted in different Cr(VI)-resistant bacteria isolated from polluted sediments, that evidenced values of resistance to Cr(VI) corresponding to 250 µg ml^{-1} [110].

Bacterial strain	MIC, µg ml^{-1}
chr2	150
chr3	250

Table 2. MIC values expressed as µg ml^{-1} of Cr(VI) evidenced in the isolated Cr(VI)-resistant strains.

The mechanism of resistance to Cr(VI) was investigated in the isolated bacterial strains, evidencing their capability to reduce Cr(VI), the most toxic and extremely soluble form of chromium, as revealed by the tests evidencing the depletion of the Cr(VI) content in cultures in conjunction with the increase of the bacterial biomass. The bacterial strain Cr(VI)-resistant *Cellulomonas* sp. chr2 evidence high levels of growth which, after 24 hours of incubation, reached an absorbance of 1.6 without Cr(VI) added, and an absorbance of 1.4 in the presence of 50 mg ml^{-1} of Cr(VI), evidencing a high adaptability of the bacterial strain to the presence of the toxic chromate (Fig. 3 A). Parallel to the growth of bacterial cells, a correspondent Cr(VI) reduction in the presence of different Cr(VI) concentrations was evidenced, with a residual 8% of Cr(VI) after 24 hours, when the toxic anion was added at a concentration of 50 mg ml^{-1} (Fig. 3 B). The behaviour of the bacterial isolate *Cellulomonas* sp. chr2 suggested its possible involvement in Cr(VI) removal from polluted sites.

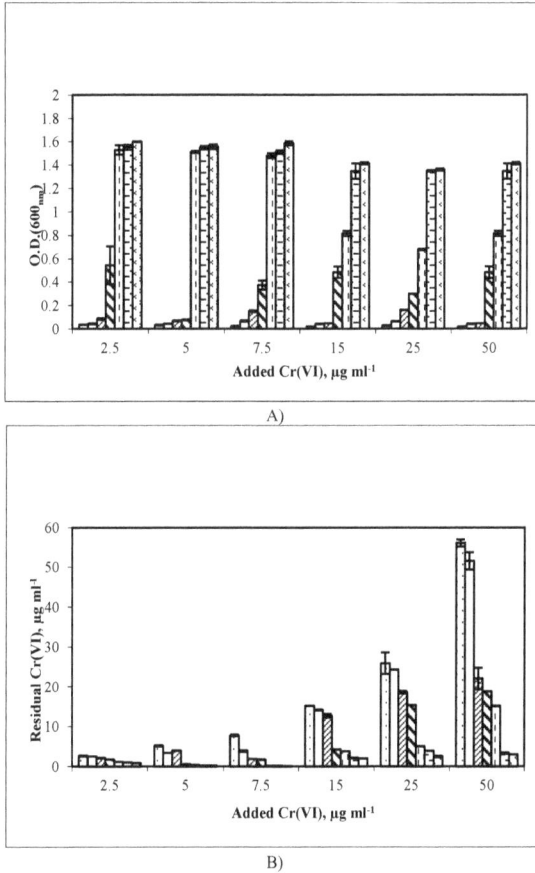

Figure 3. Growth of strain *Cellulomonas* sp. chr2 in the presence of different concentrations of Cr(VI) (A) and correspondent reduction of Cr(VI) (B). Values were detected at different times: 0 (▢), 0.5 (▨), 3 (▨), 6 (▢), 12 (▢), 18 (▢) and 24 (▢) hours.

The bacterial strain *Microbacterium* sp. chr3 evidenced higher levels of growth if compared to those of the bacterial strain *Cellulomonas* sp. chr2. In fact, after 24 hours of incubation the strain chr3 evidenced values of absorbance corresponding to 1.8 and 1.6 in the presence of 0 and 50 mg ml^{-1} of Cr(VI), respectively (Fig. 4 A). In the same experiment, the capability to reduce Cr(VI) added at different concentrations was evidenced for strain *Microbacterium* sp. chr3 in the presence of a concentration of 50 mg ml^{-1}, with a residual concentration of Cr(VI) equal to 32% after 24 hours of incubation (Fig. 4 B).

The isolated bacterial strain *Cellulomonas* sp. chr2 evidenced an higher efficiency in reducing Cr(VI) added at a concentration of 50 mg ml^{-1} as respect to the isolate *Microbacterium* sp. chr3.

In fact, at the end of the 24 hours of incubation a reduction equal to 92% of Cr(VI) was recovered in strain *Cellulomonas* sp. chr2, whereas strain *Microbacterium* sp. chr3 showed a percentage of reduction of 68%. The higher levels of Cr(VI) reduction evidenced in cultures of strain *Cellulomonas* sp. chr2 suggested that it could be a better element for possible processes of Cr(VI) bioremediation.

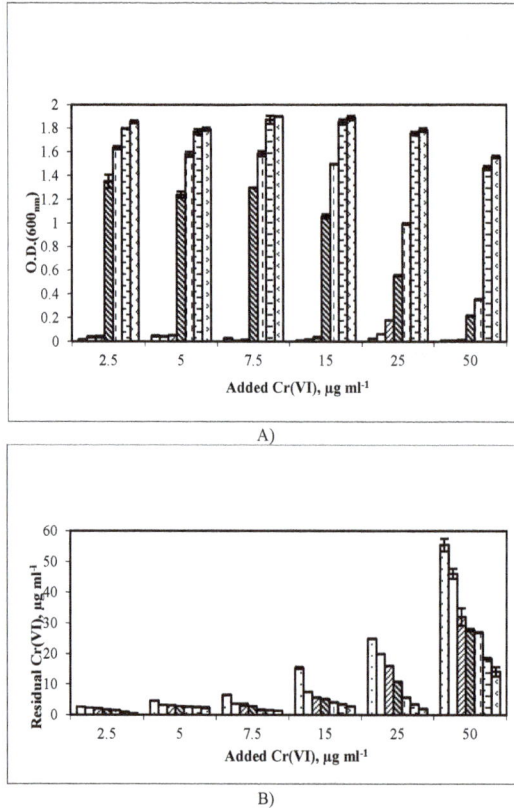

Figure 4. Growth of strain *Microbacterium* sp. chr3 in the presence of different concentrations of Cr(VI) (A) and correspondent reduction of Cr(VI) (B). Values were detected at different times: 0 (▢), 0.5 (▨), 3 (▩), 6 (▢), 12 (▢), 18 (▢) and 24 (▢) hours.

The two isolated Cr(VI)-resistant strains belonging to Actinobacteria were investigated for their capability to grow and to reduce Cr(VI) added at different concentrations, tested in a range of temperatures included from 4°C to 42°C. Concerning the strain *Cellulomonas* sp. chr2, growth was absent at 4°C, and scarce at 18 and 22°C, with an absorbance equal to 0.2 at a wavelength of 600 nm. The optima levels of growth was detected at a temperature of 28°C, with a value of absorbance of 1.1. A similar value, equal to 1.0 of absorbance was evidenced at 37°C, whereas a decrease of growth was detected at 42°C, with a value of growth corresponding to an absorbance of 0.4 (Fig. 5 A).

Cr(VI)-reduction in the same cultures of the isolate *Cellulomonas* sp. chr2 followed the pathway of bacterial growth, with a maximum level of reduction evidenced at a temperature of 28°C where a residual level of Cr(VI) of 30% was showed (Fig. 5 B).

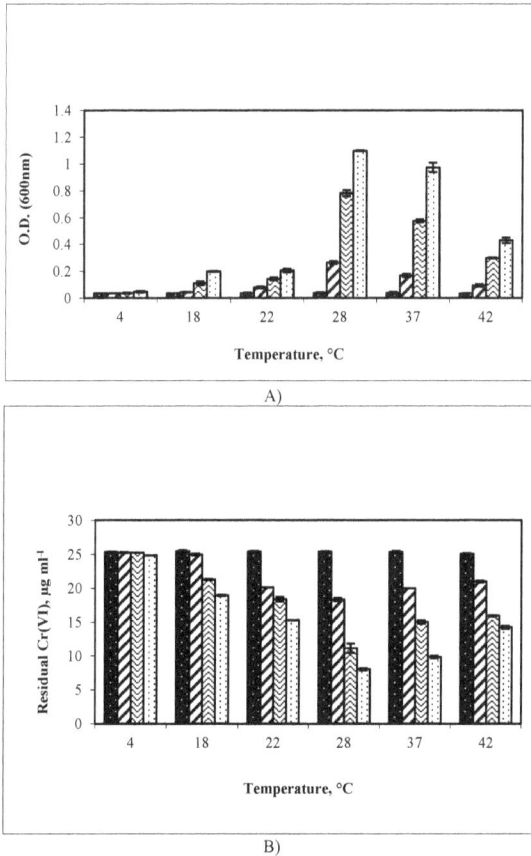

Figure 5. Growth of the strain *Cellulomonas* sp. chr2 in medium YEPG-NaCl in the presence of different temperatures, and a fixed concentration of Cr(VI) pair to 25 µg ml^{-1} (A). Related Cr(VI) reduction revealed at different temperature (B). Growth and Cr(VI)-reduction were revealed after different times of incubation: 0 (■), 6 (▨), 12 (▨) and 24 (□) hours.

The isolated strain *Microbacterium* sp. chr3 did not evidence growth at 4°C, and low levels of growth, corresponding to an absorbance equal to 0.19 and 0.22 at 18°C and 22°C, respectively. Higher levels of growth were reached at 28°C and 37°C, with values of 1.2 and 0.99, whereas at 42°C, growth of the isolate *Microbacterium* sp. chr3 evidenced an absorbance with a value of

0.75 (Fig. 6 A). Reduction of Cr(VI) by strain *Microbacterium* sp. chr3 evidenced maxima levels at temperatures of 28°C and 37°C, with residual values of Cr(VI) corresponding to percentages of 5.0% and 10%, respectively (Fig. 6 B).

This experiment was conducted in the presence of a fixed concentration of Cr(VI) corresponding to 25 mg ml^{-1}, and in this case strain *Microbacterium* sp. chr3 resulted more efficient than strain *Cellulomonas* sp. chr2 in reducing Cr(VI) at 28°C. Nevertheless, at an higher concentration of Cr(VI), equal to 50 mg ml^{-1}, strain chr2 evidenced a better capability to reduce Cr(VI) and an higher adaptability to Cr(VI), suggesting a probable better use for bioremediation applications. On the other hand it is noteworthy to note that mixed populations of bacterial strains can represent the better solutions in bioremediation, as different microorganisms, often with complementary features, can cope efficiently with contaminants in bioremediation.

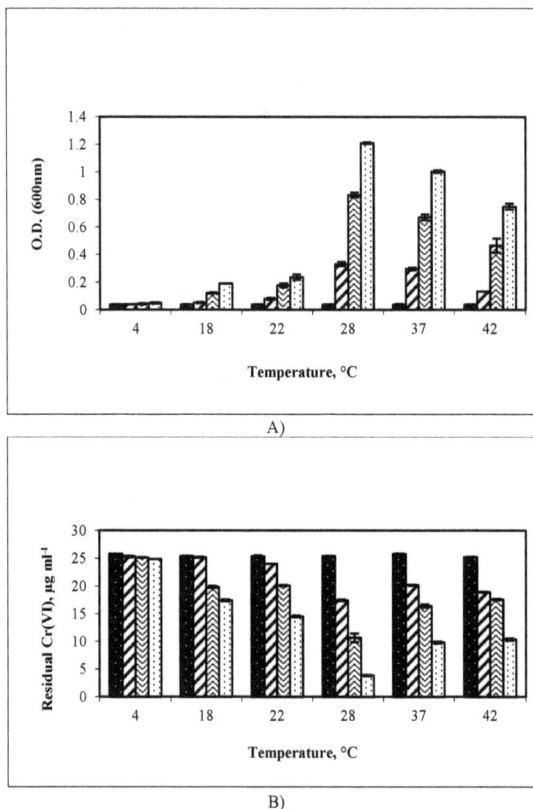

Figure 6. Growth of the strain *Microbacterium* sp. chr3 in medium YEPG-NaCl in the presence of different temperature, and a fixed concentration of Cr(VI) pair to 25 µg ml^{-1} (A). Related Cr(VI) reduction revealed at different temperature (B). Growth and Cr(VI)-reduction were revealed after different times of incubation: 0 (■), 6 (▨), 12 (▧) and 24 (☐) hours.

The capability to reduce Cr(VI) to Cr(III) was evidenced in both the bacterial strains isolated in this study. It is known that different bacterial strains were isolated from polluted sites, as a strain of *Streptomyces griseus* able to reduce Cr(VI) both in virtue of the activity of free cells and of the immobilized ones [111]. Autochthonous bacteria resistant to high levels of Cr(VI) were isolated from polluted sediments. Strain *Bacillus* sp. PB2 isolated from polluted soil evidenced an optimal growth and reduction of Cr(VI) at a temperature of 35°C, at pH values from 7.5 to 9.0, and the use of this isolate was suggested in bioremediation processes of terrestrial sites contaminated by Cr(VI) [112]. Strain *Bacillus* sp. ev3 evidenced the capability to reduce Cr(VI) to Cr(III), with an efficiency of 91% of Cr(VI) reduced in 96 hours [113]. Studies on the reduction of Cr(VI) in autochthonous bacteria in soils contaminated by Cr(VI) near stainless steel industries in the province of Hunan, in China, evidenced that in the presence of an adapt concentration of nutrients, a corresponding efficacy in reducing Cr(VI) was highlighted. A bacterial strain isolated from these polluted soils, and assigned to the genus *Bacillus* was characterized and resulted adapt for bioremediation applications [114].

Strain *Streptomyces* sp. MC1, included in the Actinobacteria, showed the capability to reduce Cr(VI) in cultures arranged in mineral medium. This strain evidenced the capability to reduce 94% of bioavailable Cr(VI) at a concentration of 1 mM in one week of incubation. Moreover the activity of the strain *Streptomyces* sp. MC1 was not inhibited by the native microbial communities resident in the native soil. Cr(VI) was almost completely removed from the polluted soil as consequence of the activity of the Cr(VI)-reducing bacteria [115]. Cr(VI)-reduction was characterized in detail in cells of the strain *Streptomyces* sp. MC1 [116]. Related chromate reductase activity in strain *Streptomyces* sp. MC1 was further evidenced and characterized [36]. The same Cr(VI)-resistant strain *Streptomyces* sp. MC1evidenced its metabolic versatility and the capability to produce a bioemulsifier in the presence of Cr(VI) [117]. Again *Microbacterium* sp. CR-07, a bacterial strain included in the Actinobacteria, showed resistance to Cr(VI) and the capability to reduce the toxic form of the metal [69].

The ability of the isolates to grow at different temperatures evidenced the range of the use of the Cr(VI)-reducing bacterial strains in eventual bioremediation processes, even in conditions of non-controlled temperature.

Recently, the bio-reduction of Cr(VI) to Cr(III) focused more attention for possible use in bioremediation processes of sites contaminated by Cr(VI). This strategy represents an environmentally friendly technology, to be applied *in situ* and acting in a selective way, and with lower costs, as respect to chemical al physical strategies. The produced Cr(III) can then precipitate as insoluble chromium-hydroxides [$Cr(OH)_3$] [118].

The two isolates evidenced the presence of plasmids when tested with primers specific for the presence of genes of Cr(VI)-resistance included into plasmids (data not shown). The bacterial strains capable of expressing both the resistance and the reduction of chromate are very useful for bioremediation. Plasmids involved both in resistance and in the reduction of Cr(VI) have been described in a strain of *Bacillus brevis* isolated from wastes of a tanning industry [119]. Plasmids of this type can be a source of genes for resistance to Cr(VI), that can be transferred *via* cloning assays, and possibly used as DNA probes for the detection of chromate-resistant bacteria in waters and soils highly contaminated with heavy metals, and the same genes can be used in Cr(VI) biosensors construction [119].

8. Conclusions

The potentiality of the Cr(VI)-resistant microorganisms in bioremediation of polluted sites was evidenced in this chapter. A case study was moreover reported with the description of bacterial strains isolated from sediments contaminated by Cr(VI), and tests of Cr(VI)-reduction were included. The isolated bacterial strains showed resistance to Cr(VI) and phylogenetic analyses of the 16S rRNA gene assigned them to the genera *Microbacterium* and *Cellulomonas*, of the order Actinomycetales, class Actinobacteria. The mechanism of Cr(VI)-resistance was due to the reduction of Cr(VI) to Cr(III) and the isolates showed the adaptability of resistance to Cr(VI) in the presence of different temperatures. These results suggested the use of the Cr(VI)-resistant isolated bacterial strains for possible bioremediation processes of contaminated sites.

Further studies including investigations on mechanisms of resistance to Cr(VI) in autochthonous microorganisms isolated from polluted sites, and on the adaptability of microorganisms to contaminants, could give insights for new researches, favoring the development of new technologies for environmental recovery.

Author details

Silvia Focardi, Milva Pepi and Silvano E. Focardi

*Address all correspondence to: silvia.focardi@unisi.it; milva.pepi@unisi.it

Department of Physical, Earth and Environmental Sciences, University of Siena, Siena, Italy

References

[1] McGrath SP, Smith S. Chromium and nickel. In: Alloway B.J. (ed.) Heavy Metals in Soils. New York: Wiley; 1990. p125–150.

[2] Shewry PR, Peterson PJ. Distribution of chromium and nickel in plants and soil from serpentine and other sites. Journal of Ecology 1976,64 195–212.

[3] Vajpayee P, Sharma SC, Tripathi RD, Rai UN, Yunus M. Bioaccumulation of chromium and toxicity to photosynthetic pigments, nitrate reductase activity and protein content of *Nelumbo nucifera* Gaertn. Chemosphere 1999;39 2159–2169.

[4] Debski B, Zalewski W, Gralak MA, Kosla T. Chromium yeast supplementation of chicken broilers in an industrial farming system. Journal of Trace Elements Medicine and Biology 2004;18 47–51.

[5] Gulan Zetic V, Stehlik Tomas V, Grba S, Lutilsky L, Kozlek D. Chromium uptake by *Saccharomyces cerevisiae* and isolation of glucose tolerance factor from yeast biomass. Journal Bioscience 200;23 217–223.

[6] Zayed AM, Terry N. Chromium in the environment: factors affecting biological remediation. Plant Soil 2003;249 139–156.

[7] Poopal AC, Laxman RS. Chromate reduction by PVA-alginate immobilized *Streptomyces griseus* in a bioreactor. Biotechnology Letters 2009;31 71–76.

[8] Gibb HJ, Lees PS, Pinsky PF, Rooney BC. Lung cancer among workers in chromium chemical production. American Journal of Industrial Medicine 2000;38 115–126.

[9] Gibb HJ, Lees PS, Pinsky PF, Rooney BC. Clinical findings of irritation among chromium chemical production workers. American Journal of Industrial Medicine 2000;38 127–131.

[10] Saxena DK, Murthy RC, Jain VK, Chandra SV. Fetoplacental-maternal uptake of hexavalent chromium administered orally in rats and mice. Bulletin of Environmental Contamination and Toxicology 1990;45 430–435.

[11] Zhou J, Xia B, Treves DS, Wu L-Y, Marsh TL, O'Neill RV, Palumbo AV, Tiedje JM. Spatial and resource factors influencing high microbial diversity in soil. Applied and Environmental Microbiology 2002;68 326–334.

[12] Shi W, Becker J, Bischoff M, Turco RF, Konopka AE. Association of microbial community composition and activity with lead, chromium, and hydrocarbon contamination. Applied and Environmental Microbiology 2002;68 3859–3866.

[13] Codd R, Rillon CT, Levina A, Lay PA. Studies on the genotoxicity of chromium: from the test tube to the cell. Coordination Chemistry Review 2001;216,217 537–582.

[14] Francisco R, Moreno A, Vasconcelos Morais P. Different physiological responses to chromate and dichromate in the chromium resistant and reducing strain *Ochrobactrum tritici* 5bvl1. BioMetals 2010;23 713-725.

[15] Ohta N, Galsworthy PR, Pardee AB. Genetics of sulfate transport by *Salmonella typhimurium*. Journal of Bacteriology 1971;105 1053–1062.

[16] Ohtake H, Cervantes C, Silver S. Decreased chromate uptake in *Pseudomonas fluorescens* carrying a chromate resistance plasmid. Journal of Bacteriology 1987;169 3853–3856.

[17] Cervantes C, Silver S. Plasmid chromate resistance and chromate reduction. Plasmid 1992;27 65–71.

[18] Silver S, Schottel J, Weiss A. Bacterial resistance to toxic metals determined by extrac-hromosomal R factors. International Biodeterioration and Biodegradation 2001;48 263–281.

[19] Pepi M, Baldi F. Modulation of chromium(VI) toxicity by organic and inorganic sul-phur species in yeast from industrial waste. BioMetals 1992;5, 179–185.

[20] Pepi M, Baldi F. Chromate tolerance in strains of *Rhodosporidium toruloides* modulated by thiosulphate and sulfur amino acids. BioMetals 1995;8 99–109.

[21] Costa M. Potential hazards of hexavalent chromium in our drinking water. Toxicolo-gy and Applied Pharmacology 2003;188 1–5.

[22] Xu XR, Li HB, Gu J-D. Reduction of hexavalent chromium by ascorbic acid in aque-ous solutions. Chemosphere 2004;57 609–613.

[23] Xu XR, Li HB, Gu J-D., Li XY. Kinetics of the reduction of chromium (VI) by vitamin C. Environmental Toxicology and Chemistry 2005;24 1310–1314.

[24] Pesti M, Gazdag Z, Belágyi J. In vivo interaction of trivalent chromium with yeast plasma membrane as revealed by EPR spectroscopy. FEMS Microbiology Letters 2000;182 375–380.

[25] Cervantes C, Campos-García J, Devars S, Gutiérrez-Corona F, Loza-Tavera H, Torres-Guzmán JC, Moreno-Sanchez R. Interactions of chromium with microorgan-isms and plants. FEMS Microbiology Reviews 2001;25 335–347.

[26] Zhitkovich Y, Song G, Quievryn V, Voitkun VX. Nonoxidative mechanisms are re-sponsible for the induction of mutagenesis by reduction of Cr(VI) with cysteine: role of ternary DNA adducts in Cr(III)-dependent mutagenesis. Biochemistry 2001;16 549–560.

[27] Raspor P, Batic M, Jamnik P, Josic D, Milacic R, Pas M, Recek M, Rezic-Dereani V, Skrt M. The influence of chromium compounds on yeast physiology. Acta Microbiol-ogy Immunology Hung 2000;47 143–173.

[28] Cervantes C, Campos-García J, Devars S, Gutiérrez-Corona F, Loza-Tavera H, Torres-Guzmán JC, Moreno-Sanchez R. Interactions of chromium with microorgan-isms and plants. FEMS Microbiology Reviews 2001;5 335–347.

[29] Plaper A, Jenko-Brinovec S, Premzl A, Kos J, Raspor P. Genotoxicity of trivalent chro-mium in bacterial cells. Possible effects on DNA topology. Chemical Research in Tox-icology 2002;15 943–949.

[30] Reynolds MF, Peterson-Roth EC, Bespalov IA, Johnston T, Gurel VM, Menard HL, Zhitkovich A. Rapid DNA double-strand breaks resulting from processing of Cr-DNA cross-links by both MutS dimers. Cancer Researches 2009;69 1071–1079.

[31] Valko M, Morris H, Cronin MT. Metals, toxicity and oxidative stress. Current Medi-cal Chemistry 2005;12 1161–1208.

[32] Quievryn G, Messer J, Zhitkovich A. Carcinogenic chromium(VI) induces cross-link-ing of vitamin C to DNA in vitro and in human lung A549 cells. Biochemistry 2002;41 3156–3167.

[33] Quievryn G, Peterson E, Messer J, Zhitkovich A. Genotoxicity and mutagenicity of chromium(VI)/ascorbate-generated DNA adducts in human and bacterial cells. Bio-chemistry 2003;42 1062–1070.

[34] Costa M, Klein CB. Toxicity and carcinogenicity of chromium compounds in hu-mans. Critical Review Toxicology 2006;36 155-63.

[35] Czakó-Vér K, Batié M, Raspor P, Sipiczki M, Pesti M. Hexavalent chromium uptake by sensitive and tolerant mutants of *Schizosaccharomyces pombe*. FEMS Microbiol. Let-ters 1999;178 109–115.

[36] Polti MA, Amoroso MJ, Abate CM. Chromate reductase activity in *Streptomyces* sp. MC1. Journal of General and Applied Microbiology 2010;56 11obio.

[37] Wong PT, Trevors JT. Chromium toxicity to algae and bacteria. In: Nriagu JO., Nie-boer E. (eds.), Chromium in the Natural and Human Environments. New York: Wi-ley; 1988. p305–315.

[38] Katz SA, Salem H. The toxicology of chromium with respect to its chemical specia-tion: a review. Journal of Applied Toxicology 1993;13 217–224.

[39] Stern R.M. Chromium compounds: production and occupational exposure. In: Lan-gard S. (ed.) Biological and Environmental Aspects of Chromium. Amsterdam: Elsevier; 1982 p5–47.

[40] Saha R, Nandi R, Saha B. Sources and toxicity of hexavalent chromium. Journal of Coordination Chemistry 2011;64 1782–1806.

[41] Barceloux DG, Barceloux D. Chromium. Clinical Toxicology 1999;37 173-194.

[42] Langard S. One hundred years of chromium and cancer: a review of epidemiological evidence and selected case reports. American Journal of Industrial Medicine 1990;17 189–214.

[43] James BR. Chemical transformations of chromium in soils relevance to mobility, bio-availability and remediation: The Chromium File 2002;8 8 p.

[44] United States Environmental Protection Agency Contaminated sediment remedia-tion guidance for hazardous waste sites. Office of Solid Waste and Emergency Re-sponse. Revised Version. 2005; OSWER 9355.0-85. EPA/540/R-05/012. December. Available at: http://www.epa.gov/superfund/resources/sediment/guidance.htm

[45] Villegas LB, Fernández PM, Amoroso MJ, de Figueroa LIC. Chromate removal by yeasts isolated from sediments of a tanning factory and a mine site in Argentina. Bio-metals 2008;21 591–600.

[46] United States Environmental Protection Agency Ambient water quality criteria for chromium. 1984; EPA 440/5-84-029. Washington, DC.

[47] Hansel CM, Wielinga BW, Fendorf S. Fate and stability of Cr following reduction by microbially generated Fe(II). SSRL (Stanford Synchrotron Radiation Laboratory) Science Highlights, a digital monthly publication. 2003; Available at: http://www.ssrl.slac. stanford.edu/research/highlights_archive/cr_contamination.pdf as of 5/29/07

[48] Schmieman EA, Yonge DR, Rege MA, Petersen JN, Turick CE, Johnstone DL, et al. Comparative kinetics of bacterial reduction of chromium. Journal of Environmental Engineering 1998;124 449–455.

[49] Schroeder DC, Lee GF. Potential transformations of chromium in natural waters. Water Air and Soil Pollution 1975;4 355–365.

[50] Saleh FY, Parkerton TF, Lewis RV, Huang JH, Dickson KL. Kinetics of chromium transformations in the environment. Science of Total Environment 1989;86 25–41.

[51] Eary LE, Rai D. Kinetics of chromate reduction by ferrous ions derived from hematite and biotite at 25 degrees C. American Journal Science 1989;289 180–213.

[52] Weaver RM, Hochella MF. The reactivity of seven Mn-oxides with $Cr^{3+}aq$: A comparative analysis of a complex, environmentally important redox reaction. American Mineralogist 2003;88 2016–2027.

[53] Fendorf SE. Surface reactions of chromium in soils and waters. Geoderma 1995;57-65–71.

[54] Martello L, Fuchsman P, Sorensen M, Magar V, Wenning RJ. Chromium geochemistry and bioaccumulation in sediments from the lower Hackensack River, New Jersey. Archive Environmental Contamination and Toxicology 2007;53 337–350.

[55] Romanenko VI, Koren'kov VN. A pure culture of bacteria utilizing chromates and bichromates as hydrogen acceptors in growth under anaerobic conditions. Mikrobiologiya 1977;46 414–417.

[56] Lebedeva EV, Lyalikova NN. Reduction of chrocoite by Pseudomonas chromatophila sp. nov. Mikrobiologya 1979;48 517–522.

[57] Tebo BM, Obratzsova AY. Sulfate-reducing bacterium grows with Cr(VI), U(VI), Mn(VI), and Fe(III) as electron acceptors. FEMS Microbiology Letters 1998;162 193–198.

[58] Park CH, Keyhan M, Wielinga B, Fendorf S, Matin A. Purification to homogeneity and characterization of a novel Pseudomonas putida chromate reductase. Applied and Environmental Microbiology 2000;66 1788–1795.

[59] Megharaj M, Avudainayagam S, Naidu R. Toxicity of hexavalent chromium and its reduction by bacteria isolated from soil contaminated with tannery waste. Current Microbiology 2003;47 51–54.

[60] Sarangi A, Krishnan C. Comparison of in vitro Cr(VI) reduction by CFEs of chromate resistant bacteria isolated from chromate contaminated soil. Bioresource Technology 2008;99 4130-4137.

[61] Camargo FA, Okeke BC, Bento FM, Frankenberger WT. In vitro reduction of hexavalent chromium by a cell-free extract of *Bacillus* sp. ES 29 stimulated by Cu^{2+}. Applied Microbiology and Biotechnology 2003;62 569–573.

[62] He M, Xiangyang L, Liu H, Miller SJ,Wang G, Rensing C. Characterization and genomic analysis of a chromate resistant and reducing bacterial strain *Lysinibacillus fusiformis* ZC1. Journal of Hazardous Materials 2011;185 682–688.

[63] Shen H, Wang Y. Characterization of enzymatic reduction of hexavalent chromium by *Escherichia coli* ATCC 33456. Applied and Environmental Microbiology 1993;59 3171-3777.

[64] Ishibashi Y, Cervantes C, Silver S. Chromium reduction in *Pseudomonas putida*. Applied Microbiology and Biotechnology 1990;56 2268-2270.

[65] Mabbett AN, Macaskie LE. A novel isolates of *Desulfovibrio* sp. with the enhanced ability to reduce Cr (VI). Biotechnology Letters 2001;23 683-687.

[66] Liu Y, Xu W, Zeng G, Li X, Goa H. Cr(CI) reduction by *Bacillus* sp. isolated from chromium landfill. Process Biochemistry 2006;41 1981–1986.

[67] Myers CR, Carstens BP, Antholine WE, Myers JM. Chromium(VI) reductase activity is associated with the cytoplasmic membrane of anaerobically grown *Shewanella putrefaciens* MR-1. Journal of Applied Microbiology 2000;88 98-106.

[68] Asatiani NV, Abuladze MK, Kartvelishvili TM, Bakradze NG, Sapojnikova NA, et al. Effect of chromium(VI) action on *Arthrobacter oxydans*. Current Microbiology 2004;49 321-326.

[69] Liu Z, Wu Y, Lei C, Liu P, Gao M. Chromate reduction by a chromate-resistant bacterium, *Microbacterium* sp. World Journal of Microbiology and Biotechnology 2012;28 1585–1592.

[70] Branco R, Chung A-P, Veríssimo A, Morais PV. Impact of chromium-contaminated wastewaters on the microbial community of a river. FEMS Microbiology Ecology 2005;54 35–46.

[71] Focardi S, Pepi M, Landi G, Gasperini S, Ruta M, Di Biasio P, Focardi SE. Hexavalent chromium reduction by whole cells and cell free extract of the moderate halophilic bacterial strain *Halomonas* sp. TA-04. International Biodeterioration and Biodegradation 2012;66 63–70.

[72] Ramírez-Díaz MI, Díaz-Pérez C, Vargas E, Riveros-Rosas H, Campos-García J, Cervantes C. Mechanisms of bacterial resistance to chromium compounds. BioMetals 2007;21 321–332.

[73] Losi ME, Frankenberger WT. Chromium-resistant microorganisms isolated from evaporation ponds of a metal processing plant. Water Air and Soil Pollution 1994;74 405–413.

[74] Wang Y-T. Microbial reduction of chromate. In: Lovley DR. (Ed.), Environmental Microbe–Metal Interactions. American Society for Microbiology Press 2000; Washington DC.

[75] Cheung KH, Gu J-D. Mechanism of hexavalent chromium detoxification by microorganisms and bioremediation application potential: A review. International Biodeterioration and Biodegradation 2007;59 8–15.

[76] Komori K, Rivas A, Toda K, Ohtake H. A method for removal of toxic chromium using dialysis-sac cultures of a chromate-reducing strain of Enterobacter cloacae. Applied Microbiology and Biotechnology 1990;33 91-121.

[77] Jeyasingh J, Ligy P. Bioremediation of chromium contaminated soil: optimization of operating parameters under laboratory conditions. Journal of Hazardous Materials 2005;118 113–120.

[78] Kratochvil D, Pimentel P, Volesky B. Removal of trivalent and hexavalent chromium by seaweed biosorbent. Environmental Science and Technology 1998;32 2693–2698.

[79] Patterson JW. Industrial wastewater treatment technology. Stoneham, MA: Butterworth Publishers; 1985.

[80] Ganguli A, Tripathi AK. Bioremediation of toxic chromium from electroplating effluent by chromate-reducing Pseudomonas aeruginosa A2Chr in two bioreactors. Applied Microbiology and Biotechnology 2002;58 416–420.

[81] Ohtake H, Cervantes C, Silver S. Decreased chromate uptake in Pseudomonas fluorescens carrying a chromate resistance plasmid. Journal of Bacteriology 1987;169 3853–3856.

[82] Cervantes C, Campos-García J, Devars S, Gutiérrez-Corona F, Loza-Tavera H, Torres-Guzmán JC, Moreno-Sánchez R. Interactions of chromium with microorganisms and plants. FEMS Microbiology Reviews 2001;25 335–347.

[83] Kiran B, Kaushik A, Kaushik CP Biosorption of Cr (VI) by native isolate of Lyngbya putealis (HH-15) in the presence of salts Journal of Hazarous Materials 2007;141 662–667.

[84] Xu L, Luo M, Li W, Wei X, Xie K, Liu L, Jiang C, Liu H. Reduction of hexavalent chromium by Pannonibacter phragmitetus LSSE-09 stimulated with external electron

donors under alkaline conditions. Journal of. Hazardous Materials 2011;185
1169-1176.

[85] Jain P, Amatullah A, Alam Rajib S, Mahmud Reza H. Antibiotic resistance and chromium reduction pattern among Actinomycetes. American Journal of Biochemistry
and Biotechnology 2012;8 111-117.

[86] Olguín EJ, Sánchez-Galván G. Heavy metal removal in phytofiltration and phycoremediation: the need to differentiate between bioadsorption and bioaccumulation.
New Biotechnology 2012;30 Number 1.

[87] Poljsak B, Pócsi I,, Raspor P, Pesti M. Interference of chromium with biological systems in yeasts and fungi: a review. Journal of Basic Microbiology 2010;50 21–36.

[88] Monachese M, Burton JP, Reid G. Bioremediation and tolerance of humans to heavy
metals through microbial processes: a potential role for probiotics? Applied and Environmental Microbiology 2012;78 6397-404.

[89] Lebeau T, Braud A, Jézéquel K. Performance of bioaugmentation-assisted phytoextraction applied to metal contaminated soils: A review. Environmental Pollution
2008;153 497-522.

[90] Focardi S, Pepi M, Ruta M, Marvasi M, Bernardini E, Gasperini S, Focardi SE. Arsenic
precipitation by an anaerobic arsenic-respiring bacterial strain isolated from polluted
sediments of the Orbetello Lagoon, Italy. Letters in Applied Microbiology 2010;51
578-585.

[91] Cheng Y, Holman H-Y, Lin Z. Minerals, microbes, and remediation: Remediation of
chromium and uranium contamination by microbial activity. Elements 2012;8
107-112.

[92] United States Environmental Protection Agency Sewer sediment and control. 2004;
Office of Research and Development, Edison, NJ. Available at: http://
www.epa.gov/ORD/NRMRL/pubs/600r04059/600r04059.pdf. Accessed: March 2006.

[93] Smibert RM, Krieg NR. General characterization In: Gerhardt P, Murray RGE, Costilow RN, Nester EW, Wood WA, Krieg NR, Philips GB, (eds.) Manual of methods for
general bacteriology. Washington: American Society for Microbiology; 1981. p411–
442.

[94] Pepi M, Volterrani M, Renzi M, Marvasi M, Gasperini S, Franchi E, Focardi SE. Ar-
senic-resistant bacteria isolated from contaminated sediments of the Orbetello Lagoon, Italy, and their characterization. Journal of Applied Microbiology 2007;103
2299–2308.

[95] Altschul SF, Madden TL, Schäffer AA, Zhang J, Zhang Z, MillerW, Lipman DJ. Gap-
ped BLAST and PSI-Blast: a new generation of protein database search programs.
Nucleic Acids Research 1997;25 3389–3402.

[96]   Cole JR, Wang Q, Cardenas E, Fish J, Chai B, Farris RJ, Kulam-Syed-Mohideen AS,
       McGarrell DM, Marsh T, Garrity GM, Tiedje JM. The Ribosomal Database Project:
       improved alignments and new tools for rRNA analysis. Nucleic Acids Researches
       2009;37 (Database issue): D141–D145; doi:10.1093/nar/gkn879.

[97]   Kumar S, Dudley J, Nei M, Tamura K. MEGA: A biologist-centric software for evolu-
       tionary analysis of DNA and protein sequences. Briefings in Bioinformatics 2008;9
       299–306.

[98]   Saitou N, Nei M. The neighbour-joining method: a new method for reconstructing
       phylogenetic trees. Molecular Biology and Evolution 1987;4 406–425.

[99]   Kimura M. A simple method for estimating evolutionary rates of base substitutions
       through comparative studies of nucleotide sequences. Journal of Molecular Evolu-
       tion 1980;16 111–120.

[100] Bartlett RJ, James BR. Chromium. Part 3. Chemical methods. In: Sparks DL. (ed.)
       Methods of soil analysis. Madison, Wisconsin: ASA and SSSA; 1996. p683–701.

[101] Sambrook J, Fritsch EF, Maniatis T. Molecular cloning. A laboratory manual, Second
       Edition. Cold Spring Harbour: Laboratory Press; 1989.

[102] Viamajala S, Smith WA, Sani RK, Apel WA, Petersen JN, Neal AL, Roberto FF, New-
       by DT, Peyton BM. Isolation and characterization of Cr(VI) reducing *Cellulomonas*
       spp. from subsurface soils: implications for long term chromate reduction. Biore-
       source Technology 2007;98 612–622.

[103] Molokwane PE, Meli KC, Nkhalambayansi-Chirwa E. Chromium(VI) reduction in
       activated sludge bacteria exposed to high chromium loading: Brits culture (South Af-
       rica). Water Researches 2008;42 4538–4548.

[104] Sani RK, Peyton BM, Smith WA, Apel WA, Petersen JN. Dissimilatory reduction of
       Cr(VI), Fe(III), and U(VI) by *Cellulomonas* isolates. Applied Microbiology and Bio-
       technology 2002;60 192–199.

[105] Molokwane PE, Meli KC, Nkhalambayansi-Chirwa E. Chromium (VI) reduction in
       activated sludge bacteria exposed to high chromium loading. Water Research 2008;42
       4538–4548.

[106] Pattanapipitpaisal P, Brown NL, Macaskie LE. Chromate reduction by *Microbacterium
       liquefaciens* immobilized in polyvinyl alcohol. Biotechnology Letters 2001;23 61–65.

[107] You JL, Cao LX, Liu GF, Zhou SN, Tan HM, et al. Isolation and characterization of
       actinomycetes antagonistic to pathogenic *Vibrio* spp. from nearshore marine sedi-
       ments. World Journal Microbiology and Biotechnology 2005;21 679-682.

[108] Megharaj M, Avudainayagam S, Naidu R. Toxicity of hexavalent chromium and its
       reduction by bacteria isolated from soil contaminated with tannery waste. Current
       Microbiology 2003;47 51–54.

[109] Okeke BC, Layman J, Grenshaw S, Oji C. Environmental and kinetic parameters for Cr(VI) bioreduction by a bacterial monoculture purified from Cr(VI)-resistant consortium. Biology Trace Elements Researches 2008;123 229–241.

[110] Luli GW, Telnagi JW, Strohl WR, Pfister RM. Hexavalent chromium-resistant bacteria isolated from river sediments. Applied and Environmental Microbiology 1983;46 846–854.

[111] Poopal AC, Laxman RS. Chromate reduction by PVA-alginate immobilized *Streptomyces griseus* in a bioreactor. Biotechnology Letters 2009;31 71–76.

[112] Rehman A, Zahoor A, Muneer B, Hasnain S. Chromium tolerance and reduction potential of a *Bacillus* sp. ev3 isolated from metal contaminated wastewaters. Bullettin of Environmental Contamination and Toxicology 2008;81 25–29.

[113] Chai L, Huang S, Yang Z, Peng B, Huang Y, Chen Y. Cr(VI) remediation by indigenous bacteria in soils contaminated by chromium-containing slag. Journal of Hazardous Materials 2009;167 516–522.

[114] Polti MA, García RD, Amoroso MJ, Abate CM. Bioremediation of chromium (VI) contaminated soil by *Streptomyces* sp. MC1. Journal of Basic Microbiology 2009;49 285–292.

[115] Rai D, Sass BM, Moore DA. Chromium(III) hydrolysis constants and solubility of chromium(III) hydroxide. Inorganic Chemistry 1987;26 345–349.

[116] Polti MA, Amoroso MJ, Abate CM, Chromate reductase activity in *Streptomyces* sp. MC1. Journal of General and Applied Microbiology 2009; 56 11–18.

[117] Colin VL, Pereira CE, Villegas LB, Amoroso MJ, Abate CM. Production and partial characterization of bioemulsifier from a chromium-resistant actinobacteria. Chemosphere 2013;90 1372–1378.

[118] Verma T, Garg SK, Ramtake PW. Genetic correlation between chromium resistance and reduction in *Bacillus brevis* isolated from tannery effluent. Journal of Applied Microbiology 2009;107 1425–1432.

[119] Adeel SS, Wajid A, Hussain S, Malik F, Sami Z, Ul Haq I, Hameed A, Channa RA. Recovery of chromium from the tannery wastewater by use of *Bacillus subtilis* in Gujranwala, Pakistan. Journal of Pharmacy and Biological Sciences 2012;2 2278-3008

Permissions

The contributors of this book come from diverse backgrounds, making this book a truly international effort. This book will bring forth new frontiers with its revolutionizing research information and detailed analysis of the nascent developments around the world.

We would like to thank Rolando Chamy and Francisca Rosenkranz, for lending their expertise to make the book truly unique. Thy have played a crucial role in the development of this book. Without their invaluable contribution this book wouldn't have been possible. They have made vital efforts to compile up to date information on the varied aspects of this subject to make this book a valuable addition to the collection of many professionals and students.

This book was conceptualized with the vision of imparting up-to-date information and advanced data in this field. To ensure the same, a matchless editorial board was set up. Every individual on the board went through rigorous rounds of assessment to prove their worth. After which they invested a large part of their time researching and compiling the most relevant data for our readers. Conferences and sessions were held from time to time between the editorial board and the contributing authors to present the data in the most comprehensible form. The editorial team has worked tirelessly to provide valuable and valid information to help people across the globe.

Every chapter published in this book has been scrutinized by our experts. Their significance has been extensively debated. The topics covered herein carry significant findings which will fuel the growth of the discipline. They may even be implemented as practical applications or may be referred to as a beginning point for another development. Chapters in this book were first published by InTech; hereby published with permission under the Creative Commons Attribution License or equivalent.

The editorial board has been involved in producing this book since its inception. They have spent rigorous hours researching and exploring the diverse topics which have resulted in the successful publishing of this book. They have passed on their knowledge of decades through this book. To expedite this challenging task, the publisher supported the team at every step. A small team of assistant editors was also appointed to further simplify the editing procedure and attain best results for the readers.

Our editorial team has been hand-picked from every corner of the world. Their multi-ethnicity adds dynamic inputs to the discussions which result in innovative outcomes. These outcomes are then further discussed with the researchers and contributors who give their valuable feedback and opinion regarding the same. The feedback is then collaborated with the researches and they are edited in a comprehensive manner to aid the understanding of the subject.

Apart from the editorial board, the designing team has also invested a significant amount of their time in understanding the subject and creating the most relevant covers. They scrutinized every image to scout for the most suitable representation of the subject and create an appropriate cover for the book.

The publishing team has been involved in this book since its early stages. They were actively engaged in every process, be it collecting the data, connecting with the contributors or procuring relevant information. The team has been an ardent support to the editorial, designing and production team. Their endless efforts to recruit the best for this project, has resulted in the accomplishment of this book. They are a veteran in the field of academics and their pool of knowledge is as vast as their experience in printing. Their expertise and guidance has proved useful at every step. Their uncompromising quality standards have made this book an exceptional effort. Their encouragement from time to time has been an inspiration for everyone.

The publisher and the editorial board hope that this book will prove to be a valuable piece of knowledge for researchers, students, practitioners and scholars across the globe.

List of Contributors

Elisa Tamariz
Biomedical Department, Health Science Institute, Veracruzana University, Veracruz, México

Ariadna Rios-Ramírez
Neurobiology Institute, National University of México, Querétaro, México

Encarnación Jurado, Mercedes Fernández-Serrano, Francisco Ríos and Manuela Lechuga
Department of Chemical Engineering, University of Granada, Granada, Spain

R.S. Reis
University of Sydney, School of Molecular Biology, NSW, Australia

D.M.G. Freire
Department of Biochemistry, Chemistry Institute, Federal University of Rio de Janeiro, RJ, Brazil

G.J. Pacheco
Department of Biochemistry, Chemistry Institute, Federal University of Rio de Janeiro, RJ, Brazil
Laboratory of Toxinology, IOC, Oswaldo Cruz Foundation, RJ, Brazil

A.G. Pereira
Department of Cellular Biology, University of Brasília, DF, Brazil

Stefania Gheorghe, Irina Lucaciu, Iuliana Paun, Catalina Stoica and Elena Stanescu
National Research and Development Institute for Industrial Ecology — ECOIND Bucharest, Romania

Babak Ghanbarzadeh and Hadi Almasi
University of Tabriz, Iran

Taisuke Banno
Department of Basic Science, Graduate School of Arts and Sciences, The University of Tokyo, Japan

Taro Toyota
Department of Basic Science, Graduate School of Arts and Sciences, The University of Tokyo, Japan
Precursory Research of Embryonic Science and Technology (PRESTO), Japan Science and Technology Agency (JST), Japan

Shuichi Matsumura
Department of Applied Chemistry, Faculty of Science and Technology, Keio University, Japan

Vladimir Sedlarik
Centre of Polymer Systems, University Institute, Tomas Bata University in Zlin, Zlin, Czech Republic
Polymer Centre, Faculty of Technology, Tomas Bata University in Zlin, Zlin, Czech Republic

Olusola Abayomi Ojo-Omoniyi
Department of Microbiology, Faculty of Science, Lagos State University, Ojo. Lagos, Nigeria

Marie Zarevúcka
Institute of Organic Chemistry and Biochemistry, AS CR, Flemingovo nám, Prague, Czech Republic

Ma. Laura Ortiz-Hernández, Enrique Sánchez-Salinas, Edgar Dantán-González and María Luisa Castrejón-Godínez
Biotechnology Research Center, Autonomous University of State of Morelos, Col. Chamilpa, Cuernavaca, Morelos, C.P., Mexico

Nezha Tahri Joutey, Wifak Bahafid, Hanane Sayel and Naïma El Ghachtouli
Microbial Biotechnology Laboratory, Sidi Mohammed Ben Abdellah University, Faculty of Science and Technics, Fez, Morocco

M. S. Nascimento
Departamento de Antibióticos, Universidade Federal de Pernambuco, UFPE, Recife, PE, Brazil

A. L. B. D. Santana
Departamento de Química, Universidade Federal Rural de Pernambuco, UAST, Serra Talhada, PE, Brazil

C. A. Maranhão
Instituto Federal Educação, Ciência e Tecnologia de Pernambuco, IFPE, Campus Recife, Recife, PE, Brazil

L. S. Oliveira and L. Bieber
Departamento de Química Fundamental, Universidade Federal de Pernambuco, UFPE, Recife, PE, Brazil, Brazil

Silvia Focardi, Milva Pepi and Silvano E. Focardi
Department of Physical, Earth and Environmental Sciences, University of Siena, Siena, Italy

www.ingramcontent.com/pod-product-compliance
Lightning Source LLC
Chambersburg PA
CBHW070716190326
41458CB00004B/995